Nature-Inspired Structured Functional Surfaces

Nature-Inspired Structured Functional Surfaces

Design, Fabrication, Characterization, and Applications

Zhiwu Han

WILEY-VCH

Author

Prof. Zhiwu Han
Jilin University
Key Laboratory of Bionic Engineering
No.5988 Renmin Street Nanguan
130012 Changchun
China

Cover Image: © nono57/Getty Images

Library of Congress Card No.: applied for

British Library Cataloguing-in-Publication Data
A catalogue record for this book is available from the British Library.

Bibliographic information published by the Deutsche Nationalbibliothek
The Deutsche Nationalbibliothek lists this publication in the Deutsche Nationalbibliografie; detailed bibliographic data are available on the Internet at <http://dnb.d-nb.de>.

Print ISBN: 978-3-527-35021-6
ePDF ISBN: 978-3-527-83654-3
ePub ISBN: 978-3-527-83655-0
oBook ISBN: 978-3-527-83656-7

Typesetting Straive, Chennai, India

C118487_040722

Contents

Preface

In the long history of mankind, learning from the all-encompassing nature and drawing inspiration from the nature creatures has been an effective way for people to create new inventions and new things since ancient times. It is also a wise way for human beings to get rid of the predicament of survival and to develop forever. As a new and comprehensive interdisciplinary subject, bionics has maintained astrong vitality since its birth in the 1960s. It means that the people study the principle of the structure and function of living organisms, and invent new equipment, tools and technology according to these principles, and create advanced technology suitable for production, learning and living. In terms of engineering technology, bionics, based on the study of biological systems,provides new principles, methods and approaches for the design and construction of new technical equipment. The glorious mission of bionics is to provide the most reliable, the most flexible, the most efficient and the most economical technology system close to the biological system for the benefit of mankind.

Biological surface is the first interface between the organism and outside world. All the exchange and interaction of information and energy firstly occur on these surfaces. So, the evolution of organisms also firstly occurs on the surface of living organisms. As a result, the biological surfaces are endowed with nearly perfect multifunctional characteristics to adapt to their diverse harsh environments. For example, the lotus leaf rejects liquid adhesion through the hierarchical structures on its surface, which is known as the lotus leaf effect. The micro/nanostructures on surface of butterfly wings can interact with light and give some brilliant colors. The nanocone structures on surface of cicada wings can effectively inhibit the reflection of external visible light. The micro/nanocomposite structures of insect compound eyes can effectively reduce reflection and make it have excellent antifogging properties. It is not difficult to draw a conclusion from the above examples. The multifunctional characteristics of these insects are inseparable from its surface structures and material properties. Therefore, it can be conjecture the conclusion that there are almost no non-functional structures and no non-structural functions on the biological surface. In fact, research on nature-inspired structural surfaces is infull swing at home and abroad. The relevant achievements have sprung up. Nature-inspired functional surfaces are becoming new research hotspots in the field of bionic engineering and showing fire-new developments.

The scope of this book is introducing a wide variety of nature-inspired structured functional surfaces, taking several innovative, typical and major progresses finished by Bionic Engineering Research Team of Jilin University as examples. It includes seven kinds of nature-inspired surfaces with different structural features: structural antireflective surfaces, structural light trapping surfaces, structural antifogging surfaces, structurally colored surfaces, oil-water separationmaterials, underwater superhydrophobic surfaces and structural responsive surfaces toward multiple gases/vapors. It mainly focuses on the the oretical and technical progress about the basic principles, synthetic design and fabrication strategies, along with the advanced characterization methods and composition analysis techniques. These latest progress and technologies are playing important roles in inspiring science and technology innovation. Engineers inrelated fields can design and manufacture nature-inspired surfaces and materials to solve key engineering problems inspired by the biological structured functional surfaces. On the other hand, taking inspiration from the diversified biological surfaces in nature, they can start from the diversified engineering and technical problems faced by human beings, to reveal new principles, form innovative technologies, overcome major engineering technical problems, and finally benefit all of mankind.

It is hoped that this book will give the readers enough background information to begin to solve critical engineering and materials problems, or provide enough information and resources to spring board them to generate new theory, technologies, methods and equipment.

China, 25 February 2022 *Luquan Ren*

Acknowledgments

I am deeply indebted to many fundings and support toward the research in this book, including the National Key Research and Development Program of China (No. 2018YFA0703300), the Foundation for Innovative Research Groups of the National Natural Science Foundation of China (No. 52021003), the National Natural Science Foundation of China (No. 51835006), and the Jilin University Science and Technology Innovative Research Team (No. 2020TD-03).

On behalf of my research group, I would like to especially thank our academic mentor Prof. Luquan Ren for guiding us into the exciting and inspiring interdisciplinary area of bionics. He firstly established the subject of Bionic Science and Engineering and exploited the research field of terrain-machine bionics. He proposed the Bionic Non-Smooth Theory and Coupling Bionics Theory. I am grateful to the Key Laboratory of Bionic Engineering (KLBE) of Ministry of Education, Jilin University, which provided a solid and broad academic platform as well as long-term strong support for our research.

In addition, I would like to acknowledge the efforts of many postdoctoral research fellows and graduate students in my research group, including Drs. Zhengzhi Mu, Bo Li, Xiaoming Feng, Zhibin Jiao, Ze Wang, Hanliang Ding, You Chen, and Hao Xue for their initial editing. I also thank the efforts of many research staffs of KLBE, including Prof. Shichao Niu and Prof. Junqiu Zhang who contributed to the proofreading and revision of this book. Finally, I would like to extend my sincere thanks to the help, patience, and encouragement provided by our editors from Wiley Press, Lifen Yang, Maceda, Katrina, N. Kiruthigadevi, and Yoganandh Rajadurai.

1

Introduction of Nature-Inspired Functional Structural Surface

1.1 Advanced Materials Boosted by Bionics

Learning from the all-encompassing nature and drawing inspiration from natural creatures has been always an effective way for people to invent and create new things since ancient times. It is also a wise way for human beings to get rid of survival predicaments and to develop sustainably. After nearly four billion years of evolution and optimization, natural creatures have possessed many excellent properties that are far beyond human beings. The study of typical biometric organs and structures in the nature can not only help us understand nature better but also provide useful references and inspiration to solve current scientific puzzles and technical dilemmas [1–8]. As a new and comprehensive interdisciplinary subject, bionics has maintained strong vitality since its birth in the 1960s. In fact, the ideology of bionics has been existing in the world for thousands of years. The origin of bionics could be derived that nature has been the source of all kinds of technological ideas, engineering principles and major inventions since ancient times. In other words, bionics acts a bridge to link biology in the nature with technologies developed by human beings. As known to all, a wide variety of natural creatures can adapt to harsh environments through a long period of evolutionary processes, so that they can get survival and development. Surprisingly, some typical natural creatures possess outstanding properties that even precede humanmade delicate products in the fields of optics, mechanics, dynamics, and so on. Thus, taking inspiration from biology in the nature to develop new materials and technologies is a wise choice for scientists and engineers. It has flourished to this day and has gradually integrated into biology, materials science, mechanics, optics, and many other disciplines including electronics and electromagnetism. In recent years, with the rapid development of micro-/nanomanufacturing technologies, related instruments and equipment, scientists and engineers have turned to high-performance organisms in nature. They have been trying to reproduce the organisms by artificially copying the complex micro-/nanostructures with excellent performance. With the performance of traditional materials gradually entering the plateau stage, a breakthrough is needed for the flourishing research of new functional materials. The introduction of bionics ideas will provide a new pathway to break the deadlock of functional materials research.

Nature-Inspired Structured Functional Surfaces: Design, Fabrication, Characterization, and Applications, First Edition. Zhiwu Han.

In fact, research on bioinspired structural surfaces is in full swing at home and abroad. The relevant achievements have sprung up. Bioinspired functional surfaces are becoming new research hotspots in bionic engineering and are showing fire-new developments. At present, from the imitation of biological prototypes, the imitation objects range from biomolecules [9–11] (DNA molecules, RNA molecules, etc.), microorganisms [12–16] (viruses, bacteria, fungi, and algae) to plants [17, 18] (wood, leaves, etc.), animals [19–44] (beetles, butterfly wings, moth eyes, bird feathers, shells, teeth, marine life, gecko feet, mosquito, leeches, polar bear fur, etc.), and even the entire biological system [45–48], as shown in Table 1.1. Most of the imitation objects are concentrated on the creature's body surface, and the corresponding artificial replicas are the bioinspired structural surfaces. From the material point of view, the material types of bioinspired structural surfaces have gradually evolved from simple organic materials and inorganic materials to broad

Table 1.1 Typical examples of structure–function correspondence in biological systems.

	Biology	Feature structures	Functions
Biomolecules	DNA	Nanostructures	Miniaturization
Microorganisms	Virus, bacteria, fungi, yeast	Various nanostructures	Self-assembly, miniaturization
	Algae (diatom, coccolithophore)	Periodic porous structures/hierarchical microstructures	Chemical energy conversion, particular optical functions
Plants	Wood	Periodic porous structures	High mechanical strength
	Leaves	Hierarchical structures	Chemical energy conversion, superhydrophobicity, self-cleaning
Animals	Insects (beetles, butterfly wings, etc.)	Periodic porous structures/hierarchical structures	Structural color, superhydrophobicity
	Compound eyes	Periodic structures	Antireflection
	Feathers	Periodic structures	Structural color, superhydrophobicity
	Seashells, teeth	Periodic structures	Structural color, high mechanical strength
	Marine animals (sea urchin exoskeleton)	Periodic structures	Particular optical functions
	Gecko feet	Hierarchical structures	Strong adhesive force
	Mosquito's legs	Hierarchical structures	Water-supporting ability
	Fur and skin of polar bear	Hollow structures	Thermal insulation
Biological systems			Self-repair, self-heating, sensory-aid devices

Source: Reproduced with permission from Ref. [45]. Copyright © 2011 WILEY-VCH Verlag GmbH & Co. KGaA, Weinheim.

material systems such as hybrid materials and composite materials. Nature-inspired functional structural surfaces (NIFSS) are present in a variety of different material states and structural forms. As expected, they also show remarkable functional characteristics.

Here are some typical examples. A series of multidimensional biomimetic silicon-based nanocomposites were prepared by DNA origami [49]. Bioinspired photo-controlled nanochannels based on DNA molecules can be used for drug sustained release, optical information storage, and logic networks [50]. The shell-like ordered layered structure material exhibits ultra-high mechanical properties [51]. By mimicking the unique topology of plant viruses, nano-optical antennas can be prepared for molecular fingerprinting [52]. The membrane of bacteria *Bacillus subtilis* exhibits durability against liquid wetting and gas permeation and is expected to provide an example for the study of antibacterial and biomimetic drainage surfaces [53]. The natural photonic crystals with opal-like structures of algae *Cystoseira tamariscifolia* cells can produce vivid structural colors by reversibly changing the stacking state of the internal structure in response to external environmental conditions, thereby exhibiting light manipulation ability in addition to visual signals [54]. By using the natural structure of wood anisotropy and the cellulose component therein, material scientists design and manufacture low-cost, lightweight, and high-performance structured "super wood," all-wood supercapacitors, "transparent timber" with mechanical and transparent properties [55–57]. A series of hydrophobic, oleophobic, and amphiphobic "lotus effect" inspired self-cleaning surfaces [58–61] and so on. It can be seen that the surface structures of organisms and their excellent performance interdependent to form an integration of structure and function. It should be noted that structures lay the foundation for superior performance. Performance also reflects the extension of the structures. Therefore, excellent functional properties of biological surfaces are revealed. On this basis, the design and manufacture of the NIFSS that meet the requirements have become a hotspot and a challenge in bionic engineering.

Many unique functional properties of biology in nature are inextricably linked to ultrafine 3D micro-/nanostructures. Taking the most common species of plants and animals as examples, studies have shown that the self-cleaning effect of the lotus leaf, also known as the "leaf effect," benefits from the convex-packed structures densely distributed on leaf surfaces [62]. The magical phenomenon of continuous directed transport of liquid film on the surface of the *Nepenthes alata* rim is related to the multistage groove structures of the lip and the blind-hole structures with a one-way wedge angle in some grooves [63]. The gecko can freely climb on the vertical wall, which mainly relies on the rich microvilli structures of the sole to provide strong adhesion [64, 65]. Another similar case is the adhesion phenomenon of the sacral bristles of the ladybug *Coccinella septempunctata* to the rough surfaces [66]. The antireflective effect of the moth eye is closely related to the conical array of the outer surface of the eyes [64]. The directional water-collecting effect of the spider silk is realized by its unique periodic spindle knots [63]. The single scale of the chafer *Cyphochilus* wings can be dazzling white, and the optical performance is closely related to the filamentous network microstructure of the wing scale surface [67, 68].

1.2 Definition and Classification of NIFSS

Bioinspired structured surfaces are referred to as NIFSS in this book. Due to the excellent functional properties of biological surfaces in many respects, materials scientists and engineers have long focused on biomimetic structures of biological surfaces and artificial reproduction of their excellent functions. The so-called NIFSS is a general term for all kinds of artificial structural surfaces with various materials at different scale levels inspired by biological surfaces.

In this book, according to the different sources of the original micro-/nanostructures of the biological surfaces, the NIFSS can be divided into two categories. One is biology-based structure surfaces with similar or enhanced functional properties, which are designed and developed using the biological surface itself as a raw material or by chemical modification and physical evaporation. The other one is biology-inspired structural surfaces with similar or enhanced functional properties, which are designed and developed in combination with existing micro-/nanomanufacturing processes. The two types of NIFSS are explained one by one in the following text.

The NIFSS can be further subdivided into two subclasses: (i) the biological surface itself is an original structural material to get the desired functional surface with the natural functional characteristics. Various types of biological surfaces with excellent functional characteristics, previously reported, can be regarded as natural structural surfaces. Since the biological surfaces have been separated from the biological body, the excised biological surfaces themselves can be considered a special kind of NIFSS; (ii) denatured biomimetic structural surfaces based on biological surfaces are obtained by chemical modification. Since the main component of the biological surfaces itself is organic, it is still difficult to meet the stringent requirements for practical applications. Therefore, it has become a kind of modification and enhancement treatment of the natural structure surfaces.

NIFSS mainly refer to artificial biomimetic materials or devices that mimic the excellent functional properties of the biological surface or the internal mechanism, and the final realized functions can be similar to or different from original biological surfaces. The biology itself is only used as a source of inspiration and imitation, and it is not used as the original material to participate in the design and manufacture of the NIFSS.

1.3 Typical Prototypes with Structural Surfaces

1.3.1 Butterfly Wings

Among these outstanding research examples, butterfly is undoubtedly one of the most diverse and well-known biological prototypes. A variety of butterflies is biologically and geographically diverse (Figure 1.1). The collection channel of butterfly samples is convenient, providing a stable sample source and a huge database of biological structures for in-depth study of typical butterfly wings.

Figure 1.1 Different butterflies with colorful wings in the nature.

In recent decades, there have been numerous research cases based on the butterfly wings or inspired by the micro-/nanostructures of the wing scales, and the research content is rich enough. For example, in a study related to the structures of butterfly wings, the main research content includes the microscopic characterization of the micro-/nanostructures on the butterfly wing surface and the intrinsic formation mechanism of the brilliant structural colors [26, 69–82]. A quantitative study on the contribution of single wing scales to interference and diffraction in the structures of butterfly wings was also carried out [83]. A bioengineering method of butterfly wing structural colors is also an emerging hotspot [84–90]. In terms of wettability research, related studies have shown that the micro-/nanostructures of the butterfly wing scales endow the wing surface with higher roughness. They can regulate surface wettability and control the bounce behavior of droplets on the surface [91, 92]. Based on this, directional wet super-slip fibers [93] and structured waterproof surfaces [94] are developed. In the study of responsive materials, the scales of the *Morpho sulkowskyi* make it selectively optically react to different vapors. This optical response is derived from the polarity gradient of the micro-/nanostructure material itself and is superior to the performance of existing nanophotonic sensors [24, 95]. The micro-/nanostructures of the *Greta oto* butterfly wing surface have piezoelectric response characteristics, which are expected to provide a reference for the development of new optoelectronic devices. It can be applied to the field of electro-stealth [96]. Photonic crystal-type micro-/nanostructures and pigment-type micro-/nanostructures on the butterfly *Polyommatus icarus* can exhibit differential response characteristics to cold stress [97]. Photonic crystal structures of *Papilio ulysses* wing scales can follow the change in the external refractive index to produce a reversible thermochromic reaction [98]. *Morpho* wing scales can be slightly deformed by external thermal radiation. Inspired by this, researchers have proposed a new thermal imaging technology [99]. In catalytic research, the main research work includes photocatalysis induced by structural colors of butterfly wings [90] and chemical catalysis using butterfly or its imitation as support materials [100]. In addition, in the oil/water separation research, researchers have used butterfly wings as the osmosis membrane to imitate the artificial filter membrane for oil/water separation [101].

Butterfly wings have shown great research value and the application potential in various research fields, such as micro-/nano-optics, water transportation, sensing detection, optical catalysis, and even oil/water separation. They have very broad and potential application prospects.

1.3.2 Cicada Wings

Similarly, cicada is another typical biological prototype for bionic research. Especially, its wing has attracted intensive research interest in the field of antireflective materials. Cicada wings have typical periodic micro-/nanostructure arrays. Huang et al. [102] characterized the micro-/nanostructures of cicada wings by scanning electron microscope (SEM) and measured total reflectance of cicada wings in the wavelength range of 400–800 nm, which is as low as 1%. A three-dimensional (3D) array model based on the micro-/nanostructures of its wing surface was established. The simulation results were in good agreement with the experimental results (Figure 1.2). It has been confirmed that cicada wings have ultra-low reflectivity and exhibit excellent antireflective properties.

1.3.3 Moth Eyes

In the insect world, compound eyes present an attractive physiological optical performance in terms of optical sensitivity and antireflection [103–105]. Compound eyes usually contain thousands of small eyes (ommatidia) [106], as shown in Figure 1.3. The eyes are usually neatly distributed along a spherical or hemispherical surface in a hexagonal pattern. The surface of these small eyes is not smooth. It is tightly covered by hemispherical nanoscale bumps, forming a grating that enhances the ability of small eyes to absorb light [107]. Taking nocturnal moths as an example, the cornea of the subwavelength structures has an optical antireflective function, which can provide stealth help for its nighttime activities [108, 109]. It has been confirmed that the antireflective function of the moth eye is caused by the micro-/nanostructures, which makes a gradient change in the refractive index between the air and the cornea, achieving the inhibition of light reflection [64].

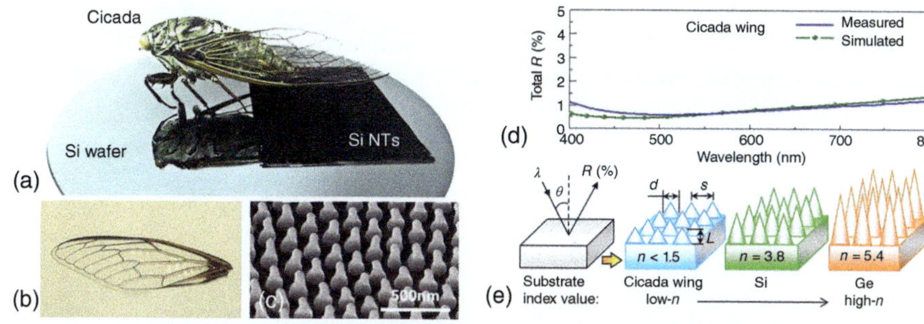

Figure 1.2 Transparent cicada wings with antireflective nanostructure arrays. (a) A cicada specimen was placed partially on a polished silicon wafer and a piece of Si nanotips. (b) Photographic image of a singing cicada wing. (c) SEM image of the cicada wing surface. (d) Comparison of measured and simulated total reflectance (total *R*%) spectrum as a function of wavelength for the cicada wing. (e) Schematics of reflectance reduction of biomimetic nanostructures with feature parameters compared to planar surfaces. Abbreviations: λ, incident wavelength; θ, angle of incidence; *d*, diameter; *S*, spacing; *L*, length; *n*, bulk refractive index; Si, silicon; Ge, germanium. Source: Reproduced with permission from Liimatainen et al. [102]. Copyright © 2015 American Chemical Society.

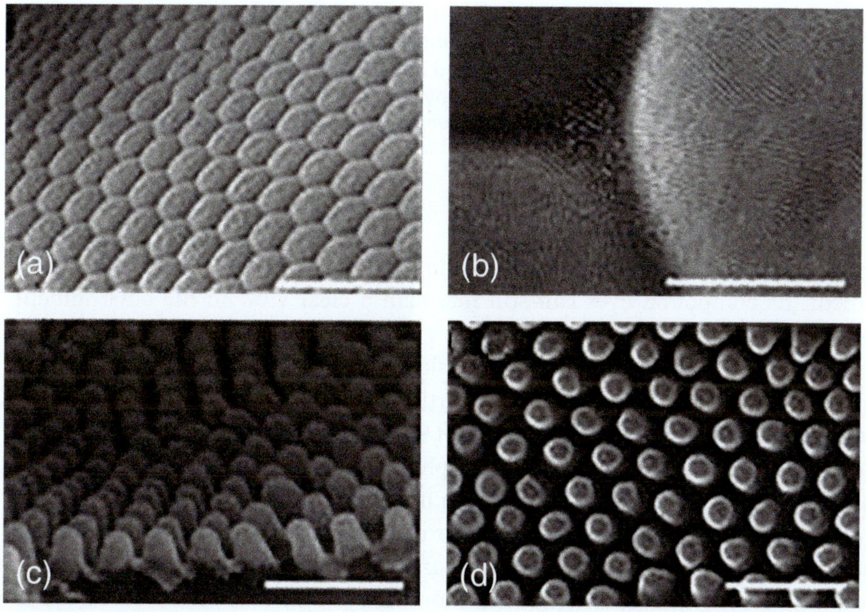

Figure 1.3 SEM images of the *Attacus atlas* moth eye showing the compound eye structures. Scale bar: (a) 100 μm, (b) 5 μm, (c) and (d) 500 nm. Source: Reproduced with permission from Wang et al. [106]. Copyright © 2011 The Royal Society.

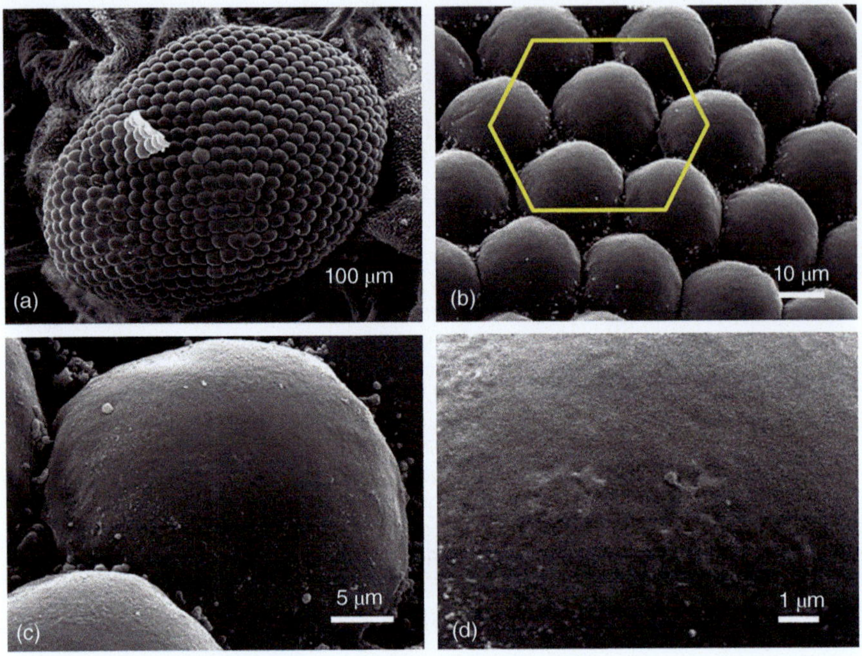

Figure 1.4 SEM images of compound eyes. (a) The compound eyes are ellipsoidal, composed of hundreds of ommatidia. (b) Ommatidia are uniform and tightly arranged. (c) The upper part of ommatidia is spherical, and the lower is cylindrical, with a diameter of 20 μm or so. (d) There is no more tiny structure on the ommatidia surface. Source: Reproduced with permission from Han et al. [110]. Copyright © 2014 Science China Press and Springer-Verlag Berlin Heidelberg.

1.3.4 Mayfly Eyes

Mayfly *Ephemera pictiventris* is a kind of insect that lives in the near water environment. Its compound eyes can still maintain a clear view in the environment where water vapor is concentrated [110]. The main component of the cornea of the compound eyes is chitin, whose intrinsic contact angle is about 100°. The eyes exhibit excellent superhydrophobic properties. The top of the small eye and the diameter of the base is not equal. The small eye can be seen as the upper and lower parts. The upper part is approximately spherical. The lower part is a truncated cone shape, which is closely arranged in a hexagonal shape. The overall height of the small eye is about 11 μm; the diameter of the base is about 22 μm. The upper spherical surface is straight. The compound eye size is consistent and closely arranged. Further enlargement results show that the surface of the small eye is relatively smooth and has no tiny nanoscale structures (Figure 1.4).

1.3.5 Mosquito Eyes

Other insects like mosquitoes *Culex pipiens* also exhibit excellent superhydrophobic properties [111], which keep mosquitoes live in extremely humid environments

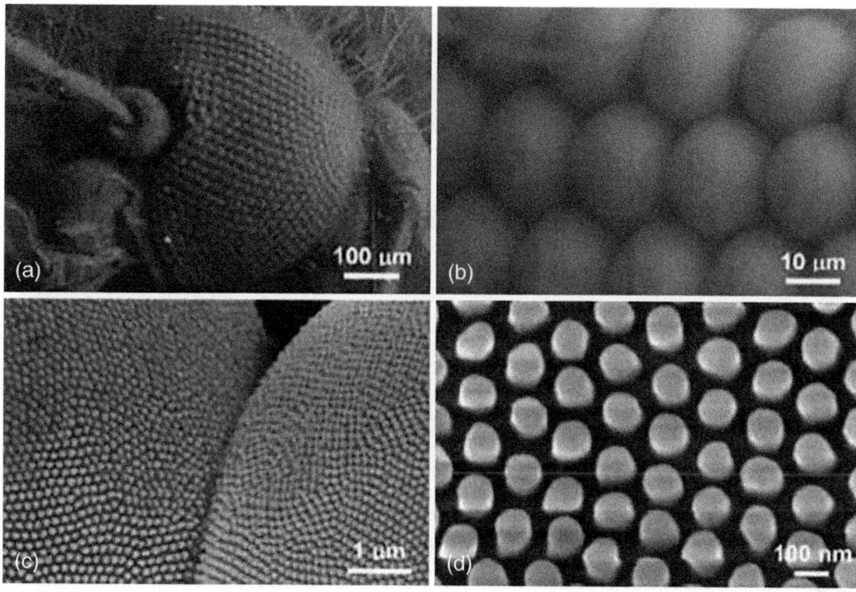

Figure 1.5 Complex hierarchical micro-/nanostructures of the compound eyes of mosquitoes. (a) SEM image of a single mosquito eye. (b) Numerous ommatidia forming a hexagonally close-packed micro-hemisphere. (c) Two neighboring ommatidia with nanonipple arrays. (d) Hexagonally non-close-packed nanonipples covering an ommatidial surface. Source: Reproduced with permission from Tadepalli et al. [111]. Copyright © 2007 WILEY-VCH Verlag GmbH & Co. KGaA, Weinheim.

with clear vision. For mosquitoes, the complex hierarchical micro-/nanostructures of the compound eyes provide a structural basis for the realization of this function as shown in Figure 1.5. Similarly, this compound eye is also composed of a large number of small eyes. On the one hand, the microlevel protrusion structure of the compound eye is uniformly arranged in close-packed hexagons, which can effectively prevent larger droplets from staying in the gap of the small eyes. On the other hand, the nanosized mastoid in the small eyes plays a key role in avoiding small-scale water vapor condensation. With the synergistic effect of these two features, hydrophobicity and anti-fogging of the mosquito compound eye are finally realized.

1.3.6 Water Striders' Legs

It is well known that water striders can float on water surface and they can be propelled rapidly with their superhydrophobic hairy legs by transporting the momentum [112, 113] waves to propel themselves across the water surface [112, 114–116]. Water strider *Gerris remigis* (Figure 1.6a) living at the water surface in a highly humid environment. Without any external force, tiny, condensed droplets in the range of femtoliters (fl) to microliters (μl) are removed from the strider's legs, owing to the presence of oriented conical setae. The leg of *Gerris* is a centimeter-sized cylinder (of typical diameter 150 μm) decorated by an array

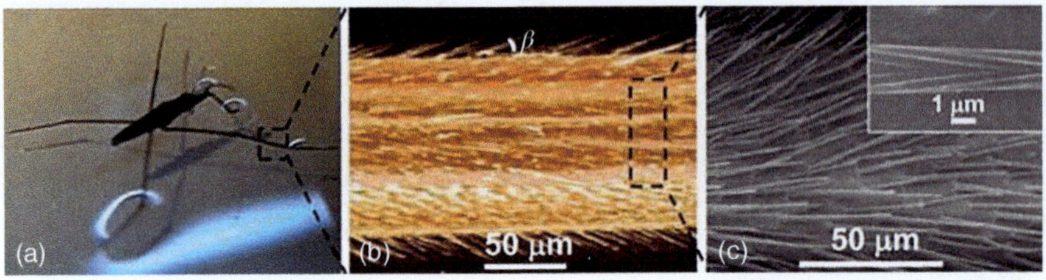

Figure 1.6 (a) *Gerris remigis* lives at the water surface in a highly humid environment. (b, c) Micro-XCT and SEM images of a water strider's leg showing typical hierarchical structures. Source: Reproduced with permission from Wang et al. [112]. Copyright © 2015 National Academy of Sciences.

of inclined tapered hairs (Figure 1.6b,c) characterized by micro X-ray computed tomography (XCT) and SEM. Individual setae have a length L = 40–50 μm, a maximum diameter of ~3 μm, and an apex angle of ~5°. They make regular arrays with a mutual distance of 5–10 μm and are tilted by an angle β = 25–35° to the base of the leg (Figure 1.6b,c). In addition, longitudinal or quasi-helicoidal nanogrooves are found on the setae surface, as shown in Figure 1.6c (*inset*).

1.3.7 Scorpion Back

For creatures surviving in deserts, the abrasion of the body surface by wind and sand is main challenge. Abrasion is also undesirable, which can cause catastrophic failures in most industrial applications [117]. In nature, some animals such as desert lizards and scorpions live in a gas–solid mixed medium environment such as sand. They perform in this environment through the synergy of special surface morphology, internal microstructure, and biological flexibility. The back of the scorpion can resist abrasion and protect it from damage (Figure 1.7). Han and Zhang et al. [118–120] showed the erosion resistance mechanism of scorpion back, which is the result of multiple coupling effects. The surface morphology, material, and elasticity of the back of the desert scorpion are important biological coupling elements to resist the erosion. According to their analysis, the scorpion can form special protrusions and grooves on the back through adaptation to the living environment and its evolution, thereby changing the flow state of the surface boundary layer and reducing surface erosion. On the other hand, the elastic internodal membranes and side membranes play the role of energy release and help reduce erosion.

1.3.8 Gecko's Feet

Gecko is the largest animal known to support its weight by producing high (dry) adhesion [121]. The ability of geckos (Figure 1.8a) to climb on vertical walls has been noticed in ancient times. However, it was not until the invention of the electron microscope in the 1950s that it was possible to observe the skin on the gecko's feet (Figure 1.8b) and toes (Figure 1.8c). The observed skin has a complex fibrous structure composed of lamellae, setae, branches, and spatula (Figure 1.8d) [64, 122–129].

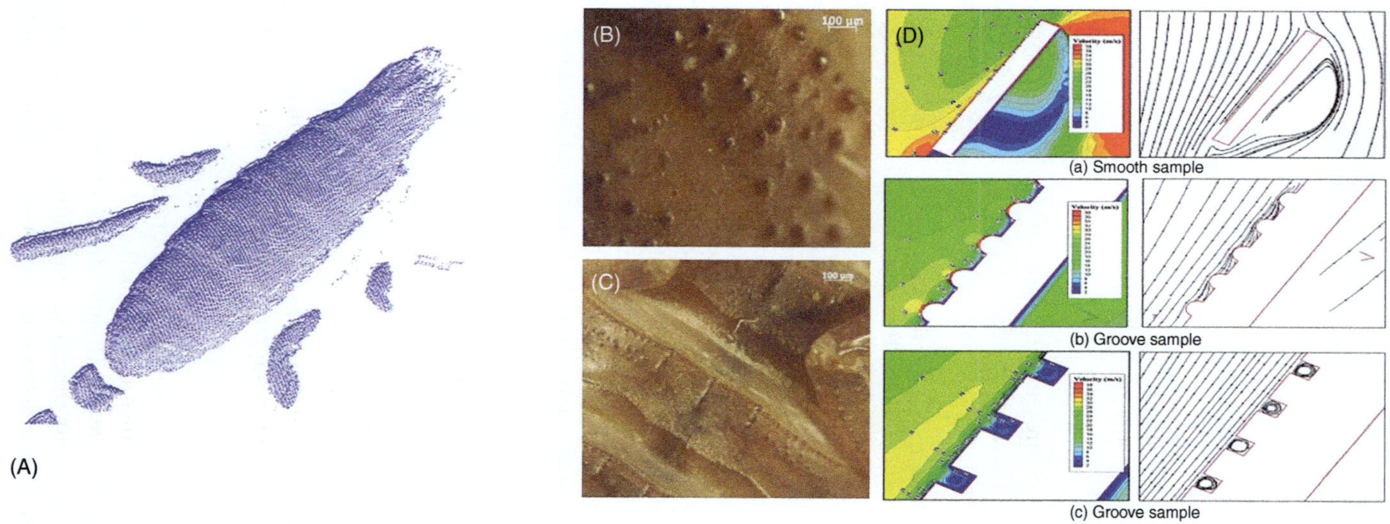

Figure 1.7 The dorsal surface of the scorpion. (a) Scanned data using a laser scanner. (b) The convex hull of scorpion back. (c) The groove of scorpion back. (d) Mechanism of the anti-wear surface of the scorpion: the air is rotating in the groove channel, forming a stable low-speed reverse flow zone. Source: Adapted from Han et al. [118] with permission from the American Chemical Society.

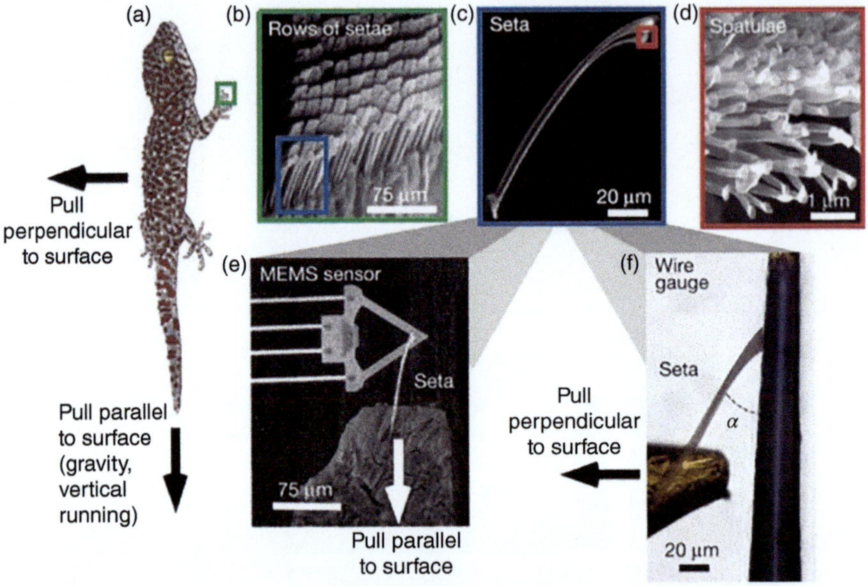

Figure 1.8 Gecko setae and apparatus for force measurement. (a) Photo of the Tokay gecko (*Gekko gecko*). (b) SEM image of arrays of setae from a toe. (c) A single seta. (d) The finest terminal spatula of a seta. (e) Single seta attached to a microelectromechanical system (MEMS) cantilever capable of measuring force production during attachment parallel and perpendicular to the surface. (f) Single seta attached to an aluminum bonding wire capable of measuring force production during detachment perpendicular to the surface. Source: Reproduced from Autumn et al. [64] with permission from Nature Publishing Group.

This hierarchical structure allows the gecko to attach to or detach from the surfaces at will. One explanation for the gecko's ability to control adhesion is that it can adapt to surface roughness and achieve a very large actual contact area between its feet and the surface [64, 125–130]. Also, compliance and adaptability of the setae contribute to high adhesion. This could inspire the innovative design of high-sensitive sensors for force measurement (Figure 1.8e and f).

1.3.9 Underwater Animals

Underwater animals, such as carp and shark, can swim freely owing to their special surface structures. For carp, the fan-shaped scales are covered by oriented nanostructured micropapillae (Figure 1.9), which not only has a drag reduction function but also has super lipophilicity in air and super oleophobicity in water [131, 132]. The surface of super-oleophobic fish originates from the micro–nano hierarchical structure of the water phase. Sharkskin is a natural low-resistance surface model. It is covered by very small individual tooth-like scales called dermal denticles (little skin teeth), with prismatic longitudinal grooves (parallel to the direction of local water flow). These grooved scales reduce the formation of vortices present on a smooth surface, resulting in water moving efficiently over their surface [133, 134].

Figure 1.9 Scale structure on a shark. Source: Bechert et al. [133].

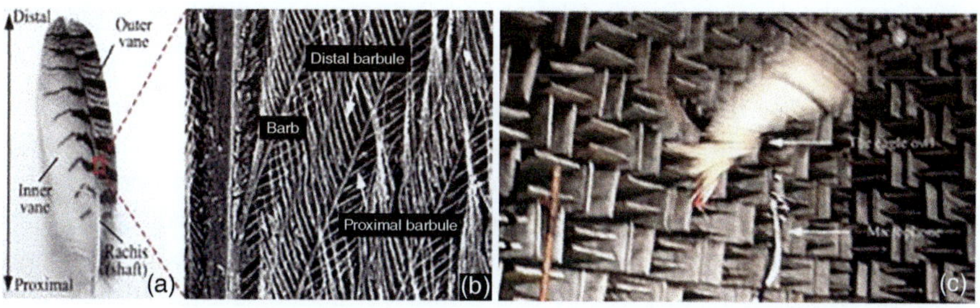

Figure 1.10 The wing feather of the eagle owl. (a) The rachis of the feather. (b) The barbules grow in different directions. (c) The eagle owl flight noise measurement. Source: Adapted from Chen et al. [135] with permission from Springer Nature.

1.3.10 Eagle Owl

Many species of owls can fly quietly. Acoustic measurements and microscopic observations on owls (*Bubo bubo*) [135] show that owls produce lower sound intensity and low-frequency flight noise, and owls' wing feathers have greater sound absorption characteristics. The microscopic structures of three special characteristics of feathers help to improve the pressure fluctuation of turbulence boundary and suppress the generation of vortex noise (Figure 1.10).

1.3.11 Desert *Stenocara* Beetle

In areas with limited water resources, such as the Namib Desert, nature has developed elegant solutions to collect water from the atmosphere. The superhydrophobic pattern on the back of the *Stenocara* beetle in the Namib Desert is a good example of micro-condensation of water [136]. The Stenocara beetle in the Namib Desert uses the hydrophilic/superhydrophobic patterned surface on its wings (Figure 1.11) to collect drinking water from the fog-filled wind.

The back of this beetle is composed of hydrophilic hills and super-hydrophobic channels. The former can collect water from the fog in the desert atmosphere, and the latter can help the collected water droplets flow into the beetle's mouth. After these small droplets converge into larger droplets, they roll into the beetle's

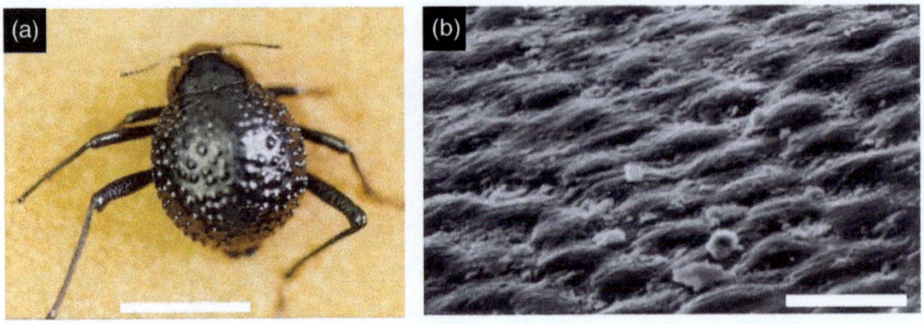

Figure 1.11 The water-capturing surface of the fused-over wings (elytra) of the desert beetle *Stenocara* sp. (a) Adult female, dorsal view, peaks and valleys are evident on the surface of the elytra. (b) SEM image of the textured surface of the depressed areas. Scale bars: (a) 10 mm and (b) 10 μm. Source: Adapted from Parker and Lawrence [136] with permission from Nature Publishing Group.

mouth and provide a fresh breakfast drink for the beetle [136, 137]. Research has shown that the formation of these large droplets is due to the uneven surface of the insect, which is composed of alternating hydrophobic, wax-coated and hydrophilic, non-wax areas. This fog-collecting structure design can be cheaply replicated on a commercial scale and can be applied to water-collecting tents and building coverings [136]. Inspired by this wonderful natural design, Rubner and coworkers produced a superhydrophobic/hydrophilic patterned surface to mimic the structure of the beetle's back [137]. The water sprayed on the superhydrophobic pattern only forms small spherical water droplets, which are mainly concentrated on the hydrophilic pattern. Later, Garrod et al. also demonstrated the preparation of superhydrophobic/hydrophilic patterned surfaces to collect water [138]. The water collection capacity of different superhydrophobic/hydrophilic ratios on the surface has been studied in detail. Through the above examples, the application prospect of superhydrophobic/hydrophilic pattern coatings in actual water collection devices can be predicted.

References

1 Bhushan, B. (2009). Biomimetics: lessons from nature – an overview. *Philosophical Transactions of the Royal Society A: Mathematical, Physical and Engineering Sciences* 367 (1893): 1445–1486.

2 Liu, K., Yao, X., and Jiang, L. (2010). Recent developments in bio-inspired special wettability. *Chemical Society Reviews* 39 (8): 3240–3255.

3 Huebsch, N. and Mooney, D.J. (2009). Inspiration and application in the evolution of biomaterials. *Nature* 462 (7272): 426–432.

4 Lee, L.P. and Szema, R. (2005). Inspirations from biological optics for advanced photonic systems. *Science* 310 (5751): 1148–1150.

5 Liu, M., Zheng, Y., Zhai, J., and Jiang, L. (2010). Bioinspired super-antiwetting interfaces with special liquid–solid adhesion. *Accounts of Chemical Research* 43 (3): 368–377.

6 Xia, F. and Jiang, L. (2008). Bio-inspired, smart, multiscale interfacial materials. *Advanced Materials* 20 (15): 2842–2858.

7 Li, Y., Zhang, J., and Yang, B. (2010). Antireflective surfaces based on biomimetic nanopillared arrays. *Nano Today* 5 (2): 117–127.

8 Liu, K. and Jiang, L. (2011). Metallic surfaces with special wettability. *Nanoscale* 3 (3): 825–838.

9 Raveendran, M., Lee, A.J., Sharma, R. et al. (2020). Rational design of DNA nanostructures for single molecule biosensing. *Nature Communications* 11 (1): 1–9.

10 Gao, Z.F., Liu, R., Wang, J. et al. (2020). Manipulating the hydrophobicity of DNA as a universal strategy for visual biosensing. *Nature Protocols* 15 (2): 316–337.

11 Tortora, M.M.C., Mishra, G., Prešern, D., and Doye, J.P.K. (2020). Chiral shape fluctuations and the origin of chirality in cholesteric phases of DNA origamis. *Science Advances* 6 (31): eaaw8331.

12 Zhang, P., Liu, G., and Chen, X. (2017). Nanobiotechnology: cell membrane-based delivery systems. *Nano Today* 13: 7–9.

13 Lv, Z., Zhou, Y., Han, S.-T., and Roy, V.A.L. (2018). From biomaterial-based data storage to bio-inspired artificial synapse. *Materials Today* 21 (5): 537–552.

14 Cressiot, B., Greive, S.J., Mojtabavi, M. et al. (2018). Thermostable virus portal proteins as reprogrammable adapters for solid-state nanopore sensors. *Nature Communications* 9 (1): 1–7.

15 Romero, E., Novoderezhkin, V.I., and van Grondelle, R. (2017). Quantum design of photosynthesis for bio-inspired solar-energy conversion. *Nature* 543 (7645): 355–365.

16 Wangpraseurt, D., You, S., Azam, F. et al. (2020). Bionic 3D printed corals. *Nature Communications* 11 (1): 1–8.

17 Chen, C., Kuang, Y., Zhu, S. et al. (2020). Structure–property–function relationships of natural and engineered wood. *Nature Reviews Materials* 5 (9): 642–666.

18 Cai, G., Ciou, J.-H., Liu, Y. et al. (2019). Leaf-inspired multiresponsive MXene-based actuator for programmable smart devices. *Science Advances* 5 (7): eaaw7956.

19 Rivera, J., Hosseini, M.S., Restrepo, D. et al. (2020). Toughening mechanisms of the elytra of the diabolical ironclad beetle. *Nature* 586 (7830): 543–548.

20 Chen, P.-Y. (2020). Tough lessons from diabolical beetles. *Nature* 586 (7830): 502–504.

21 Phan, H.V. and Park, H.C. (2020). Mechanisms of collision recovery in flying beetles and flapping-wing robots. *Science* 370 (6521): 1214–1219.

22 Siddique, R.H., Gomard, G., and Hölscher, H. (2015). The role of random nanostructures for the omnidirectional anti-reflection properties of the glasswing butterfly. *Nature Communications* 6 (1): 1–8.

23 Kolle, M., Salgard-Cunha, P.M., Scherer, M.R.J. et al. (2010). Mimicking the colourful wing scale structure of the *Papilio blumei* butterfly. *Nature Nanotechnology* 5 (7): 511–515.

24 Potyrailo, R.A., Ghiradella, H., Vertiatchikh, A. et al. (2007). *Morpho* butterfly wing scales demonstrate highly selective vapour response. *Nature Photonics* 1 (2): 123–128.

25 Sweeney, A., Jiggins, C., and Johnsen, S. (2003). Polarized light as a butterfly mating signal. *Nature* 423 (6935): 31–32.

26 Vukusic, P., Sambles, J.R., and Lawrence, C.R. (2000). Colour mixing in wing scales of a butterfly. *Nature* 404 (6777): 457.

27 Tao, P., Shang, W., Song, C. et al. (2015). Bioinspired engineering of thermal materials. *Advanced Materials* 27 (3): 428–463.

28 Goerlitzer, E.S.A., Klupp Taylor, R.N., and Vogel, N. (2018). Bioinspired photonic pigments from colloidal self-assembly. *Advanced Materials* 30 (28): 1706654.

29 Xiao, M., Hu, Z., Wang, Z. et al. (2017). Bioinspired bright noniridescent photonic melanin supraballs. *Science Advances* 3 (9): e1701151.

30 Wegst, U.G.K., Bai, H., Saiz, E. et al. (2015). Bioinspired structural materials. *Nature Materials* 14 (1): 23–36.

31 Gu, G.X., Takaffoli, M., and Buehler, M.J. (2017). Hierarchically enhanced impact resistance of bioinspired composites. *Advanced Materials* 29 (28): 1700060.

32 Connors, M., Yang, T., Hosny, A. et al. (2019). Bioinspired design of flexible armor based on chiton scales. *Nature Communications* 10 (1): 1–13.

33 Huang, W., Restrepo, D., Jung, J. et al. (2019). Multiscale toughening mechanisms in biological materials and bioinspired designs. *Advanced Materials* 31 (43): 1901561.

34 Tan, G., Zhang, J., Zheng, L. et al. (2019). Nature-inspired nacre-like composites combining human tooth-matching elasticity and hardness with exceptional damage tolerance. *Advanced Materials* 31 (52): 1904603.

35 Velasco-Hogan, A., Deheyn, D.D., Koch, M. et al. (2019). On the nature of the transparent teeth of the deep-sea dragonfish, *Aristostomias scintillans*. *Matter* 1 (1): 235–249.

36 Amini, S., Razi, H., Seidel, R. et al. (2020). Shape-preserving erosion controlled by the graded microarchitecture of shark tooth enameloid. *Nature Communications* 11 (1): 1–11.

37 Wang, X., Yang, B., Tan, D. et al. (2020). Bioinspired footed soft robot with unidirectional all-terrain mobility. *Materials Today* 35: 42–49.

38 Liimatainen, V., Drotlef, D., Son, D., and Sitti, M. (2020). Liquid-superrepellent bioinspired fibrillar adhesives. *Advanced Materials* 32 (19): 2000497.

39 Nakata, T., Phillips, N., Simões, P. et al. (2020). Aerodynamic imaging by mosquitoes inspires a surface detector for autonomous flying vehicles. *Science* 368 (6491): 634–637.

40 Dou, S., Xu, H., Zhao, J. et al. (2021). Bioinspired microstructured materials for optical and thermal regulation. *Advanced Materials* 33 (6): 2000697.

41 Shen, Q., Luo, Z., Ma, S. et al. (2018). Bioinspired infrared sensing materials and systems. *Advanced Materials* 30 (28): 1707632.

42 Chen, C. and Hu, L. (2019). Super elastic and thermally insulating carbon aerogel: go tubular like polar bear hair. *Matter* 1 (1): 36–38.
43 Zhan, H., Wu, K., Hu, Y. et al. (2019). Biomimetic carbon tube aerogel enables super-elasticity and thermal insulation. *Chem* 5 (7): 1871–1882.
44 Wang, L., Chen, D., Jiang, K., and Shen, G. (2017). New insights and perspectives into biological materials for flexible electronics. *Chemical Society Reviews* 46 (22): 6764–6815.
45 Zhou, H., Fan, T., and Zhang, D. (2011). Biotemplated materials for sustainable energy and environment: current status and challenges. *ChemSusChem* 4 (10): 1344–1387.
46 Faber, J.A., Arrieta, A.F., and Studart, A.R. (2018). Bioinspired spring origami. *Science* 359 (6382): 1386–1391.
47 Artero, V. (2017). Bioinspired catalytic materials for energy-relevant conversions. *Nature Energy* 2 (9): 1–6.
48 Fu, T., Liu, X., Gao, H. et al. (2020). Bioinspired bio-voltage memristors. *Nature Communications* 11 (1): 1–10.
49 Liu, X., Zhang, F., Jing, X. et al. (2018). Complex silica composite nanomaterials templated with DNA origami. *Nature* 559 (7715): 593–598.
50 Li, P., Xie, G., Kong, X. et al. (2016). Light-controlled ion transport through biomimetic DNA-based channels. *Angewandte Chemie* 128 (50): 15866–15870.
51 Cheng, Q., Jiang, L., and Tang, Z. (2014). Bioinspired layered materials with superior mechanical performance. *Accounts of Chemical Research* 47 (4): 1256–1266.
52 Hong, S., Lee, M.Y., Jackson, A.O., and Lee, L.P. (2015). Bioinspired optical antennas: gold plant viruses. *Light: Science & Applications* 4 (3): e267.
53 Epstein, A.K., Pokroy, B., Seminara, A., and Aizenberg, J. (2011). Bacterial biofilm shows persistent resistance to liquid wetting and gas penetration. *Proceedings of the National Academy of Sciences of the United States of America* 108 (3): 995–1000.
54 Lopez-Garcia, M., Masters, N., O'Brien, H.E. et al. (2018). Light-induced dynamic structural color by intracellular 3D photonic crystals in brown algae. *Science Advances* 4 (4): eaan8917.
55 Song, J., Chen, C., Zhu, S. et al. (2018). Processing bulk natural wood into a high-performance structural material. *Nature* 554 (7691): 224–228.
56 Chen, C., Zhang, Y., Li, Y. et al. (2017). All-wood, low tortuosity, aqueous, biodegradable supercapacitors with ultra-high capacitance. *Energy & Environmental Science* 10 (2): 538–545.
57 Zhu, M., Song, J., Li, T. et al. (2016). Highly anisotropic, highly transparent wood composites. *Advanced Materials* 28 (26): 5181–5187.
58 Chen, L., Guo, Z., and Liu, W. (2016). Biomimetic multi-functional superamphiphobic FOTS-TiO_2 particles beyond lotus leaf. *ACS Applied Materials & Interfaces* 8 (40): 27188–27198.
59 Ueda, E. and Levkin, P.A. (2013). Emerging applications of superhydrophilic–superhydrophobic micropatterns. *Advanced Materials* 25 (9): 1234–1247.

60 Yang, S., Ju, J., Qiu, Y. et al. (2014). Peanut leaf inspired multifunctional surfaces. *Small* 10 (2): 294–299.

61 Bixler, G.D. and Bhushan, B. (2013). Fluid drag reduction and efficient self-cleaning with rice leaf and butterfly wing bioinspired surfaces. *Nanoscale* 5 (17): 7685–7710.

62 Barthlott, W. and Neinhuis, C. (1997). Purity of the sacred lotus, or escape from contamination in biological surfaces. *Planta* 202 (1): 1–8.

63 Zheng, Y., Bai, H., Huang, Z. et al. (2010). Directional water collection on wetted spider silk. *Nature* 463 (7281): 640–643.

64 Autumn, K., Liang, Y.A., Hsieh, S.T. et al. (2000). Adhesive force of a single gecko foot-hair. *Nature* 405 (6787): 681–685.

65 Geim, A.K., Dubonos, S.V., Grigorieva, I.V. et al. (2003). Microfabricated adhesive mimicking gecko foot-hair. *Nature Materials* 2 (7): 461–463.

66 Peisker, H., Michels, J., and Gorb, S.N. (2013). Evidence for a material gradient in the adhesive tarsal setae of the ladybird beetle *Coccinella septempunctata*. *Nature Communications* 4 (1): 1661.

67 Luke, S.M., Benny, T.H., and Vukusic, P. (2010). Structural optimization for broadband scattering in several ultra-thin white beetle scales. *Applied Optics* 49 (22): 4246–4254.

68 Cortese, L., Pattelli, L., Utel, F. et al. (2015). Anisotropic light transport in White Beetle Scales. *Advanced Optical Materials* 3 (10): 1337–1341.

69 Kinoshita, S., Yoshioka, S., and Kawagoe, K. (2002). Mechanisms of structural colour in the *Morpho* butterfly: cooperation of regularity and irregularity in an iridescent scale. *Proceedings of the Royal Society B: Biological Sciences* 269 (1499): 1417–1421.

70 Werner, T., Koshikawa, S., Williams, T.M., and Carroll, S.B. (2010). Generation of a novel wing colour pattern by the Wingless morphogen. *Nature* 464 (7292): 1143–1148.

71 Liu, F., Shi, W., Hu, X., and Dong, B. (2013). Hybrid structures and optical effects in *Morpho* scales with thin and thick coatings using an atomic layer deposition method. *Optics Communications* 291: 416–423.

72 Dhungel, B. and Otaki, J.M. (2014). Morphometric analysis of nymphalid butterfly wings: number, size and arrangement of scales, and their implications for tissue-size determination. *Entomological Science* 17 (2): 207–218.

73 Wu, W., Shi, T., Liao, G., and Zuo, H. (2011). Research on spectral reflection characteristics of nanostructures in *Morpho* butterfly wing scale. *Journal of Physics Conference Series* 276 (1): 012049.

74 Liao, G., Zuo, H., Cao, Y., and Shi, T. (2010). Optical properties of the micro/nano structures of *Morpho* butterfly wing scales. *Science China Technological Sciences* 53 (1): 175–181.

75 Zhu, D., Kinoshita, S., Cai, D., and Cole, J.B. (2009). Investigation of structural colors in *Morpho* butterflies using the nonstandard-finite-difference time-domain method: effects of alternately stacked shelves and ridge density. *Physical Review E – Statistical, Nonlinear, and Soft Matter Physics* 80 (5): 1–12.

76 Jiang, X., Shi, T., Zuo, H. et al. (2012). Investigation on color variation of *Morpho* butterfly wings hierarchical structure based on PCA. *Science China Technological Sciences* 55 (1): 16–21.

77 Van Hooijdonk, E., Barthou, C., Vigneron, J.P., and Berthier, S. (2011). Detailed experimental analysis of the structural fluorescence in the butterfly *Morpho sulkowskyi* (Nymphalidae). *Journal of Nanophotonics* 5 (1): 053525.

78 Giraldo, M.A. and Stavenga, D.G. (2016). Brilliant iridescence of *Morpho* butterfly wing scales is due to both a thin film lower lamina and a multilayered upper lamina. *Journal of Comparative Physiology A: Neuroethology, Sensory, Neural, and Behavioral Physiology* 202 (5): 381–388.

79 Zhu, D. (2017). Reflection characterization of nano-sized dielectric structure in *Morpho* butterfly wings. *Journal of Nanophotonics* 11 (4): 043503.

80 Giraldo, M.A., Yoshioka, S., Liu, C., and Stavenga, D.G. (2016). Coloration mechanisms and phylogeny of *Morpho* butterflies. *Journal of Experimental Biology* 219 (24): 3936–3944.

81 Vukusic, P., Sambles, R., Lawrence, C., and Wakely, G. (2001). Sculpted-multilayer optical effects in two species of *Papilio* butterfly. *Applied Optics* 40 (7): 1116.

82 Plattner, L. (2004). Optical properties of the scales of *Morpho rhetenor* butterflies: theoretical and experimental investigation of the back-scattering of light in the visible spectrum. *Journal of the Royal Society Interface* 1 (1): 49–59.

83 Vukusic, P., Sambles, J.R., Lawrence, C.R., and Wootton, R.J. (1999). Quantified interference and diffraction in single *Morpho* butterfly scales. *Proceedings of the Royal Society B: Biological Sciences* 266 (1427): 1403–1411.

84 Liu, Y., Huang, L., and Shi, W. (2012). Structural color bio-engineering by replicating *Morpho* wings. *Advanced Materials Research* 391, 392: 409–417.

85 Zhang, S. and Chen, Y. (2015). Nanofabrication and coloration study of artificial *Morpho* butterfly wings with aligned lamellae layers. *Scientific Reports* 5: 1–10.

86 Song, B., Johansen, V.E., Sigmund, O., and Shin, J.H. (2017). Reproducing the hierarchy of disorder for *Morpho*-inspired, broad-angle color reflection. *Scientific Reports* 7 (1): 46023.

87 Siddique, R.H., Hünig, R., Faisal, A. et al. (2015). Fabrication of hierarchical photonic nanostructures inspired by *Morpho* butterflies utilizing laser interference lithography. *Optical Materials Express* 5 (5): 996–1005.

88 Song, B., Eom, S.C., and Shin, J.H. (2014). Disorder, and broad-angle iridescence from *Morpho*-inspired structures. *Optics Express* 22 (16): 19386–19400.

89 Mohri, Y., Kobashi, J., Yoshida, H., and Ozaki, M. (2017). *Morpho*-butterfly-inspired patterning of helical photonic structures for circular-polarization-sensitive, wide-angle diffuse reflection. *Advanced Optical Materials* 5 (7): 1601071.

90 Rodriguez, R.E., Agarwal, S.P., An, S. et al. (2018). Biotemplated *Morpho* butterfly wings for tunable structurally colored photocatalysts. *ACS Applied Materials & Interfaces* 10 (5): 4614–4621.

91 Wanasekara, N.D. and Chalivendra, V.B. (2011). Role of surface roughness on wettability and coefficient of restitution in butterfly wings. *Soft Matter* 7 (2): 373–379.

92 Zheng, Y., Gao, X., and Jiang, L. (2007). Directional adhesion of superhydrophobic butterfly wings. *Soft Matter* 3 (2): 178–182.

93 Cao, M., Jin, X., Peng, Y. et al. (2017). Unidirectional wetting properties on multi-bioinspired magnetocontrollable slippery microcilia. *Advanced Materials* 29 (23): 1606869.

94 Liimatainen, V., Vuckovac, M., Jokinen, V. et al. (2017). Mapping microscale wetting variations on biological and synthetic water-repellent surfaces. *Nature Communications* 8: 1798.

95 Potyrailo, R.A., Starkey, T.A., Vukusic, P. et al. (2013). Discovery of the surface polarity gradient on iridescent *Morpho* butterfly scales reveals a mechanism of their selective vapor response. *Proceedings of the National Academy of Sciences of the United States of America* 110 (39): 15567–15572.

96 Binetti, V.R., Schiffman, J.D., Leaffer, O.D. et al. (2009). The natural transparency and piezoelectric response of the *Greta oto* butterfly wing. *Integrative Biology* 1 (4): 324–329.

97 Kertesz, K., Piszter, G., Horvath, Z.E. et al. (2017). Changes in structural, and pigmentary colours in response to cold stress in *Polyommatus icarus* butterflies. *Scientific Reports* 7: 1118.

98 Wang, W., Wang, G.P., Zhang, W., and Zhang, D. (2018). Reversible thermochromic response based on photonic crystal structure in butterfly wing. *Nanophotonics* 7 (1): 217–227.

99 Grujić, D., Vasiljević, D., Pantelić, D. et al. (2018). Infrared camera on a butterfly's wing. *Optics Express* 26 (11): 14143–14158.

100 Fang, J., Gu, J., Liu, Q. et al. (2018). Three-dimensional CdS/Au butterfly wing scales with hierarchical rib structures for plasmon-enhanced photocatalytic hydrogen production. *ACS Applied Materials & Interfaces* 10 (23): 19649–19655.

101 Han, Z., Li, B., Mu, Z. et al. (2017). Energy-efficient oil–water separation of biomimetic copper membrane with multiscale hierarchical dendritic structures. *Small* 13 (34): 1701121.

102 Huang, Y.F., Jen, Y.J., Chen, L.C. et al. (2015). Design for approaching cicada-wing reflectance in low- and high-index biomimetic nanostructures. *ACS Nano* 9 (1): 301–311.

103 Tadepalli, S., Slocik, J.M., Gupta, M.K. et al. (2017). Bio-optics and bio-inspired optical materials. *Chemical Reviews* 117 (20): 12705–12763.

104 Bernhard, C.G. and Miller, W.H. (1962). A corneal nipple pattern in insect compound eyes. *Acta Physiologica Scandinavica* 56: 385–386.

105 Wilson, S. and Hutley, M. (1982). The optical properties of 'moth eye' antireflection surfaces. *Acta Ophthalmologica* 29 (7): 993–1009.

106 Ko, D.H., Tumbleston, J.R., Henderson, K.J. et al. (2011). Biomimetic microlens array with antireflective "moth-eye" surface. *Soft Matter* 7 (14): 6404–6407.

107 Stavenga, D.G., Foletti, S., Palasantzas, G., and Arikawa, K. (2006). Light on the moth-eye corneal nipple array of butterflies. *Proceedings of the Royal Society B: Biological Sciences* 273 (1587): 661–667.

108 Sun, C., Jiang, P., and Jiang, B. (2008). Broadband moth-eye antireflection coatings on silicon. *Applied Physics Letters* 92 (6): 23–25.

109 Clapham, P. and Hutley, M. (1973). Reduction of lens reflexion by the "moth eye" principle. *Nature* 244 (5414): 281–282.

110 Han, Z., Guan, H., Cao, Y. et al. (2014). Antifogging properties and mechanism of micron structure in *Ephemera pictiventris* McLachlan compound eyes. *Chinese Science Bulletin* 59 (17): 2039–2044.

111 Gao, X., Yan, X., Yao, X. et al. (2007). The dry-style antifogging properties of mosquito compound eyes and artificial analogues prepared by soft lithography. *Advanced Materials* 19 (17): 2213–2217.

112 Wang, Q., Yao, X., Liu, H. et al. (2015). Self-removal of condensed water on the legs of water striders. *Proceedings of the National Academy of Sciences of the United States of America* 112 (30): 9247–9252.

113 Gao, X. and Jiang, L. (2004). Water-repellent legs of water striders. *Nature* 432 (7013): 36–36.

114 Hu, D. and Bush, J.W.M. (2010). The hydrodynamics of water-walking arthropods. *Journal of Fluid Mechanics* 644: 5–33.

115 Gao, P. and Feng, J. (2011). A numerical investigation of the propulsion of water walkers. *Journal of Fluid Mechanics* 668: 363–383.

116 Suter, R.B., Rosenberg, O., Loeb, S. et al. (1997). Locomotion on the water surface: propulsive mechanisms of the fisher spider *Dolomedes triton*. *Journal of Experimental Biology* 200 (19): 2523–2538.

117 Nosonovsky, M. and Bhushan, B. (2010). Green tribology: principles, research areas and challenges. *Philosophical Transactions of the Royal Society A: Mathematical, Physical and Engineering Sciences* 368 (1929): 4677–4694.

118 Han, Z., Zhang, J., Ge, C. et al. (2012). Erosion resistance of bionic functional surfaces inspired from desert scorpions. *Langmuir* 28 (5): 2914–2921.

119 Han, Z., Zhang, J., Ge, C. et al. (2010). Anti-erosion function in animals and its biomimetic application. *Journal of Bionic Engineering* 7 (4): S50–S58.

120 Meng, F., Zhao, R., Zhan, Y., and Liu, X. (2011). Design of thorn-like micro/nanofibers: fabrication and controlled morphology for engineered composite materials applications. *Journal of Materials Chemistry* 21 (41): 16385–16390.

121 Bhushan, B. and Jung, Y.C. (2011). Natural and biomimetic artificial surfaces for superhydrophobicity, self-cleaning, low adhesion, and drag reduction. *Progress in Materials Science* 56 (1): 1–108.

122 Gao, H., Wang, X., Yao, H. et al. (2005). Mechanics of hierarchical adhesion structures of geckos. *Mechanics of Materials* 37 (2, 3): 275–285.

123 Bhushan, B., Peressadko, A.G., and Kim, T.-W. (2006). Adhesion analysis of two-level hierarchical morphology in natural attachment systems for 'smart adhesion'. *Journal of Adhesion Science and Technology* 20 (13): 1475–1491.

124 Autumn, K. (2006). How gecko toes stick – the powerful, fantastic adhesive used by geckos is made of nanoscale hairs that engage tiny forces, inspiring envy among human imitators. *American Scientist* 94 (2): 124–132.

125 Bhushan, B. and Sayer, R.A. (2006). Surface characterization and friction of a bio-inspired reversible adhesive tape. *Microsystem Technologies* 13 (1): 71–78.

126 Kim, T.W. and Bhushan, B. (2007). Adhesion analysis of multi-level hierarchical attachment system contacting with a rough surface. *Journal of Adhesion Science and Technology* 21 (1): 1–20.

127 Kim, T.W. and Bhushan, B. (2007). Effect of stiffness of multi-level hierarchical attachment system on adhesion enhancement. *Ultramicroscopy* 107 (10, 11): 902–912.

128 Fratzl, P. (2007). Biomimetic materials research: what can we really learn from nature's structural materials. *Journal of the Royal Society Interface* 4 (15): 637–642.

129 Filippov, A.E. and Gorb, S.N. (2015). Spatial model of the gecko foot hair: functional significance of highly specialized non-uniform geometry. *Interface Focus* 5 (1): 20140065.

130 Bhushan, B. and Sayer, R.A. (2007). Gecko feet: natural attachment systems for smart adhesion. In: *Applied Scanning Probe Methods VII* (ed. B. Bhushan and H. Fuchs), 41–76. Springer.

131 Liu, M., Wang, S., Wei, Z. et al. (2009). Bioinspired design of a superoleophobic and low adhesive water/solid interface. *Advanced Materials* 21 (6): 665–669.

132 Liu, K. and Jiang, L. (2011). Bio-inspired design of multiscale structures for function integration. *Nano Today* 6 (2): 155–175.

133 Bechert, D.W., Bruse, M., and Hage, W. (2000). Experiments with three-dimensional riblets as an idealized model of shark skin. *Experiments in Fluids* 28 (5): 403–412.

134 Dean, B. and Bhushan, B. (2010). Shark-skin surfaces for fluid-drag reduction in turbulent flow: a review. *Philosophical Transactions of the Royal Society A – Mathematical Physical and Engineering Sciences* 368 (1929): 4775–4806.

135 Chen, K., Liu, Q., Liao, G. et al. (2012). The sound suppression characteristics of wing feather of owl (*Bubo bubo*). *Journal of Bionic Engineering* 9 (2): 192–199.

136 Parker, A.R. and Lawrence, C.R. (2001). Water capture by a desert beetle. *Nature* 414 (6859): 33–34.

137 Zhai, L., Berg, M.C., Cebeci, F.C. et al. (2006). Patterned superhydrophobic surfaces: toward a synthetic mimic of the Namib Desert beetle. *Nano Letters* 6 (6): 1213–1217.

138 Garrod, R.P., Harris, L.G., Schofield, W.C.E. et al. (2007). Mimicking a *Stenocara* beetle's back for microcondensation using plasmachemical patterned superhydrophobic–superhydrophilic surfaces. *Langmuir* 23 (2): 689–693.

2

Characterization, Analysis, Modeling, and Fabrication of NIFSS

2.1 Characterization Techniques and Analysis Methods of NIFSS

2.1.1 Preparation of Biological Prototypes

Since biological surfaces are our object of in-depth investigation, how to collect the biological prototypes is a primary and important issue. Creatures living in harsh environments should be valuable choices for bionic study due to their spectacular surfaces evolved after millions of years of struggle with brutal nature. However, it should be noted that the conditions where these special creatures live are also being unfriendly to human beings. Although capturing live creatures like butterflies in the wild could be the most common approach to collect biological prototypes, professional practices are necessary to capturers to protect the live status or organ integrity of the biological prototypes. Here we have listed several commonly used approaches to collect target biological prototypes.

As mentioned above, collecting biological prototypes in the wild is a big challenge for fresh researchers; sometimes it's unexpected and dangerous, especially the creatures living in harsh environments that are far away from the activity space of human beings and full of uncertainties. Moreover, capturers should be very familiar with the habits of biological prototypes such as snakes, spiders, scorpions, to avoid possible fatal risks. Also, some special creatures are distributed in certain gathering places that are not easy to access due to distance, climates, altitude, safety, etc. For example, *Morpho* butterflies are mainly distributed in rainforests around the world. Thus, collecting in the wild often applies to those common creatures in daily life. For some common biological prototypes, we can find special online stores offering biological specimens. To some degree, these biological specimens are adequate for scientific research of NIFSS, especially for those biological surfaces that are not related to the living state of the original biological prototypes. Museums are good places to acquire information about biological prototypes due to the abundant inventory of specimens including various animals and plants. Although it is hard to get permission to occupy the biological specimens for study, taking photos is generally allowed and is helpful to identify the feature structures of target biological prototypes.

Nature-Inspired Structured Functional Surfaces: Design, Fabrication, Characterization, and Applications, First Edition. Zhiwu Han.

Preservation of biological prototypes is essential because surface research is usually time-consuming. Thus, specimens should be preserved for a relatively long period for further study. For the living biological prototypes, understanding their habits and creating an analogous environment are obligatory to keep them alive and even help them multiply. Specimens are usually preserved under low temperature, dry, and ventilating conditions to keep them fresh as much as possible.

Before starting surface characterization of the target biological prototypes, surface pretreatment is the key step to obtain satisfactory characterization results. Certainly, various physical and chemical methods can be applied to pretreat the surfaces based on the features of the selected biological prototypes. For instance, since the size of a butterfly is too large, it is not suitable to take the whole butterfly as a research object. To facilitate subsequent test characterization, butterfly wings need to be selected as an initial test material. The wings must be cleaned after the test and characterization because the wings of the butterfly sample may contaminants. These interference components may have originated from impurities such as environmental dust around the butterfly and may also be some biological components carried by the butterfly. Specifically, the cleaning process of the entire butterfly wing can be performed as follows:

First, using the scalpel the forewing of the butterfly is cut neatly from the root of the wing. Then, the removed fore wing entirely is placed in a glass petri dish, an appropriate amount of preconfigured ethanol solution is poured into the petri dish, and it should be ensured that the wings are fully immersed in the ethanol solution for 10–15 minutes to do the preliminary cleaning. Secondly, the remaining ethanol solution in the culture dish is taken out by a plastic dropper, and the residual ethanol solution on the surface of the wings is volatilized, so that the wings are naturally dried in the air. Then, an appropriate amount of ether reagent is used to soak the dried butterfly wings again for 10–15 minutes, and the wings are cleaned twice to remove the remaining ether. Next, by repeating the washing process, on the one hand, the ether reagent on the wing surface can be washed away, and on the other hand, the wings are also subjected to secondary cleaning and drying treatment. Finally, the area with uniform color and flat fins is selected, and the green plaque area of the wings is cut using the scalpel in the direction of the wing veins for further characterization.

2.1.2 Characterization Techniques of NIFSS

2.1.2.1 Optical Microscopy (OM) Technique

Optical microscope (OM) is an optical instrument that uses optical principles to magnify and image small objects that cannot be distinguished by the human eye to extract fine structural information. The optical system of the microscope mainly includes four parts: an objective lens, an eyepiece, a mirror, and a concentrator. The mechanical mechanism of the microscope is the important part. Its role is to fix and adjust the optical lens, fixed and moving specimens. It mainly consists of a mirror base, a mirror arm, a stage, a lens barrel, an objective lens converter, and a focusing device. Researchers usually adopt OM for the elementary characterization of the

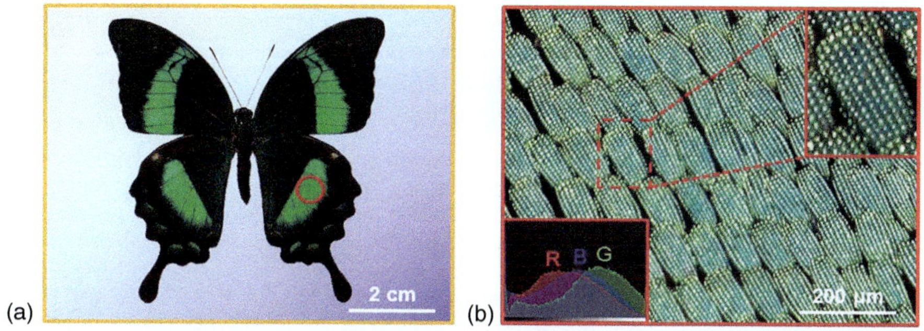

Figure 2.1 (a) A photograph of an intact male *Papilio palinurus* butterfly exhibiting brilliant axisymmetric green scale regions. (b) 3D ultradepth stereoscopic microscope image of the green scale region and its three primary colors (RGB) distribution. Source: Reproduced with permission from Han et al. [1]. Copyright © 2017 American Chemical Society.

NIFSS. Han et al. used a 3D ultradepth stereoscopic microscope to obtain the image of the green scale region of male *Papilio palinurus* butterfly [1] (Figure 2.1).

2.1.2.2 Field Emission Scanning Electron Microscope (FESEM) Technique

Field emission scanning electron microscope (FESEM) is a type of electron microscope. The instrument has ultrahigh resolution and can perform secondary electron image, reflection electron image observation, and image processing of various solid sample surface topography. The instrument utilizes the principle of secondary electron imaging to obtain a loyal and original 3D appearance by observing biological samples such as tissues, cells, microorganisms, and biological macromolecules at the nanoscale on a plated or uncoated basis. Ultrafine topography structural information of the sample surface can also be obtained. It has a high-performance X-ray energy spectrometer, which can simultaneously perform qualitative, semiquantitative, and quantitative analysis of micro-area line elements on the surface of the sample, and has comprehensive analysis capabilities of morphology and chemical composition. This technique is preferred for most surface characterizations, especially characterizations of feature structures. Han et al. used this method to characterize the ultrafine multiscale hierarchical micro-/nanostructures of *Morpho menelaus terrestris* butterfly [2] (Figure 2.2), and more details of feature structures can be identified clearly under FESEM.

2.1.2.3 Scanning Electron Microscope (SEM) Technique

Scanning electron microscope (SEM) is an observation method between transmission electron microscope (TEM) and optical microscope. It uses a narrow-focused high-energy electron beam to scan the sample, through the interaction between the beam and the material, to stimulate a variety of physical information, to collect, amplify, and reimage the information, for micromorphology characterization of a material. The resolution of the new SEM can reach 1 nm and the magnification can reach 3×10^5 times or more and can be adjusted continuously. In addition, the combination of SEM and other analytical instruments can be used to observe

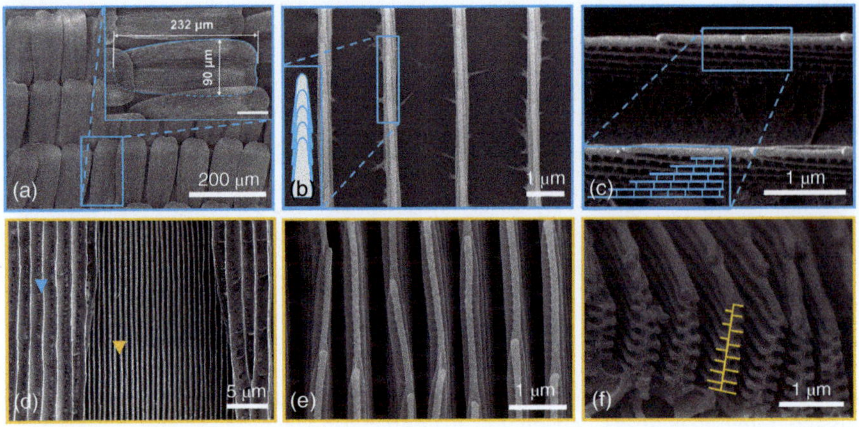

Figure 2.2 FESEM images of *Morpho* butterfly wings demonstrating feature structures of the original cover scales (a–c) and ground scales (d–f). (a) Shapes and distributions of cover scales on the wing substrate and a single scale (inset) showing its feature sizes. (b) Periodic parallel ridges arrange along the length of the wing scales. The overlapped lamellae (inset) constitute narrow ridges. (c) A side view of the ridges shows the ultrafine window-like structures (inset) between two adjacent lamellae. (d) Overlapping region of cover scales (blue triangle) and ground scales (yellow triangle). (e) Denser ridges with multiscale hierarchical structures. (f) Cross section of the ridges demonstrating pagoda-like structures (yellow profile). Source: Reproduced with permission from Han et al. [2]. Copyright © 2016 American Chemical Society.

the microscale morphology and analyze the microscale composition of the material at the same time. SEM is widely used in the research of geotechnical, graphite, ceramics, and nanomaterials. Therefore, SEM plays an important role in the field of characterization and analysis of NIFSS.

2.1.2.4 Transmission Electron Microscope (TEM) Technique

TEM can be used to see fine structures of less than 0.2 μm that are invisible under an optical microscope. These structures are called submicroscopic structures or ultrastructures. To see these structures, a light source with a shorter wavelength should be chosen to increase the resolution of the microscope. In 1932, Ruska invented a TEM using an electron beam as a light source. The wavelength of the electron beam is much shorter than that of visible light and ultraviolet light, and the wavelength of the electron beam is inversely proportional to the square root of the voltage of the emitted electron beam, that is, the higher the voltage, the shorter the wavelength. At present, the resolution of TEM can reach 0.2 nm. Han et al. applied the TEM technique to obtain the cross-sectional structural information of the cover scales of *Parnassius nomion* butterfly [3] as shown in Figure 2.3.

2.1.2.5 X-ray Diffraction (XRD) Technique

X-ray diffraction (XRD) of the target material is used to analyze the diffraction pattern to obtain information about the material composition and the structure or morphology of the atoms or molecules inside the material. XRD is generally used to

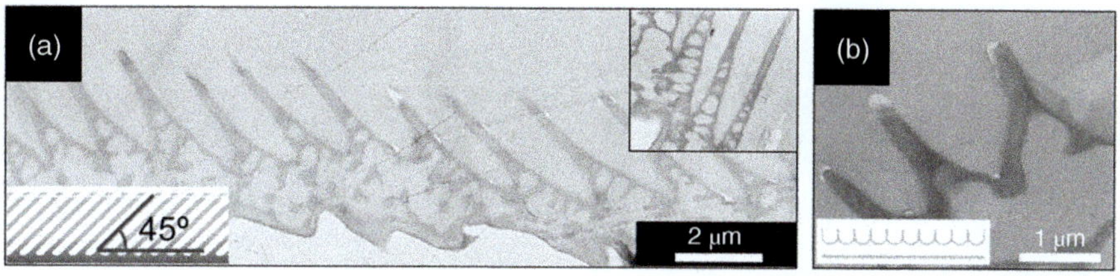

Figure 2.3 (a) The cross-sectional images of cover scales of butterfly wings along their ridges. The inclination angle of each comb teeth relative to their flat base plane is about 45°. (b) The U-type architecture of the section is perpendicular to the ridges. Source: Reproduced with permission from Han et al. [3]. Copyright © 2013 AIP Publishing LLC.

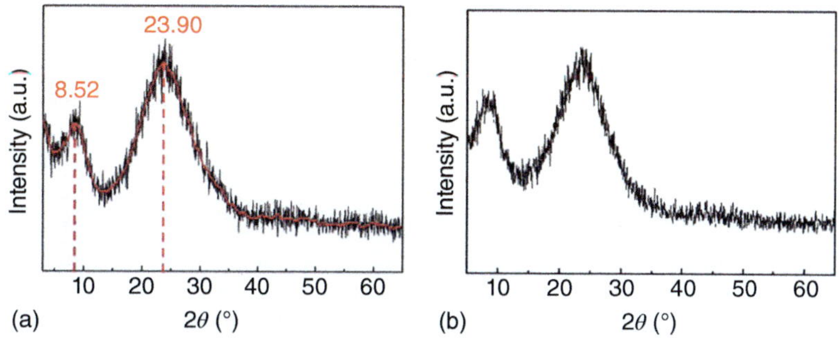

Figure 2.4 (a) XRD spectrum of multiscale hierarchical pagoda structures presents its quasi-PC structures and corresponding grazing angles. Source: Reproduced with permission from Han et al. [2]. Copyright © 2016 American Chemical Society. (b) XRD spectrum of the SiO_2 negative replica. Source: Reproduced with permission from Han et al. [4]. Copyright © 2015 Elsevier B.V.

determine the crystal structure. The crystal structure causes the incident X-ray beam to be diffracted into many specific directions. By measuring the angle and intensity of these diffracted beams, the crystallographer can produce a 3D image of the electron density within the crystal. Based on this electron density, the average position of the atoms in the crystal, as well as their chemical bonds and various other information, can be determined. To determine the periodical structures of butterfly wings, Han et al. used XRD to confirm the structural information of wing scales and their artificial replicas [2, 4] (Figure 2.4).

2.1.2.6 Atomic Force Microscope (AFM) Technique

Atomic force microscope (AFM) is a new atomic-level high-resolution instrument invented after the scanning tunneling microscope. AFM can measure various materials within nanometer range in atmospheric and liquid environments. The physical properties of the region include profiling or direct nanomanipulation. AFM is now widely used in the research and experiments of various nano-related disciplines in semiconductors, nanofunctional materials, biology, chemical, food, medical research, and scientific research institutes. It has become the basic tool for nanoscience research. For the study of highly efficient mechanoelectrical energy

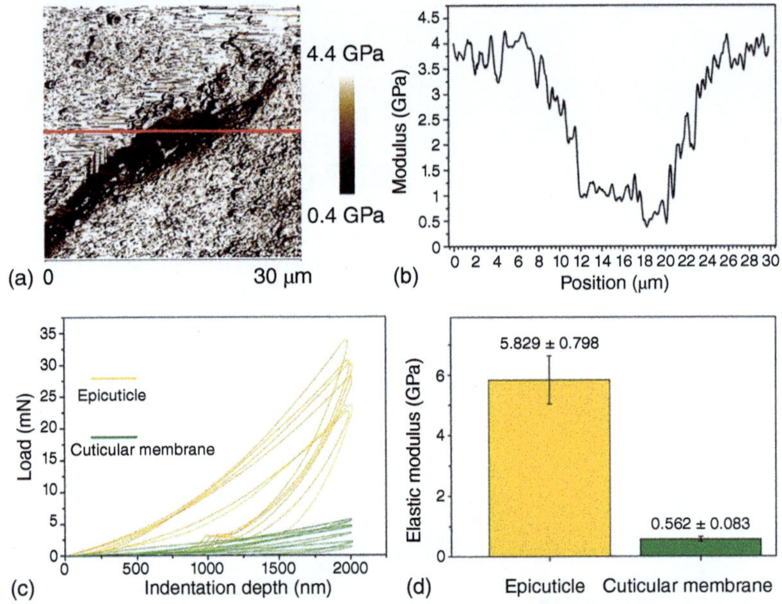

Figure 2.5 The elastic modulus distribution of the basitarsal compound slit sensilla (BCSS) characterized by using the AFM technique. (a) Elastic modulus distribution map obtained by peak-force QNM mode in the local area of the BCSS near the slit's tip. (b) Sectional profiles of elastic modulus data taken from the location denoted by the red line. (c) Representative nanoindentation load–displacement curves at 2 μm indentation depth. The yellow curves are obtained from the epicuticle above the block, and the green curves are from the cuticular membrane covering the slit. (d) Elastic modulus of epicuticle (yellow) and cuticular membrane (green), respectively. Source: Reproduced with permission from Wang et al. [5]. Copyright © 2019 WILEY-VCH Verlag GmbH & Co. KGaA, Weinheim.

conversion based on the near-tip stress field of an anti-fracture slit observed in scorpions, Wang et al. acquired high-resolution tomography imaging and conducted elastic measurements using a Dimension Icon AFM system. An AFM probe with a spring constant of $0.7\,\mathrm{N\,m^{-1}}$ was used, and the actual value was calibrated by a thermal tune method before each measurement. The AFM scanning mode based on quantitative nanomechanical mapping (QNM) and Peak Force QNM was adopted in the measurements. The elastic mapping image of samples was obtained using the Derjaguin–Müller–Toporov mode. The elastic modulus mapping image was calculated according to the bias of the force curve from the baseline. All the mapping images were obtained simultaneously with the AFM topography imaging [5] as shown in Figure 2.5.

2.1.3 Analysis Methods of NIFSS

2.1.3.1 Ultraviolet–Visible Spectroscopy (UV–vis) Method

Ultraviolet–visible absorption spectroscopy (UV–vis) is a kind of electronic spectrum, which is generated from transition of valence electrons. The ultraviolet visible spectrum and the degree of absorption generated by the absorption of ultraviolet

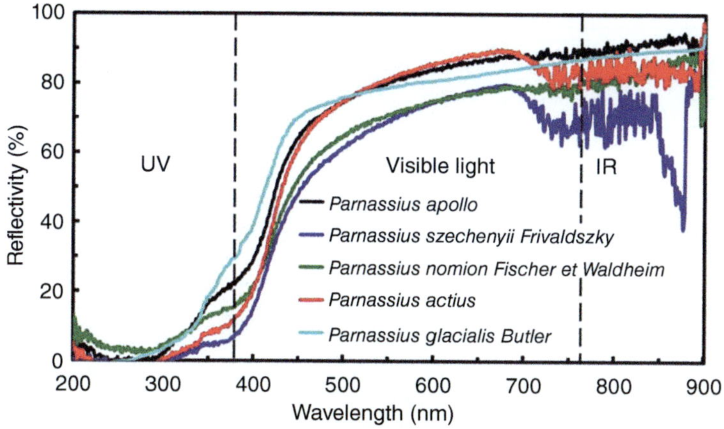

Figure 2.6 The reflectance spectra of five *Parnassius* butterflies confirmed that all these butterflies have a lower reflectivity below 20% in the range of 200–350 nm (ultraviolet region) and a higher reflectivity in the visible light. Source: Reproduced with permission from Han et al. [3]. Copyright © 2013 AIP Publishing LLC.

and visible light by molecules or ions of a substance can be analyzed, determined, and inferred for the composition, content, and structure of the substance. Han et al. performed reflectance measurements in the range of 200–900 nm using the TU-1901 spectrometer with integral ball. It is obvious that reflectivity of the range of both visible and infrared light is always higher than 60% and that of UV range is lower than 20%. Furthermore, the reflectivity of the range from 200 to 340 nm is lower than 10%. All these butterflies have an outstanding antireflective effect for UV light [3] (Figure 2.6).

2.1.3.2 Energy Dispersive Spectrometer (EDS) Method

Energy dispersive spectrometer (EDS) is used to analyze the types and contents of elements in the micro-domain of a material, combined with the use of SEM and TEM. Each element has its X-ray characteristic wavelength, and the size of the characteristic wavelength depends on the characteristic energy ΔE released during the energy level transition. The energy spectrometer is different from the X-ray photon characteristic energy of different elements for component analysis. Its range of application is abroad: (i) Analysis of inorganic or organic solid materials such as polymers, ceramics, concrete, biology, minerals and fibers; (ii) phase analysis, composition analysis, and identification of inclusions of metal materials; (iii) it can analyze the surface coating and coating of solid materials, such as the detection of the surface coating of the metalized film; (iv) identification of gold and silver jewelry, gemstone jewelry, archeological and cultural relics identification, and criminal investigation and identification; (v) qualitative and quantitative analysis of the micro-area components on the surface of the material, and the surface, line, and point distribution analysis of the elements on the surface of the material. Han et al. applied EDS for the acquisition of elemental species and distribution of various butterfly wings and corresponding artificial replicas [2, 6] as shown in Figure 2.7.

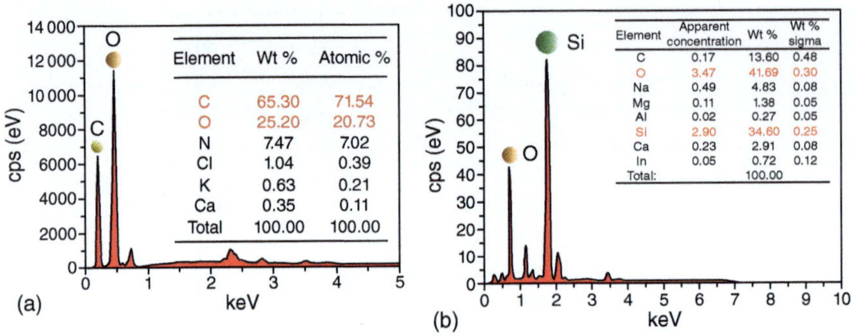

Figure 2.7 (a) The major elements included are carbon (C), hydrogen (H), oxygen (O), and nitrogen (N), which constitute the nonwetting organic framework of butterfly wings. Source: Reproduced with permission from Han et al. [6]. Copyright © 2017 Wiley-VCH Verlag GmbH & Co. KGaA, Weinheim. (b) EDS spectrum indicates the main elements of the biomimetic film: Si and O. Source: Reproduced with permission from Han et al. [2]. Copyright © 2016 American Chemical Society.

2.1.3.3 X-ray Photoelectron Spectroscopy (XPS) Method

X-ray photoelectron spectroscopy (XPS) is an advanced analytical technique used in microscopic analysis of electronic materials and components and often used in conjunction with Auger electron spectroscopy (AES). Because it can measure the inner electron binding energy and chemical shift of atoms more accurately than AES, it not only provides information about molecular structure and atomic valence state for chemical research but also facilitates research on electronic materials. It can provide abundant information about elemental content, chemical state, molecular structure and chemical bond of various compounds. When analyzing electronic materials, it provides not only general chemical information but also information about surface, micro-area, and depth distribution. In addition, since the X-ray beam incident on the surface of the sample is a photon beam, sample destruction is very small, which is very advantageous for analyzing organic materials and polymer materials.

The energy of X-ray photons is between 1000 and 1500 eV, which not only ionizes the valence electrons of the molecules but also excites the inner electrons. The energy levels of the inner electrons are slightly affected by the molecular environment. The inner electron binding energy of the same atom is very small in different molecules, so it is characteristic. Photoelectrons are excited upon photon incidence on a solid surface, and the photoelectrons are analyzed by an energy analyzer called photoelectron spectroscopy.

The principle of XPS is to use X-rays to radiate samples, so that the inner electrons of atoms or molecules are excited to emit. An electron excited by a photon is called a photoelectron. The energy of photoelectrons can be measured. The photoelectron kinetic energy/binding energy $E_b = h\nu$ (light energy) − E_k (kinetic energy) − w (work function) is plotted on the abscissa, and the relative intensity (pulse/second) on the ordinate, which gives information about the sample. X-ray photoelectron spectroscopy is a most widely used technique for surface analysis, which is also called electron spectroscopy for chemical analysis.

2.1.3.4 Fourier-Transform Infrared Spectroscopy (FTIR) Method

Fourier-transform infrared spectroscopy (FTIR) is a method of analyzing and identifying the mathematical processing of Fourier transform using computer technology and infrared spectroscopy. It mainly consists of an optical detection part and a computer part. When a sample is placed in the interferometer optical path, the absorbed interferogram intensity curve correspondingly changes due to the absorption of energy at a certain frequency. By mathematical Fourier transform technique, the time-domain interferogram can be transformed into frequency-domain interferogram. The light intensity and the entire infrared spectrum are obtained. According to the different characteristics of the spectrum, the functional group of unknowns can be determined, chemical structures can be determined, the chemical reaction history can be observed, isomers can be distinguished, and purity of analytes can be determined.

The main advantage is the multi-path transmission of signals, which can measure all the information of all frequencies, greatly improve the signal-to-noise ratio, multi-wavenumber accuracy, up to 0.01 cm^{-1}, high resolution, up to 0.1–0.005 cm^{-1}. The energy has a wide spectral range and can measure a range of 10 000 to 10 cm^{-1}. It is widely used in the fields of chemistry, physics, biology, pharmacy, etc. It also has many applications for the analysis of organic matter in the environment, such as the organic pollution of coal. FTIR has high detection sensitivity, high measurement accuracy, high resolution, fast measurement speed, low astigmatism, and wide band. With the continuous advancement of computer technology, FTIR has also been constantly evolving. The method has been widely used in organic chemistry, metal organic, inorganic chemistry, catalysis, petrochemical, materials science, biology, medicine, and environment. Han et al. adopted FTIR to achieve the functional groups and obtain chemical structure information of the original biological prototypes and biomimetic samples [6] as shown in Figure 2.8.

2.1.3.5 Ion Probe Analysis Method

Ion probe analysis, also known as ion probe microanalysis, works by accelerating and focusing inert gas or oxygen ions into small high-energy ion beams and bombards the sample surface by electronic optics, stimulating and splashing secondary ions. A mass spectrometer is used to separate the ions with different masses to load ratios (mass/charge) to detect all elements in the micro region within several atomic depths and micrometers, isotopes can be determined. The sensitivity of the method is higher than that of electron probe micro-analyzer (EPMA), and it is particularly sensitive to ultralight elements. It can detect 10^{-10} g trace elements, and its relative sensitivity is 10^{-10}–10^{-19} g. The analysis speed is fast, and the plane distribution image of elements can be easily obtained. The ion sputtering effect can also be used to analyze the element distribution in the depth of several microns under the surface. However, the quantitative analysis method of ion probes is not mature.

In 1938, the interaction between ions and solids was studied, but it was not until the 1960s that a practical ion probe analyzer was produced. The basic components of the ion probe analyzer include vacuum system, ion source, primary ion focusing optical system, mass spectrometer, detection and image display system, sample room, etc. Ion probe is suitable for the analysis of ultralight elements, trace elements,

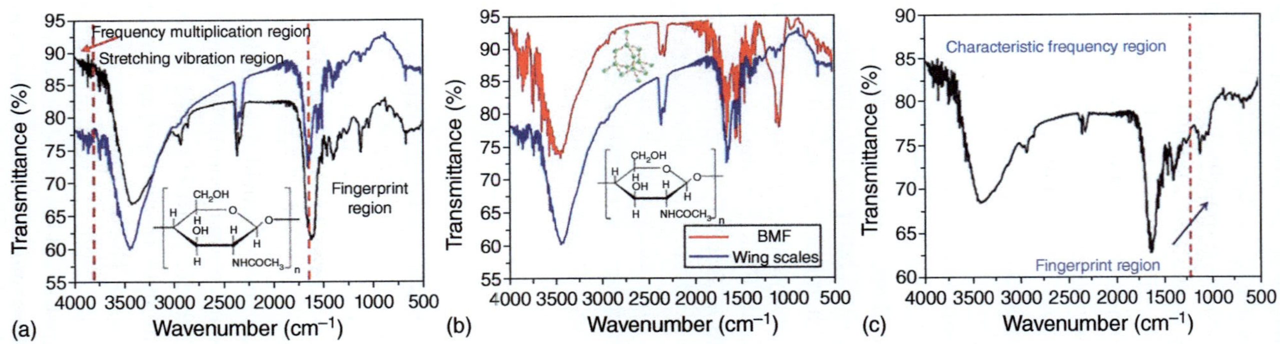

Figure 2.8 (a) FTIR spectra of original (black line) and treated (blue line) wing scales for comparison indicate the brilliant blue colors resulted from fine micro-/nanostructures in wing scales rather than pigments. (b) FTIR spectra of chitin-based wing scales and SiO_2-based biomimetic film for comparison. Source: Reproduced with permission from Han et al. [2]. Copyright © 2016 American Chemical Society. (c) FTIR spectrum indicates that butterfly wings are mainly chitin-based and primarily composed of water-insoluble organic components. Source: Reproduced with permission from Han et al. [6]. Copyright © 2017 Wiley-VCH Verlag GmbH & Co. KGaA, Weinheim.

and identification of isotopes. It is widely used in the study of oxidation, corrosion, diffusion, and precipitation of metal, especially hydrogen embrittlement, as well as in the analysis of surface coating and permeability.

2.1.3.6 Nanoindentation

The nanoindentation test can control the load continuously by the computer and monitor the indentation depth online. A complete indentation process consists of two steps: the loading process and the unloading process. During loading, an external load is applied to the indenter to press it into the surface of the sample. With the increasing load, the depth of the indenter into the sample increases. When the load reaches the maximum, there will be residual indentation marks on the sample surface after the removal of the external load.

2.2 Modeling and Simulation Methods for Bionic Design of NIFSS

2.2.1 Modeling Methods

2.2.1.1 Modeling of Self-Cleaning for Gecko Setae

Gecko setae are self-cleaning adhesives. Geckos with dirty feet recover their ability to cling to vertical surfaces after only a few steps. Self-cleaning occurs in arrays of setae isolated from the gecko. Contact mechanical models suggest that self-cleaning occurs by an energetic disequilibrium between the adhesive forces attracting a dirt particle to the substrate and those forces attracting the same particle to one or more spatulae. To predict the number of particle–spatula interactions (N) needed to achieve an energetic equilibrium with the particle–wall interaction, researchers [7, 8] use the ratio of the interaction energies:

$$N = \left(1 + \frac{R_{\mathrm{P}}}{Rs}\right) \frac{A_{\mathrm{PW}} D_{\mathrm{PS}}}{A_{\mathrm{PS}} D_{\mathrm{PW}}} \tag{2.1}$$

where R_{S} and R_{P} represent the radius of gecko spatula and aspherical dirt particle, respectively. P, W, and S refer to the particle, wall, and spatula, respectively. A is the Hamaker constant (typically $\approx 10^{-19}$ J for van der Waals interactions in the air). D is the particle-to-wall distance.

In the case of N spatulae attached to each particle, approximately half of the particles will remain attached to the wall, and self-cleaning will occur in each step, assuming a clean substrate is encountered. If less than N spatulae are attached to each particle, self-cleaning will occur rapidly due to the energetic disequilibrium; particles tend to remain attached to the wall rather than to the spatula.

2.2.1.2 Modeling of Superhydrophobic Surfaces

2.2.1.2.1 Superhydrophobic Modeling for Water Striders

Water striders (*Gerris remigis*) have remarkable non-wetting legs that enable them to stand effortlessly and move quickly on the water, a feature believed to be due to a surface tension effect caused by secreted wax [9]. But it is the special hierarchical structure of the legs, which are covered by large numbers of oriented tiny hairs (microsetae) with fine nanogrooves, which is more important in inducing this water resistance. According to Cassie's law [10] for surface wettability, such microstructures can be regarded as heterogeneous surfaces composed of solid and air. The apparent contact angle θ_1 of the legs is given by

$$\cos\theta_1 = f_1 \cos\theta_w - f_2 \tag{2.2}$$

where f_1 is the area fraction of microsetae with nanogrooves, f_2 is the fraction of air on the leg surface, and θ_w is the contact angle of the secreted wax. Using measured values of θ_1 and θ_w, the air fraction between the leg and the water surface corresponds to f_2 = 96.86%. Available air is trapped in spaces in the micro setae and nanogrooves to form a cushion at the leg–water interface that prevents the legs from being wetted [11].

2.2.1.2.2 Superhydrophobic Modeling for Gecko Feet

The high adhesive forces of gecko feet to water can be explained by the following considerations (Figure 2.9). The morphology and orientation of gecko setae are heterogeneous. It is difficult to accurately control the contact modes between gecko spatulae and water droplets. Therefore, researchers came up with two typical solid–liquid contact state, the default state that represents minimizing spatulae contact with water and the adhered state that represents maximizing spatulae contact with water. Autumn and Hansen [12] fixed isolated gecko seta arrays on a glass substrate, which can be seen as nearly ideal default state arrays, and demonstrated that this array is non-sticky to water. The adhered state, in which the setae are bent so that almost all the spatulae can contact water, contributes the main adhesive force. The origin of the high adhesive force toward water results from high-density nanopillars contacting the water. Besides the abovementioned conformational changes in the surface proteins of gecko setae upon exposure to water, the complex contact condition between water droplets and setae should also be responsible for the wide range of adhesive forces from low to high values [13].

2.2.1.2.3 Superhydrophobic Modeling of Directional Adhesion for Butterfly Wings

It is known that a droplet tends to slide on the surface of groove microstructures for butterfly wings more easily along the direction parallel to the grooves rather than perpendicular. Two reasonable hypothetical modes are shown in Figure 2.10 to clarify the distinct adhesive properties of butterfly wings. When the wing is

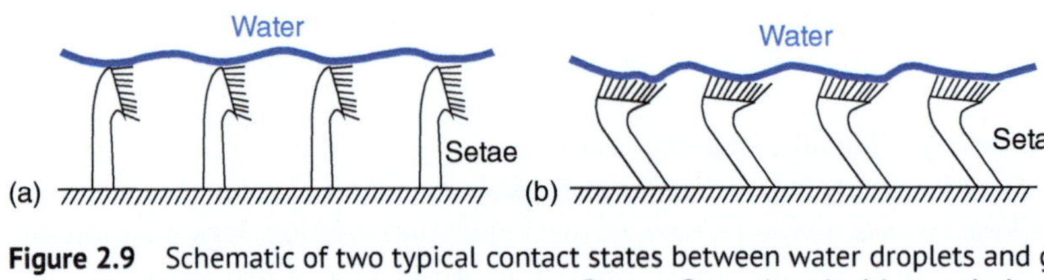

Figure 2.9 Schematic of two typical contact states between water droplets and gecko setae: (a) default state and (b) adhered state. Source: Reproduced with permission from Liu et al. [13]. Copyright © the Royal Society of Chemistry 2012.

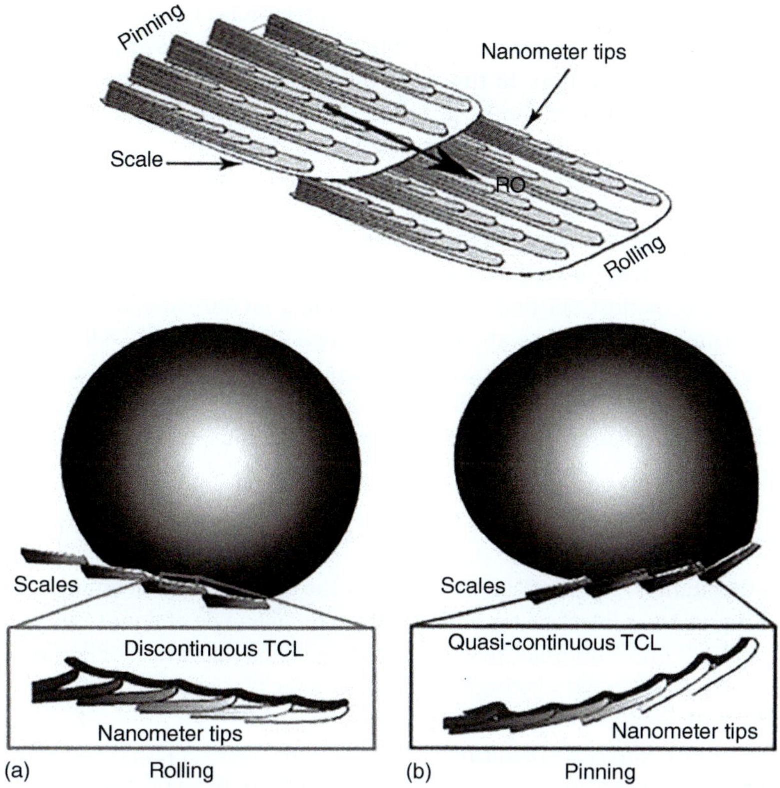

Figure 2.10 The models proposed for elucidating the potential mechanism of distinct adhesion dependent on the direction along and against the RO direction. (a) When the wing is tilted down. (b) When the wing is tilted upward. Source: Reproduced with permission from Zheng et al. [18]. Copyright © the Royal Society of Chemistry 2007.

tilted downward, the microscales with ridged nanostripes are spatially separated from each other and the oriented nanotips tend to be unwound with flexible microscales. In this case, air can be efficiently trapped in the nanoscale voids among the nanotips extended by lamellae and the ridged nanostripes and thus the droplet only touches the top of the nanotips, with a minimal contact area [10]. This ensures superhydrophobicity of the wings, with a high contact angle above 150°, which has been verified. Moreover, it is known that the ordered arrangement of the microstructures may influence the contour, length, and continuity of the three-phase (solid/liquid/gas) contact line and thus control the way a droplet tends to move [14–17]. Accordingly, the ordered arrangement of the microscales and nanostripes on the wings along the radial outward (RO) direction and the formation of the extremely discontinuous three-phase (solid/liquid/gas) contact line as illustrated in the bottom of Figure 2.10a make the droplet easily roll off the wings along the RO direction. However, when the wing is tilted upward, the flexible nanotips and microscales take on a close arrangement as shown in Figure 2.10b. The nanotips on the top of the nanostripes are raised with the flexible microscales to closely contact the droplet. As a result, a quasi-continuous three-phase (solid/liquid/gas) contact line is formed when the droplet attempts to move under the gravitational potential against the RO direction. In this case, the pinning at numerous corners (nanotips) of the steps between the neighboring lamellae on top of the ridged nanostripes may produce a very high-energy barrier, which makes the droplet pin tightly on the wing as it is tilted upward, even when it is fully upright [18].

2.2.1.3 Superhydrophobic Modeling for Fish Scales

When the fish scales come in contact with oil droplets, water molecules can be trapped in the micro-/nanostructured fish scales, forming an oil/water/solid interface. This new composite interface shows superoleophobic properties. Although Young's equation [19] was originally applied for a liquid droplet on a solid surface in air, it has also been applied to a liquid droplet on a solid surface in the presence of a second liquid [20, 21]. Using Young's equation, we can get the following equation (Figure 2.11a):

$$\cos\theta_3 = \frac{\gamma_{l1-g}\cos\theta_1 - \gamma_{l2-g}\cos\theta_2}{\gamma_{l1-l2}} \tag{2.3}$$

where $r_{11\text{-}g}$ is the liquid 1/gas interface tension, θ_1 is the contact angle of liquid 1 in air, $r_{l2\text{-}g}$ is the liquid 2/gas interface tension, θ_2 is the contact angle of liquid 2 in air, $r_{l2\text{-}l1}$ is the liquid 1/liquid 2 interface tension, and θ_3 is the contact angle of liquid 1 in liquid 2. Using Eq. (2.3), why a hydrophilic surface in the air becomes oleophobic in water can be explained.

For a rough surface composed of solid and air, Cassie and coworker [10] proposed a model describing the contact angle in a water/air/solid system. In an oil/water/solid system, on the other hand, where the rough surface is composed of solid and water (Figure 2.11b and c), the Cassie model is expressed as follows [22]:

$$\cos\theta_3{}' = f\cos\theta_3 + f - 1 \tag{2.4}$$

where f is the area fraction of solid, θ_3 is the contact angle of an oil droplet on a smooth surface in water, and θ_3 is the contact angle of an oil droplet on a rough surface in water (the area fraction f). For a micro-/nanostructured silicon surface, f is close to 0. So, θ_3 is close to 180° according to Eq. (2.4) [22].

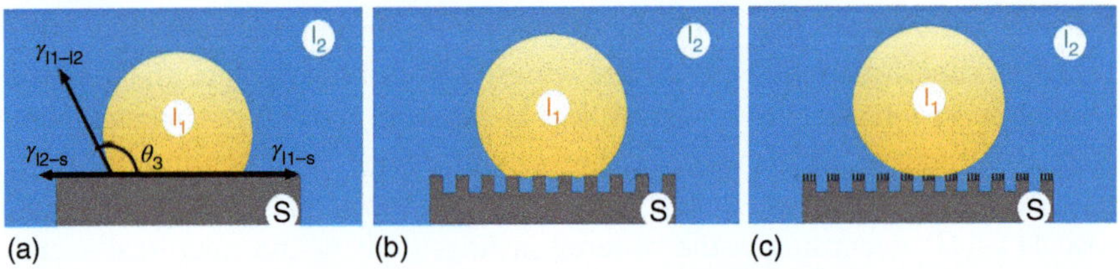

Figure 2.11 Effect of surface structure on the wetting behaviors of solid substrates in solid/oil/water three-phase systems. (a) Diagram of Young's equation at the condition of a liquid 1 droplet on a smooth surface with a contact angle θ_3 in a liquid 2 phase; (b) an oil droplet on a microstructure substrate in water, in which /1 represents oil and /2 represents water; (c) same system on a micro-/nanostructured substrate. Source: Reproduced with permission from Liu et al. [22]. Copyright © 2009 WILEY-VCH Verlag GmbH & Co. KGaA, Weinheim.

2.2.1.4 Frictional Adhesion Modeling

Geckos can climb easily and rapidly. Models of peeling tape generally treat the adhesive surface as a continuum. The force during peeling of a flexible strip of tape is given by [23]

$$F = bdE\left(\cos\alpha - 1 + \sqrt{\cos^2\alpha - 2\cos\alpha + 1 + {}^{2R}/_{dE}}\right) \tag{2.5}$$

where b is the width of the strip, d is the thickness of the strip, E is material stiffness, R is the adhesion energy, and α is the peel angle. Consider a weight suspended from a strip of tape attached to a surface of angle α vertically. Solving for α in Eq. (2.5), the angle (α^*) at the onset of peeling is obtained by Eq. (2.6):

$$\alpha^* = \cos^{-1}\left(1 - \frac{bR}{F} + \frac{F}{2bdE}\right) \tag{2.6}$$

The peeling model predicts that greater weight will initiate peeling at shallower angles. Including the elastic stretch term, maximum peeling force, which occurs at low angles, is limited by some factor in Eq. (2.7):

$$F_{\max} = \sqrt{2Rb^2dE} \tag{2.7}$$

where b is the width of the strip, d is the thickness of the strip, E is material stiffness, R is the adhesion energy, $F_{\max}$ is the maximum peeling force.

It indicated that when given the same adhesion energy, stiffer materials will peel at higher loads. Peeling mechanics can be applied at least in theory to fibrillar gecko-like materials [24]. However, in real geckos where the attachment is via a series of scissors bearing anisotropic setae, the validity of conventional peeling mechanics is less clear. Geckos hold their toes in a hyperextended position when not climbing possibly protecting the sensors from abrasion, suggesting that digital hyperextension could have functions other than a reduction of detachment force via peeling mechanics. Indeed, it has been suggested that spatulae could detach more or less simultaneously [25], due to their mechanical independence. When dragged against their natural path (against curvature), setal arrays remain compressed and do not adhere (Figure 2.12a). When dragged along their natural path (with curvature), setal arrays are compressed initially and then adhered, resulting in tensile normal forces (Figure 2.12b).

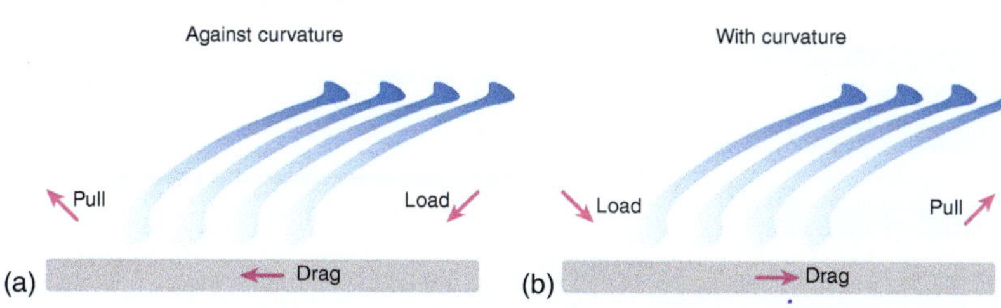

Figure 2.12 Shear and normal forces in isolated gecko setal arrays on a glass surface. (a) Setal array during load (1), drag (2), and pull (3) (LDP) against the curvature of the setal shafts exhibits Coulomb friction; (b) setal array during LDP with the curvature of the setal shafts is compressed initially and then pulled into tension as the setal tips are adhered.

2.2.1.5 Modeling of Light-Trapping Structures

A structural model was built to mimic the butterfly scale absorbing solar heat. The array of scales plays the role of a very efficient optical diffuser of the visible range. It is assumed that low reflectance in shelf microstructures is due to the alternating layers of air and chitin, which can cause reflection, transmission, and absorption at each interface (Figure 2.13a,b). The thickness of each layer determines what wavelength will be reflected and of these, what will constructively or destructively interfere at specific viewing angles [26]. In this case, the arrangement of the layers induces interference destructively, and consequently, most of the solar energy is absorbed by chitin layers after multi-absorption. As for the quasi-honeycomb-like microstructure, micro-diffraction gratings are formed with periodic parallel slits or grooves at the magnitude of the light wavelength [26], and the tiling of these micro-gratings generates a quasi-honeycomb-like structure (Figure 2.13c,d) that can diffract incident light into the scales to be absorbed. Whenever the light meets another part of the surface after it enters the scale, it is reflected back into the material, embracing the second reflection, and consequently, it will be reflected multiple times. The structural color exhibited by the wing is determined by the position of the reflectivity peak, which can be verified by the adjusted Snell's law [27]:

$$N\lambda = 2d\left(n_{\text{eff}}^{2} - \sin^{2}\theta\right)^{1/2}, N = 1, 2, 3, \ldots \tag{2.8}$$

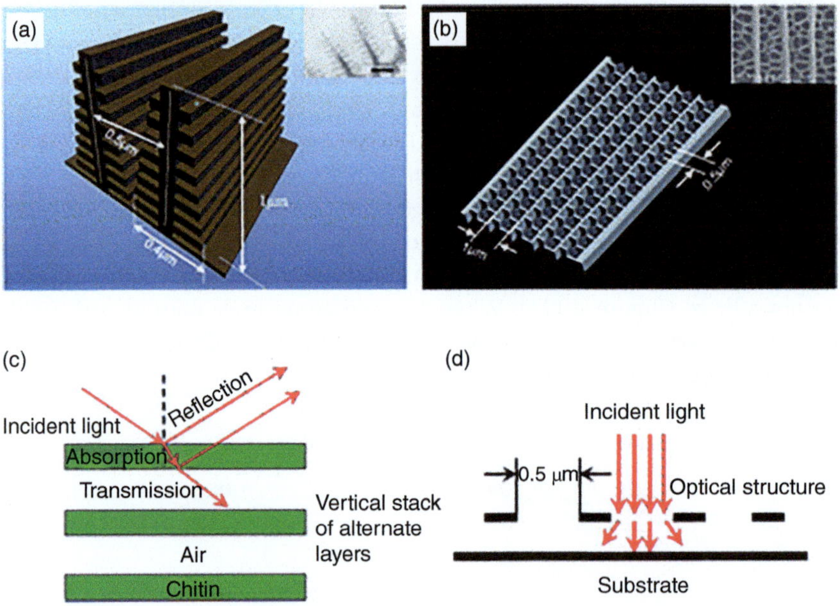

Figure 2.13 3D single-scale models of (a) shelf structure and (b) quasi-honeycomb-like structure. (c) The optimized vertical stack of the alternating layers of air and chitin is the interpretation of the mechanism of the multiple absorption, and the mechanism of the light-trapping effect of the quasi-honeycomb-like structure is explained appropriately by (d) the diffraction at the interface. Source: Reproduced from Han et al. [28]. Copyright © The Royal Society of Chemistry 2012.

where d is the single lattice constant, λ is the wavelength, θ is the incidence angle, and N is the integer. n_{eff}^2 is the effective refractive index, which is determined by the relative refractive index of different materials and shown by the following function:

$$n_{\text{eff}}^2 = n_1^2 f_1 + n_2^2(1 - f_1) \tag{2.9}$$

One lattice had two kinds of materials: cuticula and air. n_1 and n_2 represent the refractive index of cuticula and air, respectively. f_1 is the spacing ratio of the lattice, namely, $a/(a+b)$. a and b are the thickness of the cuticula and the air film, respectively. As a result, almost all the incident light is absorbed by the scales.

2.2.1.6 Fluid-Drag Reduction Modeling

The fact that ribbed surfaces (riblets) aligned in the streamwise direction do reduce turbulent skin friction has been established beyond any reasonable doubt [29]. Representative models of shark surface morphology include triangular, parabolic, and blade shapes. The triangular shape has the best characteristics of drag reduction. Luchini [30] believed when $h \geq 0.6s$ (where h and s are the height and width of a riblet, respectively), the fluid resistance is the smallest. Becher [29] used an adjustable height of the riblet in the pipeline for the test. The result showed that the best riblet height should be $h = 0.5s$. Meanwhile, the adhesion resistance of the riblet surface was reduced by 9.9% than that of the smooth surface [31]. The drag reduction effect was worse as the height of the riblet increased when $h = 0.6s$.

2.2.1.7 Erosion Resistance Modeling

Androctonus australis is a desert scorpion living in sandy deserts and may face erosive action of blowing sand at a high speed. Bionics studies show that some special morphologies of the organism surface exhibit excellent functions. The four patterns of bionic models for desert scorpions are shown in Figure 2.14. The figure shows that the convex pattern bionic model and many convex domes distribute on the material surface. The influencing factor of the erosion resistance of this model is the distribution of the convex domes, which contains the radius and the height of the dome, and position distribution of the center coordinates of the domes. In the groove pattern bionic model, many grooves are excavated on the material surface and the grooves are parallel to each other. The feature of the cross section is an arched structure. The influencing factors of the erosion resistance of this model are the depth and width of the groove and the spacing between two grooves.

The erosion resistance performance of the dorsal surface of the scorpion is not only related to the special surface morphology but also attributed to multiple factors coupling function. The domes and grooves on the surface are morphological coupling elements, and the flexible connection is a flexible coupling element. The coupling bionic model is built by combining the coupling elements (Figure 2.14c). The flexible connection of the model is removed and divided into two layers. The upper layer is a hard material and processed into bionic surface morphology, and the lower layer is a flexible material. The two-layer structure forms a soft and hard alternating composite structure (Figure 2.14d).

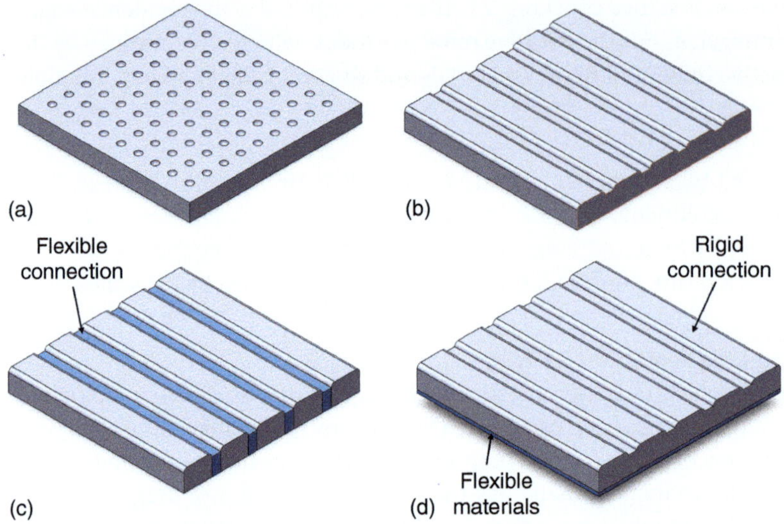

Figure 2.14 Bionic modeling of desert scorpion back. (a) Convex pattern; (b) groove pattern; (c) coupling bionic models with flexible connection; (d) coupling bionic models with soft and hard alternating composite structure. Source: Han et al., 2016 / With permission of Elsevier.

2.2.2 Simulation Methods

2.2.2.1 Translight Method

Translight is a software used for photonic bandgap calculation developed by Glasgow University in the United Kingdom. Its basic method is to calculate the reflection performance of a multilayered 1D photonic crystal (PC) by using the transfer matrix method, which is obtained by Fourier transform using Maxwell's equation. The components of the electromagnetic field in the x and y directions are used to obtain the transfer matrix and the reflection matrix. The software can adjust parameters such as the dielectric material, the thickness of the film, the number of basic periodic units, and the angle of incidence and calculate the corresponding transverse electric mode (TE) and transverse magnetic mode (TM) reflectance and transmittance, respectively. 1D, 2D, and 3D photonic crystal structures can be calculated in the software structure library. Among them, 1D structural models include *1D Bragg Stack*, *1D Omni-directional Bragg Stack,* and other modules; 2D structural models include *2D Hexagonal Crystal* and other modules; 3D structural models include *3D Opal Photonic Crystal 001* and other modules.

The transfer matrix method and the *Bragg stack in the air* module can be applied to solve the reflectance problem of photonic crystals (PCs) with the help of Translight. After selecting this module, *Cell* file of the structure needs to be set, then a new *Controlled* file can be generated. In the *Cell* file, the characteristic parameters such as the filling ratio of the structure should be set. Meanwhile, in the *Controlled* file, the unit cell constant, the number of unit cells, the incident angle, normalized frequency range, and other parameters need to be set. Finally, the calculation result is displayed in the *Crystal* folder, which needs to be opened with Excel or Origin executable program to extract the reflectance value for data analysis.

2.2.2.2 FDTD Method

As a well-known simulation method or gold standard for modeling nanophotonic devices, processes, and materials, the finite difference time domain method (FDTD), the 3D/2D Maxwell's solver, is widely used. Especially, the FDTD method is very suitable to solve optical issues. Notably, it can simulate and analyze light interactions with structured surfaces. In part of the following chapters, we will show how to solve the intractable case on optical performance prediction with the help of FDTD. More technical details on FDTD can be found in related professional reference books or tutorials.

2.2.2.3 Computational Fluid Dynamics

Computational Fluid Dynamics (CFD) is the combination of modern fluid mechanics, numerical mathematics, and computer science. It is interdisciplinary science with strong vitality. It approximates the integral and differential terms in the governing equations of fluid mechanics as discrete algebraic forms and converts them into algebraic equations. Then the discrete algebraic equations are solved by a computer, and the numerical solutions at discrete time/space points are obtained.

2.2.2.4 Other Modeling-Related Methods

Optical issues in PCs are a typical research case in the field of NIFSS, except Translight and FDTD method, many other modeling methods or simulation software can be adopted, such as PBS software, COMSOL Multiphysics, CST Design Studio, RSoft, and CUDOS MOF Utilities. Most alternative modeling methods can be selected on a case-by-case basis.

2.3 Design Principles and Fabrication Methods of NIFSS

2.3.1 Design Principles of NIFSS

An essential step toward NIFSS is to address primary design issues. Hence, the basic design principles are essential to guide the following NIFSS design. It also works as a bridge linking the original biological surfaces with final artificial products that are similar to natural biological surfaces at all scale levels.

2.3.1.1 Selection of Biological Prototypes

Biological prototypes are the origin of idea of the NIFSS design. The importance of the selection of biological prototypes is obvious. Also, which kind of biological prototype is the best choice for NIFSS design? How to dig the most useful parameters to assist us in designing desirable NIFSS? Here are several tips for the selection of biological prototypes before starting the primary design of NIFSS:

(a) **Most typical.** Although there are millions of creatures in nature, only the most typical ones with excellent properties can be selected as biological prototype candidates. The properties displayed by the target biological prototypes could be different, but they should be superior to any other contenders in a certain field. For example, structural color is an old but unfading theme for the research of natural photonic crystals, and numerous biological prototypes are gradually uncovering and emerging. However, *Morpho* butterfly is regarded as the most well-known biological prototype for structural color research due to its typical brilliant wings that are very attractive to materials scientists and engineers. Thus, "most typical" generally means unique and outstanding performance in surface functions.

(b) **Easy to bioengineering.** Bioengineering is an effective pathway to convert brittle organic biological surfaces into robust artificial structural surfaces. If the imitative object, namely the selected biological surface, is easy to bioengineering, it would be the best candidate for NIFSS design.

(c) **Living in harsh environment.** As we all know, "*Natural selection, survival of the fittest.*" After nearly four billion years of evolution and optimization, natural creatures possess many excellent properties. Especially, creatures living in harsh environments usually possess special abilities. These creatures could be the most valuable research objects for the design of NIFSS with desirable properties.

2.3.1.2 Information Extraction of Feature Structures

After the final confirmation of target biological prototypes, digging useful structural information including surface topography and the material composition becomes the key point for NIFSS designing. To clarify the surface topography of feature structures, *in situ* observation of biological surfaces using OM is a conventional and effective method. Usually, structures at the microscale level can be clearly described in this way due to the accuracy limitation of the OM. To obtain more structure details at a higher scale level, electron microscopy including SEM, TEM, and environmental scanning electron microscopy (ESEM) is a reliable method to characterize the profile, size, and shape of feature structures. Based on this, the framework of the NIFSS gradually comes out, and the "bone" of the NIFSS is built up roughly.

For the material composition of feature structures, mature modern material analysis technologies can be very helpful, such as XRD for the material information of composition, the structure or morphology of the atoms or molecules inside the material, XPS for qualitative and quantitative analysis of elements, solid surface analysis for structure of compounds, FTIR spectrometer for material structural information, especially polymer bonds. Based on the material information, the underlying mechanism of the excellent properties of the natural structure surfaces can be understood.

2.3.1.3 Coupling Design Strategy of Structures and Materials

According to actual demand, designing a series of surfaces with different structures will endow them with desirable material properties. To chase the best combination of structures and materials, numerical calculation and analogue simulation are essential to optimize the NIFSS design further. In fact, it is an effective way to save time and reduce costs before conducting irreversible fabrication. Usually, the experimental optimization design methods are used to the craft parameters optimal.

2.3.2 Fabrication Methods of NIFSS

No matter which design method is adopted, it will eventually be implemented in the manufacturing process of the designed micro-/nanostructures for NIFSS.

Micro-/nanomanufacturing is a cross-scale manufacturing process involving two different scales, micron and nano. First, two basic concepts need to be clarified, namely micron manufacturing and nanoscale manufacturing. Specifically, the so-called micron manufacturing means that the size of the object to be manufactured in at least one dimension in a line width ranging from 100 nm to 100 μm, and the range can be further subdivided into 100 nm–1 μm. In the submicron range, this size range is already in the category of precision machining. Nanoscale manufacturing refers to the size of the object being fabricated in at least one dimension with a line width in the range of 0.1–100 nm. In addition, materials smaller than 1 nm belong to the range of atomic clusters and are beyond the research of micro-/nanomanufacturing. Various NIFSS fabrication methods have been reported [33] including conventional and sophisticated methods, both of which are discussed in the following sections.

2.3.2.1 Nanoimprint Lithography

Nanoimprint lithography is an "up-to-down" method, which mainly adopts mechanical transfer to achieve the structured manufacturing of the surface. It is also one of the important processing methods in the manufacturing fields of microelectromechanical system (MEMS), materials, and microelectronics. Saison et al. from Saint-Gobain Recherche copied the micro-/nanostructures on the lotus leaf surface and *Papilionidae ulysse* butterfly wing surface with flexible polydimethylsiloxane (PDMS) as templates by nanoimprint lithography technology [34]. First, elastomer PDMS templates were prepared by casting a liquid PDMS solution against the surface of the lotus leaf and the butterfly wing. After solidification at 80 °C for several hours, the PDMS layer was peeled off, resulting in a negative structure of the original template. In the lotus leaf, the release from the mold was done easily, whereas the wing scales remained stuck to the PDMS. Due to the low Young modulus of the PDMS (around 0.8 MPa), the templates can follow the curvature of natural leaves or wings and can be used to imprint spin-coated resist films at low pressures. Sol–gel resists were prepared from the methyl triethoxy silane (MTEOS) sol mixed in an aqueous solution under acidic conditions (pH = 3.1). A high MTEOS/H_2O (1 : 14) molar ratio was used to ensure total hydrolysis of alkoxy-silane groups before imprinting. The solutions were stirred for one day at room temperature before use. MTEOS sol–gel films were deposited by spin coating on glass or silicon substrates. The MTEOS film thickness was approximately 900 nm as measured by a profilometer. The imprint pressure was kept lower than 2 bars and the imprint temperature was kept between 80 and 150 °C for about 20 minutes. All imprints were done with the same two molds (lotus and butterfly) on typical surfaces from 4 to 10 cm^2 with good homogeneity along the surface and reproducibility demonstrated for at least 10 successive imprints of each template. Final annealing was performed between 200 and 500 °C for two hours to ensure homogeneous and stable vitrification of the sol–gel films. For samples with the highest thermal treatment (500 °C), surface grafting of fluoroalkylsilane was performed by evaporation for 12 hours at 80 °C.

2.3.2.2 Micro-molding Technique

Micro-molding technique is widely used in the manufacture of polymer micro-products, generally including micro-compression molding, micro-injection molding, micro-extrusion molding, etc. It has the advantages of low cost and mass production and is more common in the system of microfluidics, biochemical analysis, etc. Based on the investigation of micro-/nanostructures of the swallowtail butterfly *Papilio karna*, Tanaka et al. [35] from the University of Tokyo fabricated a prototype butterfly wing model with a spherical shape. The method can be used for not only model wings but also for 3D plastics with both a macro-curved shape and micro details. The manufacturing process is shown in Figure 2.15.

2.3.2.3 Layer-by-Layer Sol–Gel-Based Deposition Technique

Sol–gel process involves hydrolysis of the sol and drying to form a gel. For a good reproduction of natural surface structures inspired by animals, a liquid precursor is favorable [36, 37]. Sol–gel process mainly uses the metal alkoxides to prepare inorganic materials. It provides a new way of tailoring and controlling the microstructure of materials through a low-temperature chemical method. With the sol–gel approach, multifunctional micro-/nanosurfaces can be prepared without any assistance from photolithography, laser lithography, and other expensive technologies. For example, Zhang et al. [38] developed a novel natural stamp, the outward wings of grasshoppers (OWG), for making anisotropic surfaces. The two-step replica exhibited anisotropic high hydrophobicity and light reflection. Figure 2.16 shows the schematic representation of the two-step replication process

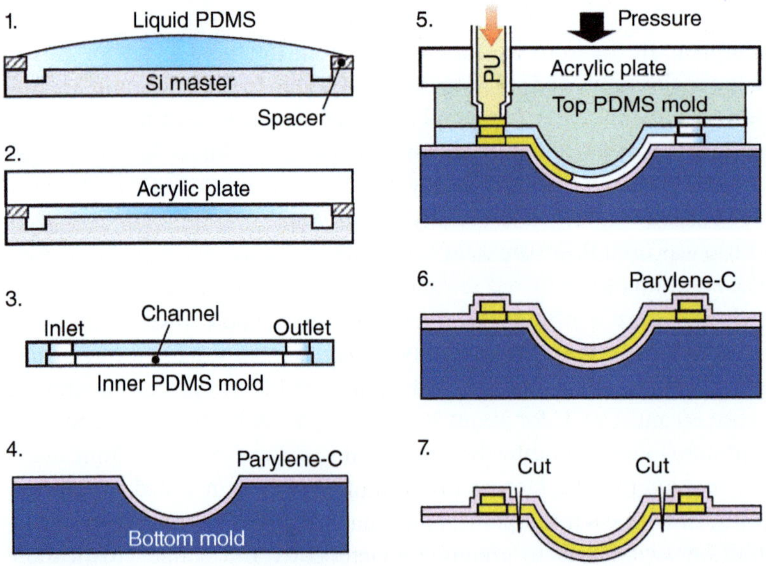

Figure 2.15 Micro-molding manufacturing process of the microstructures. Source: Reproduced with permission from Tanaka et al. [35]. Copyright © 2007 Institute of Physics Publishing.

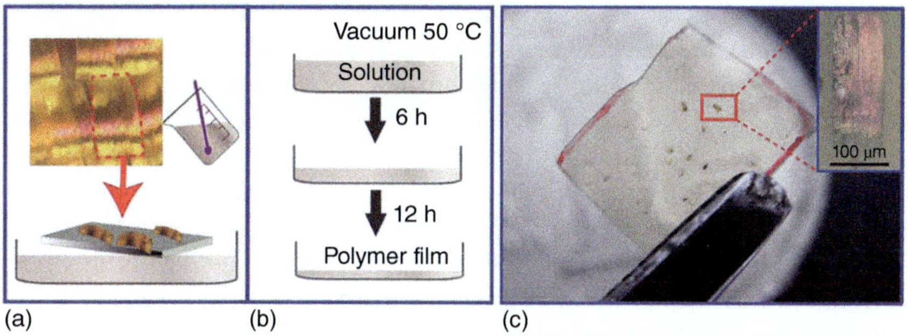

Figure 2.16 (a, b) Preparation of chitosan/polyvinyl alcohol (CTS/PVA)/wing scale composites; (c) obtained composite with wing scales buried in the CTS/PVA hydrogel. Inset: Optical microscope image of an embedded wing scale. Source: Reproduced from Zang et al. [39]. Copyright © The Royal Society of Chemistry 2011.

using sol–gel and the images of the positive and negative replicas. The results proved that anisotropic wetting surface was achieved by duplicating the surface structure of OWG.

The most useful method to coat small materials with inorganic materials is sol–gel coating carried out in aqueous media [40]. In typical sol–gel coating, inorganic sol particles undergo surface-preferred gelation to form an inorganic coat on the templates. However, examples of coating biological materials by sol–gel coating are rare [41]. This is probably because most biological materials are negatively charged at moderate conditions, which is not ideal for a sol–gel coating that is accomplished by taking advantage of electrostatic attraction between negatively charged sol particles and positively charged templates. Also, the ashes from biological materials could disturb the formation of replicas in this method. In this regard, Kim et al. have recently reported interface-selective sol–gel polymerization that was useful to coat different templates regardless of charges on them [42]. Reaction conditions in that method were so mild that living microorganisms were coated without being sacrificed [43].

Notably, some smart hydrogels, prepared by sol–gel process, have reversible swelling and deswelling properties with volume transition and have been applied in drug delivery, biosensors, and bioelectronics including tunable personal computers. With the hope to expand or shrink the natural species' inner structures, Zang et al. used an electric-field-sensitive hydrogel (EFSH) to embed and fill the wing scales of sunset moths with rich structural colors [39]. The EFSH swelled and deswelled with volume transition that modified the structures of wing scales, resulting in materials' reflectance peak shift for visible light.

In addition, few reports combine the unique multifunctional surface structures of animals with the intrinsic properties of the replica materials, even though this is one of the most important objectives of biomimetization. Zhang et al. conducted some works that ZnO replicas with photonic structures were obtained via a sol–gel bio-templating method. The dark black (DB) wing scale replicas exhibit a photonic bandgap (PBG) in the visible region, which overlaps with the visible emission range of ZnO [44, 45].

When using the sol–gel process to produce biomorphic materials, an impregnation or infiltration of the templates is an important step. For example, Cook et al. observed incompatibility between the butterfly wing and the silane coating, which led to catastrophic cracking of the coating during calcination [46]. Zhang et al. avoided this problem by choosing a zinc nitrate water-free ethanol solution as the precursor, based on the knowledge that butterfly wing surface contains chitin with functional groups that act as centers to absorb Zn(II) by hydrogen bonding, thus zinc ions could soak the template well [47]. ZnO replicas exhibiting the fine wing-scale features were obtained, due to good compatibility between the template and precursor.

In summary, the sol–gel method is very suitable considering its advantages such as chemical flexibility, facile shape control, and mild reaction conditions. When using a sol precursor to produce biomorphic materials, impregnation or infiltration of the templates is an important step. A good wetting between the sol and the template is crucial for the final product to preserve morphology, fine features, properties, and functions of the template [48]. On the other hand, there are also many problems for the sol–gel. Generally, raw materials are relatively expensive and harmful to health. There are also some large pores in the gel material. A certain amount of gas will escape out from the gel material during the drying process, resulting in structural contraction. Weatherspoon et al. [49] from the Georgia Institute of Technology first used this method to form a continuously controllable nanoscale TiO_2 coating on a fine 3D bioorganic template. In the manufacturing process, the automated surface sol–gel pumping system was used to alternately expose the chitin butterfly wing scales to the precursor solution of alcohol oxygen and water to achieve layer-by-layer controlled growth of the film. After 40 times of deposition in the mixed titanium–tin alkoxide and aqueous solution, it was calcined at 450 °C, and finally, 3D micro-/nanostructures based on nanoscale rutile titanium were obtained, which retains the butterfly wing-scale chitin. The micro-/nanostructure features of the template are shown in Figure 2.17.

2.3.2.4 Dipping Method

Zhang et al. [50] from Shanghai Jiao Tong University firstly dispersed a layer of TiO_2 colloid on the surface of conductive fluorine-doped SnO_2 by dipping method. Then, the butterfly wing was placed on the TiO_2 film and covered with a fluorine-doped

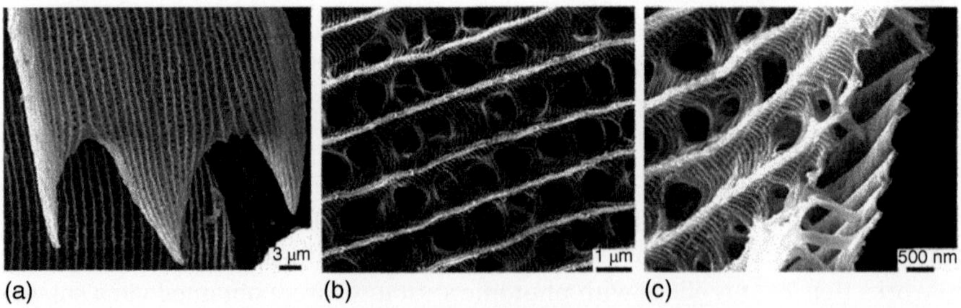

Figure 2.17 SEM images of biomimetic butterfly wing scales. Source: Reproduced with permission from Weatherspoon et al. [49]. Copyright © 2008 Wiley-VCH Verlag GmbH & Co. KGaA, Weinheim.

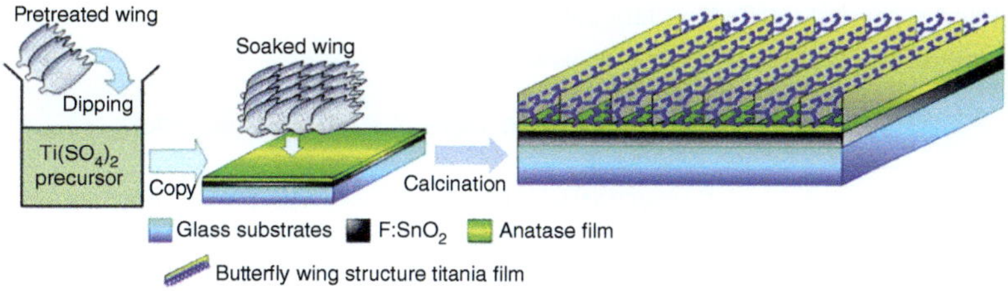

Figure 2.18 Biomimetic fabrication process using the dipping method. Source: Reproduced with permission from Zhang et al. [50]. Copyright © 2009 American Chemical Society.

layer-miscellaneous SnO_2 glass. After that, it was placed in a muffle furnace and heated to 500 °C in air at a rate of 1 °C/min for two hours to obtain an anode material based on the butterfly-wing microstructures (Figure 2.18).

2.3.2.5 Bio-template Method

The bio-template [51] method is based on the biological prototype itself or part of its organ organization as the original template, combined with various physical and chemical methods, such as wet chemical method, high-temperature sintering, chemical vapor deposition (CVD), physical vapor deposition (PVD), and atomic layer deposition (ALD) to produce functional artificial analogues. Chen et al. [52] from Shanghai Jiao Tong University used the *Euploea mulciber* butterfly wings as the original template to produce a large-area rainbow-colored ZrO_2 photonic crystal with biological morphology (Figure 2.19a). Similarly, Han et al. [53] used the *Euploea mulciber* and *Papilio paris* as the natural photonic crystal skeleton to control the synthesis of butterfly-wing/sulfurized CdS organic–inorganic composites with micro-/nanostructures (Figure 2.19b). Han et al. [54, 55] from Jilin University used *Trogonoptera brookiana* butterfly wing scales as the original template to fabricate the SiO_2-based light-trapping structural surface (Figure 2.19c).

In fact, in addition to butterfly wing scales, biological prototypes commonly used in bio-template methods include insect wings, plant leaves, and diatom cell walls.

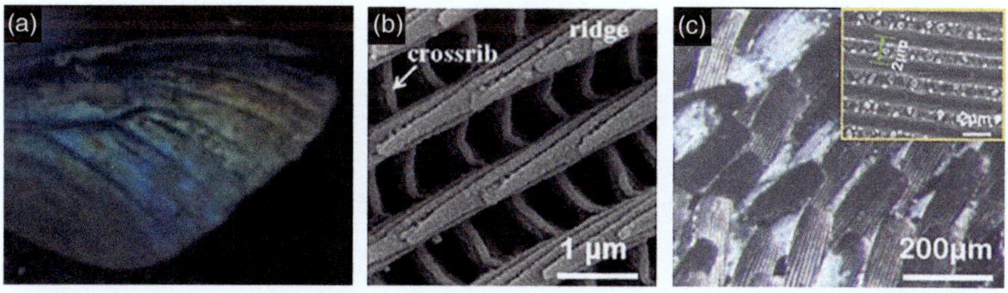

Figure 2.19 (a) ZrO_2 biomimetic wing scales with structural colors. Source: Reproduced with permission from Chen et al. [52]. Copyright © 2009 American Institute of Physics. (b) Biomimetic micro-/nanostructures based on butterfly-wing/CdS composite. Source: Reproduced with permission from Han et al. [53]. Copyright © 2009 The Royal Society of Chemistry. (c) SiO_2-based biomimetic butterfly wings. Source: Reproduced with permission from Han et al. [54]. Copyright © 2013 The Royal Society of Chemistry.

Table 2.1 Bio-templates, target materials, and fabrication routes.

Biological templates	Functional materials	Fabrication methods	Applications
Beetle elytra	SiO_2	WCM[a)] + sintering	Photonic structures
	TiO_2		
Leaves	TiO_2	WCM + sintering	Water-splitting
	ZnO		Dye-degradation
	Fe_3C	WCM + sintering	Electrodes
	SiO_2	CVD	Photonic structures
Diatom frustules[b)]	Si	Sintering (Mg-reduction)	Gas sensors
	Ag, Au, Pd	WCM	SERS substrates
	SnO_2	WCM + sintering	Gas sensors
	MOFs	WCM	Gas absorbents
Butterfly wing scales	Ag, Au	PVD	SERS substrates
	GeSeSb	CEFR	Antireflection films
	SiO_2	CVD	Photonic structures
	Al_2O_3	ALD	Photonic ICs
	TiO_2	ALD	Photonic structures
	Co, Ni, Pd, Pt	WCM	
	Ag, Au, Cu	WCM	SERS substrates
	PVA/chitosan	WCM	Stimuli-responsive materials
	Y_2O_3, TiO_2	WCM + sintering	Phosphor materials
	SiO_2, ZnO, ZrO_2	WCM + sintering	Photonic structures
	TiO_2	WCM + sintering	Photoanodes
	SnO_2	WCM + sintering	Gas sensors
	Fe_3O_4	WCM + sintering	Stimuli-responsive materials
	WO_3	WCM + sintering	Water-splitting
	Bi_2WO_6, $BiVO_4$	WCM + sintering	Photocatalysts

a) Wet-chemical methods.
b) Cell walls.
Source: Reproduced with permission from Gu et al. [77]. Copyright © 2015 WILEY-VCH Verlag GmbH & Co. KGaA, Weinheim.

The commonly used biomimetic structural materials based on the bio-template method and application examples [27, 48, 50, 56–76] are summarized in Table 2.1.

2.3.2.6 Etching

When using the etching process to produce subwavelength structures (SWSs), the parameters of etching and dewetting processes are crucial to the antireflective property of SWSs. For example, Leem et al. [78] have theoretically and experimentally investigated the influence of the process parameters on the antireflective properties of disordered Si SWSs using the Pt thin films as the etch mask and subsequent inductively coupled plasma (ICP) etching in $SiCl_4$/Ar plasma. The geometric profile and antireflective property of the fabricated Si SWSs were strongly dependent on the thermally dewetted Pt nanopatterns and various etching parameters. The Si SWS with a more tapered shape and greater height using the completely dewetted dot-like Pt nanomask pattern yielded relatively low reflectance.

Recently, antireflective structure (ARS) surfaces on silicon prepared by nanosphere lithography have been reported [79], but there are few ARS surfaces on fused silica or non-planar lenses. Li et al. have demonstrated a versatile and time-efficient method to fabricate large area fused silica cone arrays on planar fused silica substrates and plan convex lenses for high-performance antireflective and antifogging surfaces [80]. A schematic representation of the fabrication process of ARS surfaces is illustrated in Figure 2.20. The silica cone arrays were prepared by short time reactive ion etching (RIE) using 2D polystyrene (PS) colloidal crystals as masks. In another study, they also demonstrated a simple method to improve the light extraction of white organic light emitting diodes (OLEDs) by silica biomimetic antireflective surfaces. A non-close-packed hexagonal silica cone array was directly etched on the opposite side of the indium–tin-oxide (ITO)-coated fused silica substrate. This method mentioned here can be introduced in any OLED without any alteration of device structure and materials design [81]. In addition, a two-dimensional subwavelength grating (SWG) has been fabricated on a GaAlAs light emitting diode (LED). The SWG is patterned by electron beam lithography and etched by fast atom beam with Cl_2 and SF_6 gases [82].

Etching is one kind of extremely important steps of the semiconductor manufacturing process, microelectronics IC manufacturing process, and micro–nano manufacturing process. It is also one of the main crafts of graphical processing, associated with photolithography. First, through photolithography, the photoresist is exposed to lithography treatment then etching removal treatment is realized. Along with the development of the micro-fabrication process, generally speaking,

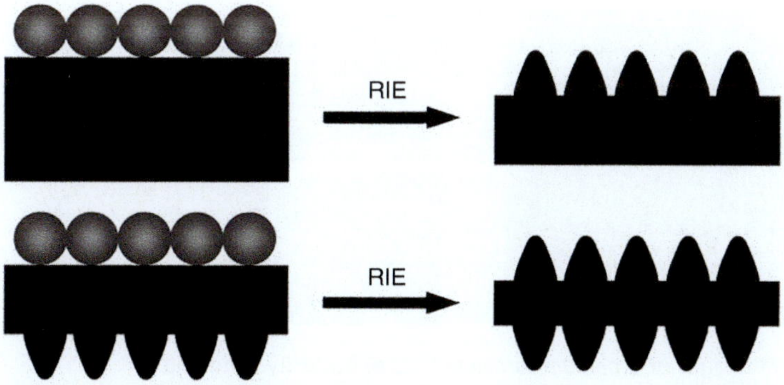

Figure 2.20 Schematic illustration of preparation of ARS surfaces. Source: Reproduced from Li et al. [80] with the permission from Wiley. Copyright © 2009 WILEY-VCH Verlag GmbH & Co. KGaA, Weinheim.

etching is a general designation of detaching or removing material through a solution, a reactive ion, or other mechanical approaches, becoming a universal way of micro processing and manufacturing. At the same time, this process has some obvious shortcomings. The thickness of the structure in nanoscale is extremely difficult to control. So, it is difficult to obtain highly ordered nanostructures when directly using etching to manufacture nanostructures.

2.3.2.7 Sonochemical Method

Sonochemical methods use the energy of sonication to promote chemical reactions for material production. For the first time, the fine hierarchical structures of butterfly wings were successfully duplicated in manganese oxide using sonochemical reduction followed by calcination [83]. The morphologies and surface structural details of the original butterfly wing and the calcined Mn_2O_3 butterfly wing are compared in Figure 2.21. New functional materials with chosen hierarchical structures of bio-templates combined with the functionality of metal oxides could be synthesized in the future by the sonochemical method [83].

This method for fabricating hollow nanostructures is simple, low cost, and energy-conserving. It can be extended to fabricate hollow nanostructures of various morphologies with bacteria of other shapes as templates, e.g. vibrio, spirillum, square bacteria, fusiform bacilli. Since sonochemical methods can produce many attractive materials, biomorphic mineralization under sonochemical conditions can be extended to these materials for the synthesis of their hollow nanostructures, multiscale structures, and so on [37].

2.3.2.8 Atomic Layer Deposition (ALD)

ALD is a thin film growth technique that uses alternating and saturating reactions between gaseous precursor molecules and a substrate to deposit films in a layer-by-layer fashion [84]. Al_2O_3 replicas of butterfly wing scales with well-controlled coating thickness were also produced through ALD [85]. Wing scales coated with a 10-, 20-, 30-, and 40-nm-thick Al_2O_3 layer were obtained by varying the cycles of deposition, and their reflected colors shifted from original blue

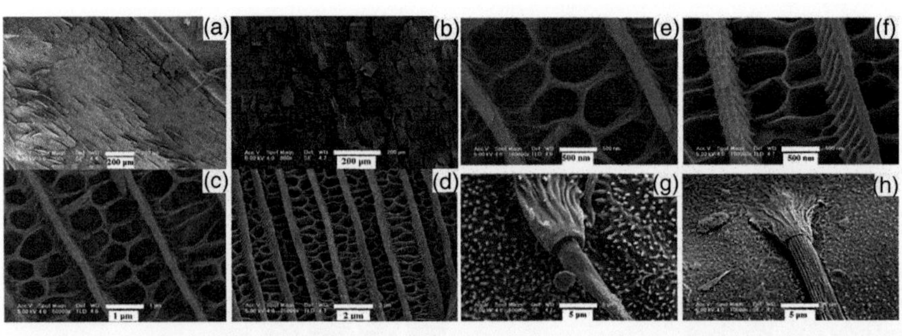

Figure 2.21 FESEM images of *Graphium sarpedon* (a, c, e) butterfly wing and (g) peg-and-socket attachments and of calcined replica Mn_2O_3 (b, d, f) butterfly wing and (h) peg-and-socket attachments. Source: Adapted from Zhu et al. [83]. Copyright © 2008 American Chemical Society.

to green, yellow, orange, and eventually pink in optical microscopic imaging. This redshift was due to the surface-film-enhanced reflection of a particular wavelength determined by the thickness and refraction index of the film. In another study, Kolle et al. used a combination of layer deposition techniques (Figure 2.22), including colloidal self-assembly, sputtering, and atomic layer deposition, to fabricate photonic structures that mimic the color mixing effect found on the wings of the Indonesian butterfly *Papilio blumei* [86]. They also showed that a conceptual variation in the natural structure leads to enhanced optical properties.

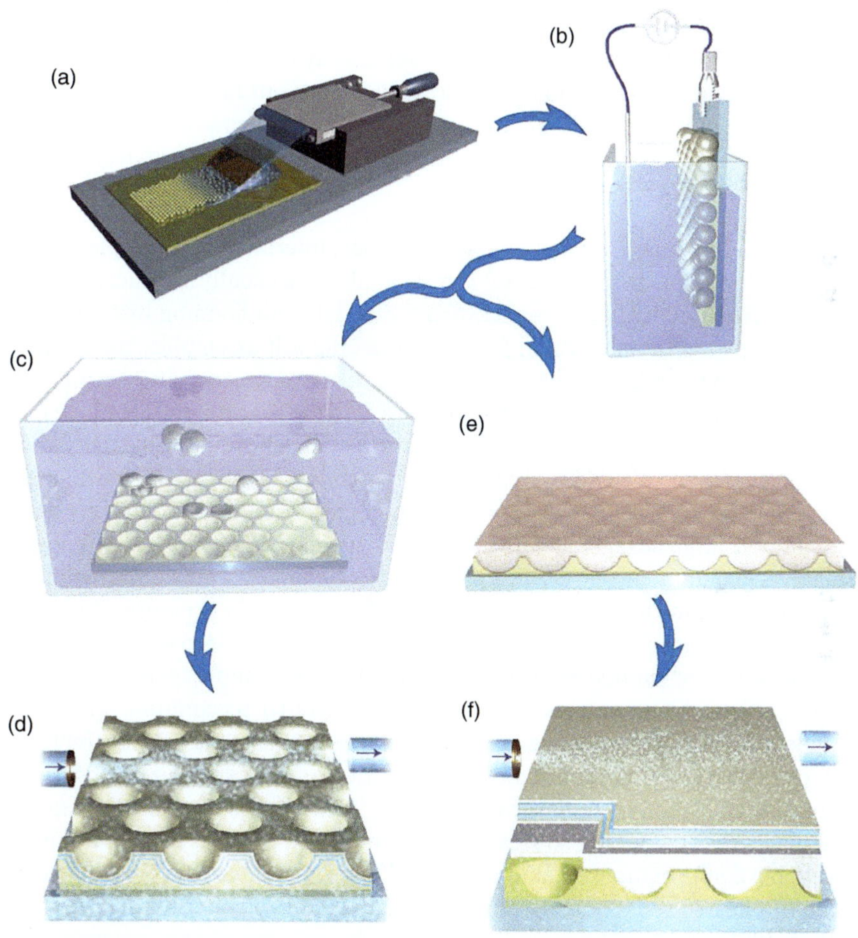

Figure 2.22 Sample fabrication. (a) Deposition of polystyrene colloids on a gold-coated silicon substrate. (b) Growth of platinum or gold in the interstices of the colloidal array by electroplating. The metal deposition is terminated when the thickness of the deposited film equals the microsphere radius. (c) Removal of the polystyrene spheres from the substrate by ultrasonication in acetone. (d) Sputtering of a thin carbon film and ALD of a stack of 11 alternating TiO_2 and Al_2O_3 layers (arrows indicate the precursor gas flow). (e, f) In a second route, the colloids are molten to cover the cavities with a homogeneous film (e), which is covered by a TiO_2–Al_2O_3 multilayer (f). Source: Reproduced from Kolle et al. [86]. Copyright © 2010 Macmillan Publishers Limited.

Among a variety of deposition methods, ALD offers a unique capability of control over the final pore size and thickness of the deposited layer with atomic layer precision [87], which makes it an attractive method for producing precise and conformal coatings over nanoporous materials. The material flexibility of ALD, as well as its capability of control over the coating thickness and hence dimension of the fine structures (like pore sizes), with atomic layer precision, makes ALD an intriguing method for fabricating materials with desired and controllable properties. Precise control of the film thickness can be achieved through controlling the response cycle number of ALD. The film thickness control can be achieved at the atomic level. Despite of many advantages, ALD is also a relatively expensive method than CVD and chemical vapor infiltration (CVI). The deposition rate of ALD is extremely slower than CVD and CVI, because deposition is carried out in atomic monolayer films.

2.3.2.9 Assembly Methods

Assembly processes involve attractive and repulsive interactions between building blocks, and building block interactions with solvents, interfaces, and templates. At the molecular level, the driving forces for the assembly are usually chemical, which can be ionic, covalent, hydrogen, noncovalent, metal–ligand bonding interactions, etc. At scales beyond the molecular, forces for material self-assemblies are usually physical, including capillary, colloidal, elastic, electric, magnetic, and shear. For example, some researchers reported a simple yet scalable self-assembly technique for fabricating efficient moth-eye antireflection coatings with adjustable reflectivity and non-close-packed microstructures, which were not easily achievable by traditional self-assembly approaches [88]. Wafer-scale, non-close-packed colloidal crystals with remarkable large hexagonal domains were created by spin-coating technology.

Assembly is generally carried out in aqueous environments, because good mobility of the preformed nanoparticles is favorable for the formation of a structure with high integrity. The preformed nanoparticles are usually prepared in the form of a colloid or a suspension. This method requires relatively high and complex external conditions and is vulnerable to outside factors such as molecular recognition, structure and number of components, nature and composition of the solvent. Self-assembly is the only practical approach for building a wide variety of nanostructures. Ensuring the components assemble themselves correctly, however, is not an easy task. Because the forces at work are so weak, self-assembling molecules can get trapped in undesirable conformations, making defects all but impossible to avoid. Any new system that relies on self-assembly must be able to either tolerate those defects or repair them [89].

2.3.2.10 Physical Evaporation and Deposition

Evaporation and deposition can reproduce fine structures of bio-templates [37]. Chung et al. fabricated a *Morpho*-mimetic thin film by depositing a dielectric multilayer on top of a silica microsphere base layer with a random size distribution. Such a combination induces a spontaneous emergence of both order and disorder across many length scales without any lithography or complicated fabrication process. This enables us to fabricate, at a large scale, thin films that not only reproduce the bright, brilliant glossy colors comparable to *Morpho rhetenor* butterflies but actually outperform both *M. rhetenor* and *Morpho didius* butterflies in maintaining their color and brightness over a wide range of viewing angles in ambient conditions [90]. Physical evaporation and deposition is a straightforward method to replicate surface features of the underlying substrates through forming a thin coat. The thickness of the coating can be controlled by the number of deposition cycles, but fine surface features may be lost if the coating is too thick.

2.3.2.11 Imprinting

Imprinting is an exciting method among structure-cloning technology of microelectronics devices. For example, Ho et al. presented an antireflective structure consisting of irregular nanopillars to increase the light extraction efficiency of flexible organic light-emitting devices. The nanopillars were made by imprinting the anodized aluminum oxide on polycarbonate substrates. With the nanoimprinted irregular and tapered nanopillars on the PC film, a high-performance antireflective (AR) layer was made. The AR layer simultaneously increased the image contrast ratio and light extraction efficiency [91]. Using inkjet printing, Kang et al. showed the ability to infuse fine droplets of silicone oils into the crystal, locally swelling it and changing the reflected color. They demonstrated the use of self-assembled colloidal photonic crystal substrates to support high resolution, multicolor, stable but erasable images printed with transparent silicon oils of varying molecular

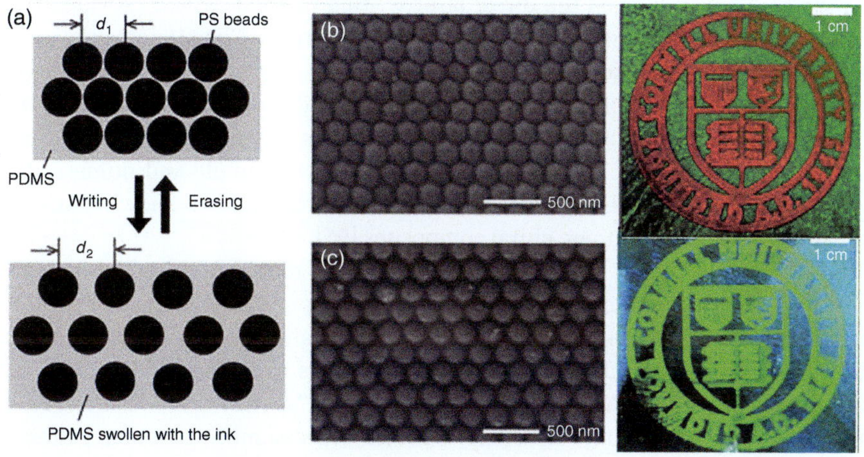

Figure 2.23 (a) Schematic of the mechanism of the color's reversible change: the lattice spacing is increased by swelling the PDMS matrix with the ink. The PDMS matrix shrinks back to its original state once the ink completely evaporates. (b) SEM image of the (111) plane of a photonic paper in a cubic close-packed lattice of 202 nm PS beads whose void spaces are completely filled with the PDMS elastomer. (c) SEM image of the same colloidal crystal after it swelled with a vinyl-terminated silicone fluid. (d) Demonstration of high-resolution multicolor images printed on green substrates. (e) Demonstration of high-resolution multicolor images printed on blue substrates. Source: Reproduced from Kang et al. [92] and Fudouzi and Xia [93]. Copyright © 2011 American Chemical Society, and copyright © 2003 WILEY-VCH Verlag GmbH & Co. KGaA, Weinheim.

weights (Figure 2.23) [92]. This method will pressure cavities or templates onto the conformal materials. Then the materials will be deformed in accordance with the template graphics. Finally, the graphics of the template will be copied onto the materials through the UV exposure or heat treatment method. Compared with most of the traditional microelectronics manufacturing processes, imprinting technology not only can copy graphics on the *XY* direction but also can press the steps of the profile of structure in the vertical direction. For the microscale fabrication technologies, imprinting is a potentially broad range of applications for manufacturing the multifunctional surface structures of animals.

2.3.2.12 Direct Laser Writing

There are many types of direct-write techniques used in science and engineering [94]. In the most classical sense, engraving or milling can be considered a direct-write process, since a tool or stylus makes contact with a surface and moves in a desired pattern to produce a feature. The coupling of a high-powered laser with direct-write processing enables similar features to be produced without requiring physical contact between a tool and the material of interest. Because of this, few techniques share the versatility of laser direct-write (LDW), in adding, subtracting, and modifying different types of materials over many different length scales, from the nanometer to the millimeter scale.

Turner et al. showed the fabrication and characterization of a novel class of biomimetic photonic chiral composites inspired by a recent finding in butterfly wing scales. They employed the direct laser writing method, which provided the ability to realize 3D microstructures with arbitrary geometry. The 3D networks were written by the 3D translation of the photoresist mounted on a piezoelectric translation stage. A key feature of the structural design is that the overall microstructure is in the shape of a pyramid with a flat top, analogous to cleaving the boundaries along the crystallographic planes. They also had experimentally and numerically characterized the transmission spectra of these microstructures showing strong circular dichroism within these chiral composites [95].

From the earliest work on laser interactions with materials, direct-write processes have been important and relevant techniques to modify, add, and subtract materials for a wide variety of systems and for applications such as metal cutting and welding. In general, direct-write processing refers to any technique that can create a pattern on a surface or volume in a serial or "spot-by-spot" fashion. This is in contrast to lithography, stamping, directed self-assembly, or other patterning approaches that require masks or preexisting patterns. At first glance, one may think that direct-write processes are slower or less important than these parallel approaches. However, direct-write allows for precise control of material properties with high resolution and enables structures that are either impossible or impractical to make with traditional parallel techniques. Furthermore, with continuing developments in laser technology providing a decrease in cost and an increase in repetition rates, there is a plethora of applications for which (LDW) method is a fast and competitive way to produce novel structures and devices.

2.3.2.13 Calcination

Calcination is an elevated-temperature treatment to burn off or decompose organic materials, as well as to crystallize and densify ceramic products, but it is not suitable for nanosized nanocrystalline products and nanostructures because elevated temperatures tend to cause heavy aggregation of nanoparticles, coarsen the crystals, and cause the collapse of nanostructures. Calcination also provides the energy for chemical reactions between the precursor and the underlying template, thus converting as well as consuming the template. During calcination, gaseous products may be released and promote the formation of porous products. Shrinkage generally occurs after calcination, but at different degrees depending on the template and previous treatments.

2.3.2.14 Selective Dissolution

Selective dissolution is another method to remove templates, which employs corrosive chemicals to dissolve the templates without damaging the products. For example, the silica inverse structure replica of butterfly wing scales was obtained through a sol–gel process and selective dissolution [96]. The butterfly wing slices were removed by acid etching, and the whole assembly was in a mixture of concentrated nitric acid and perchloric acid while heating at 130 °C for 10–15 minutes. The slides became separated and transparent when the etching was over. The etched silica inverse structure was obtained finally [97, 98]. Figure 2.24 shows the different magnification images and a TEM image obtained from the SiO_2 inverse structure replica sample. One drawback of this method is the use of corrosive chemicals, such as the concentrated nitric acid, perchloric acid and ether, which can cause pollution and are harmful to health. To avoid injury, the products can be isolated by differential dissolution using a solvent when products and byproducts present different solubilities in a solvent. This process is not necessarily corrosive. The selective dissolution could be carried out at mild temperatures, avoiding aggregation of nanoparticles or collapse of nanostructures generally caused by heat treatment. In this sense, selective dissolution for removing templates presents advantages that direct the mineralization of nanomaterials and delicate nanostructures.

2.3.2.15 Sonication

Sonication is able to disrupt biological materials with high-frequency sound waves. Sonication is also more energy conserving compared with calcination that requires elevated temperatures. Moreover, coalescence of nanomaterials and collapse of nanostructures can be avoided due to low-temperature nature of the sonication process. However, its energy is still not enough to destroy and remove hard templates like butterfly wings and sea urchin plates, which can better be removed by other methods such as calcination and dissolution.

2.3.2.16 Other Fabrication Methods for NIFSS

In addition, Kustandi et al. [99] of the A*STAR Institute of Materials Engineering in Singapore used nanoimprint lithography-pattern-shear bonding technology to mimic the micro-/nanostructures of the *Ancyluris meliboeus* butterfly, which can

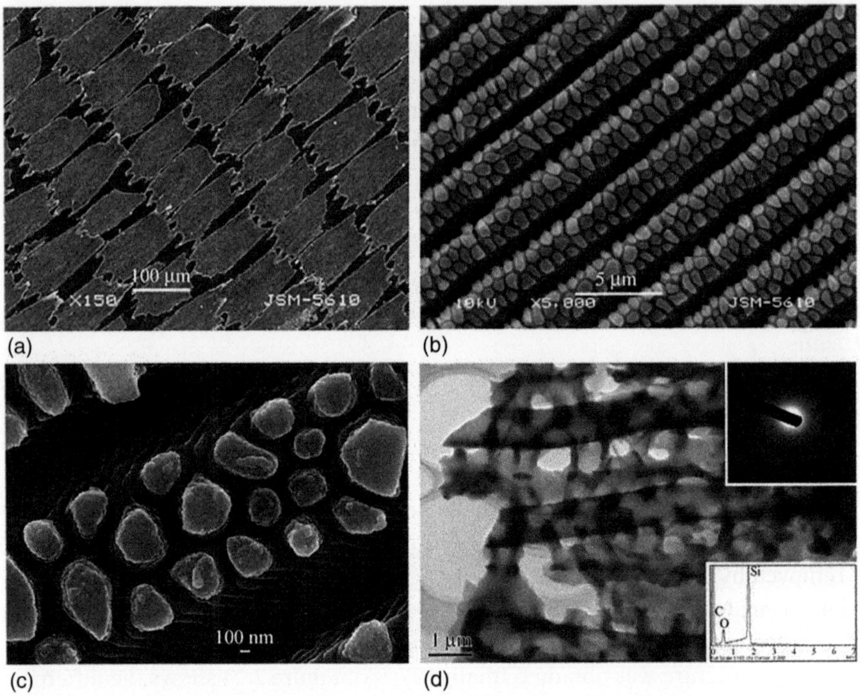

Figure 2.24 SEM images of the SiO_2 inverse structure replica. (a) Overview of the surface of the wing scale replica; (b) medium magnification image of the replica; (c) high-magnification image of the replica; (d) TEM image of the SiO_2 replica of butterfly wing scales. Inset on the right-top corner: selected area electron diffraction (SAED) pattern of the SiO_2 replica. Inset on the right-bottom corner: energy dispersive X-ray (EDX) spectrum of the sample. Source: Reproduced from Xu et al. [97]. Copyright © Tsinghua University Press and Springer-Verlag Berlin Heidelberg 2011.

be used in nano- and submicron-imprint lithography. Within the submicron range, a simpler, faster, and less expensive method of fabricating patterned materials than complex semiconductor nanoetching techniques is provided. It enables high-precision direct molding of engineered polymers and is critical to the successful formation of surface properties. Watanabe et al. [100] from the University of Hyogo in Japan successfully fabricated a similar structure of the *Morpho* butterfly by using focused ion beam-chemical vapor deposition (FIB-CVD).

2.3.3 Synthetic Design and Fabrication Strategies of NIFSS

No doubt, the design and fabrication of NIFSS are very attractive because NIFSS always exhibits striking properties in many fields. In fact, the design principles and fabrication methods should be considered together. Design principles serve as a guideline and fabrication methods work as a bridge that connects primary design ideas with final products. These structural variations can also exploit hierarchical design principles, such as morphogenesis, self-organization, and metamorphosis, which are useful to the development of new synthetic strategies in NIFSS [101].

From the basic science, a unifying design principle that underpins the advantageous properties of these NIFSS is the creation of novel micro-/nanostructures to lead to the preferential material performance without the need for external material modifications. Technically, the development of various NIFSS can be achieved through two different strategies. The first strategy is guided by fundamental knowledge and engineering implementation [102]. While the recent advances in physical understanding, as well as nanotechnology, make it possible to control the directional formation of micro-/nanostructures with a high level of specificities, these engineered devices are constrained with simple architectures and limited functionalities. Moreover, although the introduction of external sources such as electricity, light, magnetic field, and heat [103–111] can bypass these inherent limitations, it also leads to many additional drawbacks such as the reliance on the specific materials, the need for additional energy input, and sophisticated control units, all of which dramatically restrict their applications in a wide variety of settings.

The other synthetic strategy to achieve the desirable NIFSS is through mimicking nature. Many living organisms have spectacular surface properties in a myriad of ways with the minimal energy cost. More intriguingly, the preferred NIFSS is mainly ascribed to evolved topological structures with refined length scales, curvatures, shapes, densities, and array arrangements and can be preserved or retained under diverse environmental conditions, which are characterized with a temporal variation in temperature, humidity, and pressure, as well as a wide spectrum of size distributions. To some extent, nature's artistry involved in their structures and profound principles bypasses the intrinsic limitations in our imagination and provokes new design blueprints for the development of engineered devices. Remarkably, the translation of various inspirations learned from nature to practical implementations also relies on the advancements in engineering routes. Moreover, the fundamental principles for the NIFSS on both engineered and bioinspired structure design are universal. Thus, these two strategies are complementary to each other in their maneuvering and largely converging in principle as shown in Figure 2.25. As such, it is essential to formulate a general framework to guide the design, choice, optimization, and fabrication of NIFSS and then go beyond nature to develop novel structures and surfaces to satisfy specific practical applications.

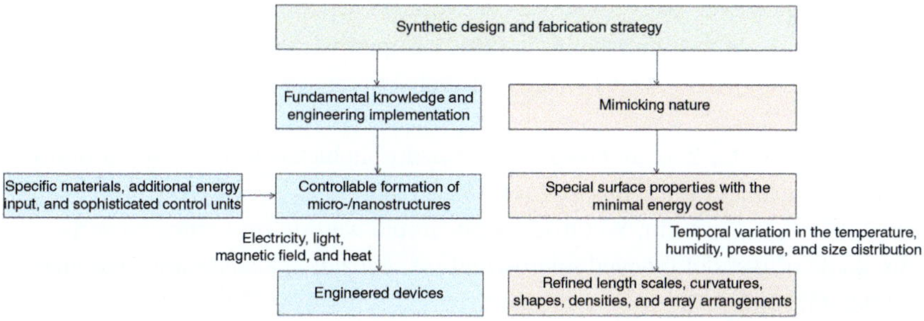

Figure 2.25 Overview of the synthetic design and fabrication strategies of NIFSS.

References

1 Han, Z., Mu, Z., Li, B. et al. (2017). Bioinspired omnidirectional self-stable reflectors with multiscale hierarchical structures. *ACS Applied Materials & Interfaces* 9 (34): 29285–29294.

2 Han, Z., Mu, Z., Li, B. et al. (2016). Active antifogging property of monolayer SiO_2 film with bioinspired multiscale hierarchical pagoda structures. *ACS Nano* 10 (9): 8591–8602.

3 Han, Z., Niu, S., Li, W., and Ren, L. (2013). Preparation of bionic nanostructures from butterfly wings and their low reflectivity of ultraviolet. *Applied Physics Letters* 102 (23): 1–5.

4 Han, Z., Li, B., Mu, Z. et al. (2015). Fabrication of the replica templated from butterfly wing scales with complex light trapping structures. *Applied Surface Science* 355: 290–297.

5 Wang, K., Zhang, J., Song, H. et al. (2019). Highly efficient mechanoelectrical energy conversion based on the near-tip stress field of an antifracture slit observed in scorpions. *Advanced Functional Materials* 29 (22): 1807693.

6 Han, Z., Li, B., Mu, Z. et al. (2017). Energy-efficient oil–water separation of biomimetic copper membrane with multiscale hierarchical dendritic structures. *Small* 13 (34): 1701121.

7 Hansen, W.R. and Autumn, K. (2005). Evidence for self-cleaning in gecko setae. *Proceedings of the National Academy of Sciences of the United States of America* 102 (2): 385–389.

8 Autumn, K. (2007). Gecko adhesion: structure, function, and applications. *MRS Bulletin* 32 (6): 473–478.

9 Hu, D., Chan, B., and Bush, J.W.M. (2003). The hydrodynamics of water strider locomotion. *Nature* 424 (6949): 663–666.

10 Cassie, A.B.D. and Baxter, S.T. (1944). Wettability of Porous Surfaces. *Faraday Society* 40: 0546–0550.

11 Gao, X. and Jiang, L. (2004). Water-repellent legs of water striders. *Nature* 432 (7013): 36–36.

12 Autumn, K. and Hansen, W. (2006). Ultrahydrophobicity indicates a non-adhesive default state in gecko setae. *Journal of Comparative Physiology. A, Neuroethology, Sensory, Neural, and Behavioral Physiology* 192 (11): 1205–1212.

13 Liu, K., Du, J., Wu, J., and Jiang, L. (2012). Superhydrophobic gecko feet with high adhesive forces towards water and their bio-inspired materials. *Nanoscale* 4 (3): 768–772.

14 Feng, L., Li, S., Li, Y. et al. (2002). Super-hydrophobic surfaces: from natural to artificial. *Advanced Materials* 14 (24): 1857–1860.

15 Tie, L., Guo, Z., and Liu, W. (2015). Anisotropic wetting properties on various shape of parallel grooved microstructure. *Journal of Colloid and Interface Science* 453: 142–150.

16 Wang, S., Liu, K., Yao, X., and Jiang, L. (2015). Bioinspired surfaces with superwettability: new insight on theory, design, and applications. *Chemical Reviews* 115 (16): 8230–8293.

17 Kwon, M.H., Shin, H.S., and Chu, C.N. (2014). Fabrication of a super-hydrophobic surface on metal using laser ablation and electrodeposition. *Applied Surface Science* 288: 222–228.

18 Zheng, Y., Gao, X., and Jiang, L. (2007). Directional adhesion of superhydrophobic butterfly wings. *Soft Matter* 3 (2): 178–182.

19 Nakajima, A., Hashimoto, K., and Watanabe, T. (2001). Recent studies on super-hydrophobic films. *Monatshefte Fur Chemie* 132 (1): 31–41.

20 Jarn, M., Granqvist, B., Lindfors, J. et al. (2006). A critical evaluation of the binary and ternary solid–oil–water and solid–water–oil interaction. *Advances in Colloid and Interface Science* 123: 137–149.

21 Gao, J., Yao, X., Zhao, Y., and Jiang, L. (2013). Lyophilic nonwettable surface based on an oil/water/air/solid four-phase system. *Small* 9 (15): 2515–2519.

22 Liu, M., Wang, S., Wei, Z. et al. (2009). Bioinspired design of a superoleophobic and low adhesive water/solid interface. *Advanced Materials* 21 (6): 665+.

23 Autumn, K., Dittmore, A., Santos, D. et al. (2006). Frictional adhesion: a new angle on gecko attachment. *Journal of Experimental Biology* 209 (18): 3569–3579.

24 Hui, C., Glassmaker, N.J., Tang, T. et al. (2004). Design of biomimetic fibrillar interfaces: 2. Mechanics of enhanced adhesion. *Journal of the Royal Society Interface* 1 (1): 35–48.

25 Gay, C. (2002). Stickiness-some fundamentals of adhesion. *Integrative and Comparative Biology* 42 (6): 1123–1126.

26 Ingram, A.L. and Parker, A.R. (2008). A review of the diversity and evolution of photonic structures in butterflies, incorporating the work of John Huxley (The natural history museum, London from 1961 to 1990). *Philosophical Transactions of the Royal Society, B: Biological Sciences* 363 (1502): 2465–2480.

27 Li, X., Fan, T., Zhou, H. et al. (2009). Enhanced light-harvesting and photocatalytic properties in morph-TiO_2 from green-leaf biotemplates. *Advanced Functional Materials* 19 (1): 45–56.

28 Han, Z., Niu, S., Shang, C. et al. (2012). Light trapping structures in wing scales of butterfly *Trogonoptera brookiana*. *Nanoscale* 4 (9): 2879–2883.

29 Bechert, D.W., Bruse, M., and Hage, W. (2000). Experiments with three-dimensional riblets as an idealized model of shark skin. *Experiments in Fluids* 28 (5): 403–412.

30 Luchini, P., Manzo, F., and Pozzi, A. (1991). Resistance of a grooved surface to parallel flow and cross-flow. *Journal of Fluid Mechanics* 228: 87–109.

31 Bechert, D., Bruse, M., Hage, W. et al. (1997). Experiments on drag-reducing surfaces and their optimization with an adjustable geometry. *Journal of Fluid Mechanics* 338: 59–87.

32 Han, Z., Zhang, J., Ge, C. et al. (2012). Erosion resistance of bionic functional surfaces inspired from desert scorpions. *Langmuir* 28 (5): 2914–2921.

33 Liu, Y., Huang, L., and Shi, W. (2012). Structural color bio-engineering by replicating *Morpho* wings. *Advanced Materials Research* 391, 392: 409–417.

34 Saison, T., Peroz, C., Chauveau, V. et al. (2008). Replication of butterfly wing and natural lotus leaf structures by nanoimprint on silica sol-gel films. *Bioinspiration & Biomimetics* 3 (4): 046004.

35 Tanaka, H., Matsumoto, K., and Shimoyama, I. (2007). Fabrication of a three-dimensional insect-wing model by micromolding of thermosetting resin with a thin elastomeric mold. *Journal of Micromechanics and Microengineering* 17 (12): 2485.

36 Guo, X., Ding, Y., and Yu, G. (2021). Anode materials: design principles and applications of next-generation high-energy-density batteries based on liquid metals. *Advanced Materials* 33 (29): 2170226.

37 Fan, T.X., Chow, S.K., and Di, Z. (2009). Biomorphic mineralization: from biology to materials. *Progress in Materials Science* 54 (5): 542–659.

38 Zhang, T., Li, M., Su, B. et al. (2011). Bio-inspired anisotropic micro/nano-surface from a natural stamp: grasshopper wings. *Soft Matter* 7 (18): 7973.

39 Zang, X., Ge, Y., Gu, J. et al. (2011). Tunable optical photonic devices made from moth wing scales: a way to enlarge natural functional structures' pool. *Journal of Materials Chemistry* 21 (36): 13913–13919.

40 Caruso, R.A. and Antonietti, M. (2001). Sol–gel nanocoating: an approach to the preparation of structured materials. *Chemistry of Materials* 13 (10): 3272–3282.

41 Teo, C.J. and Khoo, B.C. (2009). Analysis of stokes flow in microchannels with superhydrophobic surfaces containing a periodic array of micro-grooves. *Microfluidics and Nanofluidics* 7 (3): 353–382.

42 Kim, Y., Jung, E., and Yim, M. (2002). Fabrication of discrete materials by interface-selective sol–gel polymerization. *Chemistry Letters* 31 (10): 992–993.

43 Kim, Y. (2003). Small structures fabricated using ash-forming biological materials as templates. *Biomacromolecules* 4 (4): 908–913.

44 Mille, C., Tyrode, E.C., and Corkery, R.W. (2013). 3D titania photonic crystals replicated from gyroid structures in butterfly wing scales: approaching full band gaps at visible wavelengths. *RSC Advances* 3 (9): 3109–3117.

45 Zhang, Z., Yu, K., Lou, L. et al. (2012). Morphology-controlled synthesis of ZnO replicas with photonic structures from butterfly (*Papilio paris*) wing scales for tunable optical properties. *Nanoscale* 4 (8): 2606–2612.

46 Cook, G., Timms, P.L., and Göltner-Spickermann, C. (2003). Exact replication of biological structures by chemical vapor deposition of silica. *Angewandte Chemie International Edition* 42 (5): 557–559.

47 Zhang, W., Zhang, D., Fan, T. et al. (2006). Biomimetic zinc oxide replica with structural color using butterfly (*Ideopsis similis*) wings as templates. *Bioinspiration & Biomimetics* 1 (3): 89–95.

48 Silver, J., Withnall, R., Ireland, T.G. et al. (2005). Novel nano-structured phosphor materials cast from natural *Morpho* butterfly scales. *Journal of Modern Optics* 52 (7): 999–1007.

49 Weatherspoon, M.R., Cai, Y., Crne, M. et al. (2008). 3D rutile titania-based structures with *Morpho* butterfly wing scale morphologies. *Angewandte Chemie International Edition* 120 (41): 8039–8041.

50 Zhang, W., Zhang, D., Fan, T. et al. (2009). Novel photoanode structure templated from butterfly wing scales. *Chemistry of Materials* 21 (1): 33–40.

51 Mille, C., Tyrode, E.C., and Corkery, R.W. (2011). Inorganic chiral 3-D photonic crystals with bicontinuous gyroid structure replicated from butterfly wing scales. *Chemical Communications* 47 (35): 9873–9875.

52 Chen, Y., Gu, J., Zhu, S. et al. (2009). Iridescent large-area ZrO_2 photonic crystals using butterfly as templates. *Applied Physics Letters* 94 (5): 053901.

53 Han, J., Su, H., Zhang, D. et al. (2009). Butterfly wings as natural photonic crystal scaffolds for controllable assembly of CdS nanoparticles. *Journal of Materials Chemistry* 19 (46): 8741–8746.

54 Han, Z., Niu, S., Yang, M. et al. (2013). An ingenious replica templated from the light trapping structure in butterfly wing scales. *Nanoscale* 5 (18): 8500–8506.

55 Nishiyama, R. and Yoshioka, S. (2020). Detailed analysis of photonic structure in the wing scale of Rajah Brooke's birdwing butterfly. *Optics Express* 28 (11): 16782–16794.

56 Gu, J., Zhang, W., Su, H. et al. (2015). Morphology genetic materials templated from natural species. *Advanced Materials* 27 (3): 464–478.

57 Tan, Y., Gu, J., Zang, X. et al. (2011). Versatile fabrication of intact three-dimensional metallic butterfly wing scales with hierarchical sub-micrometer structures. *Angewandte Chemie International Edition* 50 (36): 8307–8311.

58 Zheng, X., Shen, G., Wang, C. et al. (2017). Bio-inspired Murray materials for mass transfer and activity. *Nature Communications* 8: 14921.

59 Galusha, J.W., Jorgensen, M.R., and Bartl, M.H. (2010). Diamond-structured titania photonic-bandgap crystals from biological templates. *Advanced Materials* 22 (1): 107–110.

60 Zhou, H., Li, X., Fan, T. et al. (2010). Artificial inorganic leafs for efficient photochemical hydrogen production inspired by natural photosynthesis. *Advanced Materials* 22 (9): 951–956.

61 Schnepp, Z., Yang, W., Antonietti, M. et al. (2010). Biotemplating of metal carbide microstructures: the magnetic leaf. *Angewandte Chemie International Edition* 49 (37): 6564–6566.

62 Seipenbusch, M. and Binder, A. (2009). Structural stabilization of metal nanoparticles by chemical vapor deposition-applied silica coatings. *The Journal of Physical Chemistry C* 113 (48): 20606–20610.

63 Bao, Z., Weatherspoon, M.R., Shian, S. et al. (2007). Chemical reduction of three-dimensional silica micro-assemblies into microporous silicon replicas. *Nature* 446 (7132): 172–175.

64 Bao, Z., Ernst, E.M., Yoo, S. et al. (2009). Syntheses of porous self-supporting metal-nanoparticle assemblies with 3D morphologies inherited from biosilica templates (diatom frustules). *Advanced Materials* 21 (4): 474–478.

65 Toster, J., Iyer, K.S., Burtovyy, R. et al. (2009). Regiospecific assembly of gold nanoparticles around the pores of diatoms: toward three-dimensional nanoarrays. *Journal of the American Chemical Society* 131 (24): 8356–8357.

66 Payne, E.K., Rosi, N.L., Xue, C. et al. (2005). Sacrificial biological templates for the formation of nanostructured metallic microshells. *Angewandte Chemie International Edition* 117 (32): 5192–5195.

67 Weatherspoon, M.R., Dickerson, M.B., Wang, G. et al. (2007). Thin, conformal, and continuous SnO_2 coatings on three-dimensional biosilica templates through hydroxy-group amplification and layer-by-layer alkoxide deposition. *Angewandte Chemie International Edition* 119 (30): 5826–5829.

68 Liu, D., Gu, J., Liu, Q. et al. (2014). Metal–organic frameworks reactivate deceased diatoms to be efficient CO_2 absorbents. *Advanced Materials* 26 (8): 1229–1234.

69 Garrett, N.L., Vukusic, P., Ogrin, F. et al. (2009). Spectroscopy on the wing: naturally inspired SERS substrates for biochemical analysis. *Journal of Biophotonics* 2 (3): 157–166.

70 Martín-Palma, R.J., Pantano, C.G., and Lakhtakia, A. (2008). Biomimetization of butterfly wings by the conformal-evaporated-film-by-rotation technique for photonics. *Applied Physics Letters* 93 (8): 083901.

71 Lakhtakia, A., Martín-Palma, R.J., Motyka, M.A. et al. (2009). Fabrication of free-standing replicas of fragile, laminar, chitinous biotemplates. *Bioinspiration & Biomimetics* 4 (3): 034001.

72 Gaillot, D.P., Deparis, O., Welch, V. et al. (2008). Composite organic–inorganic butterfly scales: production of photonic structures with atomic layer deposition. *Physical Review E* 78 (3): 031922.

73 Tan, Y., Gu, J., Xu, L. et al. (2012). High-density hotspots engineered by naturally piled-up subwavelength structures in three-dimensional copper butterfly wing scales for surface-enhanced Raman scattering detection. *Advanced Functional Materials* 22 (8): 1578–1585.

74 Mu, Z., Zhao, X., Xie, Z. et al. (2013). In situ synthesis of gold nanoparticles (AuNPs) in butterfly wings for surface enhanced Raman spectroscopy (SERS). *Journal of Materials Chemistry B* 1 (11): 1607–1613.

75 Zang, X., Tan, Y., Lv, Z. et al. (2012). Moth wing scales as optical pH sensors. *Sensors and Actuators B: Chemical* 166, 167: 824–828.

76 Peppas, N.A., Hilt, J.Z., Khademhosseini, A., and Langer, R. (2006). Hydrogels in biology and medicine: from molecular principles to bionanotechnology. *Advanced Materials* 18 (11): 1345–1360.

77 Gu, J., Zhang, W., Su, H. et al. (2015). Biomimetics: morphology genetic materials templated from natural species. *Advanced Materials* 27 (3): 394.

78 Leem, J.W., Chung, K.S., and Yu, J.S. (2012). Antireflective properties of disordered Si SWSs with hydrophobic surface by thermally dewetted Pt nanomask patterns for Si-based solar cells. *Current Applied Physics* 12 (1): 291–298.

79 Min, W., Jiang, B., and Jiang, P. (2008). Bioinspired self-cleaning antireflection coatings. *Advanced Materials* 20 (20): 3914–3918.

80 Li, Y., Zhang, J., Zhu, S. et al. (2009). Biomimetic surfaces for high-performance optics. *Advanced Materials* 21 (46): 4731–4734.

81 Li, Y., Li, F., Zhang, J. et al. (2010). Improved light extraction efficiency of white organic light-emitting devices by biomimetic antireflective surfaces. *Applied Physics Letters* 96 (15): 153305.

82 Kanamori, Y., Ishimori, M., and Hane, K. (2002). High efficient light-emitting diodes with antireflection subwavelength gratings. *IEEE Photonics Technology Letters* 14 (8): 1064–1066.

83 Zhu, S., Zhang, D., Li, Z. et al. (2008). Precision replication of hierarchical biological structures by metal oxides using a sonochemical method. *Langmuir* 24 (12): 6292–6299.

84 Elam, J.W., Routkevitch, D., Mardilovich, P.P. et al. (2003). Conformal coating on ultrahigh-aspect-ratio nanopores of anodic alumina by atomic layer deposition. *Chemistry of Materials* 15 (18): 3507–3517.

85 Huang, J., Wang, X., and Wang, Z.L. (2006). Controlled replication of butterfly wings for achieving tunable photonic properties. *Nano Letters* 6 (10): 2325–2331.

86 Kolle, M., Salgard-Cunha, P.M., Scherer, M.R.J. et al. (2010). Mimicking the colourful wing scale structure of the *Papilio blumei* butterfly. *Nature Nanotechnology* 5 (7): 511–515.

87 Ritala, M., Kukli, K., Rahtu, A. et al. (2000). Atomic layer deposition of oxide thin films with metal alkoxides as oxygen sources. *Science* 288 (5464): 319–321.

88 Linn, N.C., Sun, C.H., Jiang, P. et al. (2007). Self-assembled biomimetic antireflection coatings. *Applied Physics Letters* 91 (10): 101108.

89 Zhang, D., Zhang, W., Gu, J. et al. (2015). Inspiration from butterfly and moth wing scales: characterization, modeling, and fabrication. *Progress in Materials Science* 68: 67–96.

90 Chung, K., Yu, S., Heo, C.J. et al. (2012). Flexible, angle-independent, structural color reflectors inspired by *Morpho* butterfly wings. *Advanced Materials* 24 (18): 2375–2379.

91 Ho, Y.H., Liu, C.C., Liu, S.W. et al. (2011). Efficiency enhancement of flexible organic light-emitting devices by using antireflection nanopillars. *Optics Express* 19 (10): A295–A302.

92 Kang, P., Ogunbo, S.O., and Erickson, D. (2011). High resolution reversible color images on photonic crystal substrates. *Langmuir* 27 (16): 9676–9680.

93 Fudouzi, H. and Xia, Y. (2003). Photonic papers and inks: color writing with colorless materials. *Advanced Materials* 15 (11): 892–896.

94 Chrisey, D.B. (2000). Materials processing: the power of direct writing. *Science* 289 (5481): 879–811.

95 Turner, M.D., Schroeder-Turk, G.E., and Gu, M. (2011). Fabrication and characterization of three-dimensional biomimetic chiral composites. *Optics Express* 19 (10): 10001–10008.

96 Hardcastle, F.D., Ishihara, H., Sharma, R. et al. (2011). Photoelectroactivity and Raman spectroscopy of anodized titania (TiO_2) photoactive water-splitting catalysts as a function of oxygen-annealing temperature. *Journal of Materials Chemistry* 21 (17): 6337–6345.

97 Xu, Z., Yu, K., Li, B. et al. (2011). Optical properties of SiO_2 and ZnO nanostructured replicas of butterfly wing scales. *Nano Research* 4 (8): 737–745.

98 Han, Z., Mu, Z., Li, B. et al. (2016). A high-transmission, multiple antireflective surface inspired from bilayer 3D ultrafine hierarchical structures in butterfly wing scales. *Small* 12 (6): 713–720.

99 Kustandi, T.S., Low, H.Y., Teng, J.H. et al. (2009). Mimicking domino-like photonic nanostructures on butterfly wings. *Small* 5 (5): 574–578.

100 Watanabe, K., Hoshino, T., Kanda, K. et al. (2005). Brilliant blue observation from a *Morpho*-butterfly-scale quasi-structure. *Japanese Journal of Applied Physics* 44 (1): L48–L50.

101 Vilà-Nadal, L. and Cronin, L. (2017). Design and synthesis of polyoxometalate-framework materials from cluster precursors. *Nature Reviews Materials* 2 (10): 17054.

102 Li, J., Li, J., Sun, J. et al. (2019). Biological and engineered topological droplet rectifiers. *Advanced Materials* 31 (14): 1806501.

103 Grzybowski, B.A., Stone, H.A., and Whitesides, G.M. (2000). Dynamic self-assembly of magnetized, millimetre-sized objects rotating at a liquid–air interface. *Nature* 405 (6790): 1033–1036.

104 Ichimura, K., Oh, S.K., and Nakagawa, M. (2000). Light-driven motion of liquids on a photoresponsive surface. *Science* 288 (5471): 1624–1626.

105 Pollack, M.G., Fair, R.B., and Shenderov, A.D. (2000). Electrowetting-based actuation of liquid droplets for microfluidic applications. *Applied Physics Letters* 77 (11): 1725–1726.

106 Mugele, F. and Baret, J.C. (2005). Electrowetting: from basics to applications. *Journal of Physics: Condensed Matter* 17 (28): R705.

107 Malvadkar, N.A., Hancock, M.J., Sekeroglu, K. et al. (2010). An engineered anisotropic nanofilm with unidirectional wetting properties. *Nature Materials* 9 (12): 1023–1028.

108 He, R., Wang, Y.C., Wang, X. et al. (2014). Facile synthesis of pentacle gold-copper alloy nanocrystals and their plasmonic and catalytic properties. *Nature Communications* 5 (1): 4327.

109 Chaudhury, M.K., Chakrabarti, A., and Daniel, S. (2015). Generation of motion of drops with interfacial contact. *Langmuir* 31 (34): 9266–9281.

110 Lv, J., Liu, Y., Wei, J. et al. (2016). Photocontrol of fluid slugs in liquid crystal polymer microactuators. *Nature* 537 (7619): 179–184.

111 Kwon, G., Panchanathan, D., Mahmoudi, S.R. et al. (2017). Visible light guided manipulation of liquid wettability on photoresponsive surfaces. *Nature Communications* 8 (1): 14968.

3

Bioinspired Light-Trapping Structural Surfaces

For solar cells [1, 2], optical losses are one of the major obstacles that hinder their efficiency improvement [3]. The optical reflection loss on the surface of solar cells can be reduced in two ways: using reflection-reducing films [4] and light-trapping structures, where light-trapping structures are made by creating some surface structures to reduce the reflectivity of the surface.

Light absorption efficiency increases with the increase in the optical path length. The light-trapping surface that can effectively increase the light range will significantly improve light absorption efficiency [5]. However, the actual light-trapping performance or light-trapping efficiency of these optical structures does not reach the ideal state. Meanwhile, these optical structures have not achieved an ideal antireflective effect for ultraviolet (UV) light. Low efficiency of solar cells has become a bottleneck in the development of the solar photovoltaic industry. Therefore, the search for optimal and efficient light-trapping functional surfaces has become a hot and challenging research area in photovoltaics as well as in space exploration.

3.1 Definition and Classification of Light-Trapping Structure

Light-trapping structures increase the light range in solar cells by dispersing incident light to various angles through reflection, refraction, and scattering effects, resulting in a significant increase in the efficiency with which light is absorbed. Numerous studies have shown that the introduction of light trapping structures in solar cells is beneficial to increase the short-circuit current and conversion efficiency of solar cells. Currently, many kinds of trapped optical surfaces have been developed, such as honeycomb surfaces [6], sinusoidal grating weaving [7], dimple-shaped ordered surfaces [8], periodic pyramidal and inverted pyramidal structures [9, 10], and binary gratings [11]; these optical trapping surfaces have been well used in optical devices in many fields. However, the actual light-trapping performance or light-trapping efficiency of these optical structures does not reach the ideal state. Therefore, the search for optimal and efficient light-trapping functional surfaces has become a hot and challenging research area in photovoltaics.

Nature-Inspired Structured Functional Surfaces: Design, Fabrication, Characterization, and Applications, First Edition. Zhiwu Han.

Fortunately, nature provides a rich selection of micro-/nanostructures with high-quality potential applications for light-trapping properties. Through continuous evolution and natural selection, the best functional surface structures have been inherited from generation to generation by organisms in nature that challenge the extreme environments they are exposed to. They provide intelligent answers to scientific and engineering challenges and continue to inspire us to come up with new designs as well as high-performance structures. As a result, the study of biological micro-/nanostructures is becoming a hot topic in the bio-nanotechnology field.

3.1.1 Geometry in Light Trapping Structure

As shown in Figure 3.1, this is a typical two-dimensional (2D) model of a geometric light trapping structure, where a layer of grooves with dimensions larger than the wavelength of light is crocheted on the substrate surface, and its presence will change the structure of the optical path of the incident light. When incident light 1 vertically illuminate on the trench wall with a angle of a at point A, the first reflection and refraction occur at the same time. Then, reflected light 2 and refracted light 4 can be generated. Further, when the reflected light 2 reaches the opposite trench wall at point B, it will get a second incidence chance to generate reflected light 3 and refracted light 5, and so on.

Assuming that all incident light is perpendicular to the plane of the solar cell sheet, after geometric calculation, the size of a directly determines how many times the initial incident light has the chance to be re-incident. When a exceeds 30°, the incident light near the bottom of the trench will get a second chance to be incident on the cell surface; when a exceeds 45°, all incident light will be incident on the cell surface twice; when a exceeds 54°, the incident light near the bottom of the trench may be "thrice incident"; when a reaches 60° and above, all incident light will get a second chance to be incident on the cell surface. All incident rays will get the chance of "three incidences." Obviously, a larger trench bottom angle will always result in more incidence of incident light. If the surface of the medium has a primary reflectance of R $(0 < R < 1)$, the apparent reflectance of the medium decreases to Rn after n incidences, so the presence of a groove will make some light get multiple incidences and greatly reduce the total reflectance compared to a flat surface. Geometric trapping changes the structure of the optical path and is mainly used on surfaces that can absorb light themselves. For example, in the preparation of crystalline silicon solar cells, silicon wafers need to go through a fleece-making process before preparing the p–n junction [1], and for monocrystalline silicon, the commonly used method in industry is alkali etching to create micron-sized pyramidal structures,

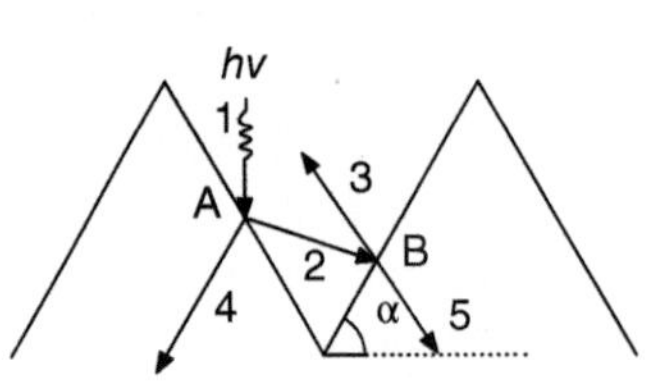

Figure 3.1 Schematic diagram of the geometric light-trapping structure.

while polycrystalline silicon is etched by acid to create micron-sized pits, both of which reduce light reflection losses by introducing geometric trap structures.

Above theoretical analysis is limited to the case when light is incident vertically. However, when incident light deviates from vertical angle, the relationship analysis between reflection number and groove angle will become somewhat blurred. Therefore, to help light get "multiple incidence" opportunity, it is necessary to prepare a transparency-enhancing film to realize reflection reduction.

3.1.2 The Principle of Light Trapping in the Film

As shown in Figure 3.2, according to the principle of optical interference phase extinction, a light-transmitting film with refractive index between the substrate and air is prepared, where the refractive index of light in air is $n_0 = 1$, the refractive index of the film is denoted as n_1, the refractive index of the substrate is denoted as n_2, and the thickness of the film is denoted as d. Assuming that incident light has the same incident intensity and wavelength throughout the film, which is recorded as λ_0 in the air and λ_1 in the film, these two satisfy the following law: $n_1/n_0 = \lambda_0/\lambda_1$. And only considering transmission and reflection of light, when light a is incident at angle i on the surface of the film and refracted into the film at angle r at point A, then reflected at point C and refracted back into the air at point B, and light b is incident from the air at the same angle i and reflected directly into the air at point B, then light a and light b are incident at the same angle i and reflected directly into the air. Then the difference in the optical range of the two incident rays a and b is calculated geometrically as

$$\delta = n_1(\text{AC} + \text{CB}) - \text{DB} \tag{3.1}$$

According to the principle of interference of light, if δ satisfies the following equation, the rays a and b will interfere and cancel at point B.

$$\delta = (2k + 1)\,\lambda/2, k = 0, 1, 2, 3 \tag{3.2}$$

Based on the geometric relationship, it follows that

$$\text{AC} = \text{CB} = d/\cos r \tag{3.3}$$

$$\text{DB} = \text{AB} \times \sin i = 2d \times \tan r \times \sin i \tag{3.4}$$

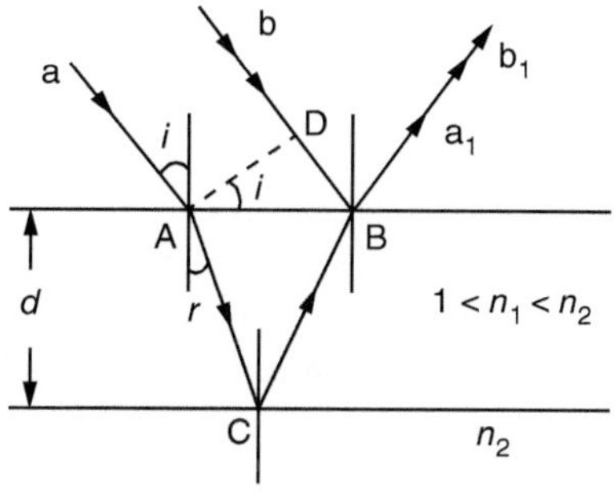

Figure 3.2 Principle diagram of reflected light interference. Source: Based on Dennler et al. [2].

And by the law of refraction: $\frac{n_1}{n_0} = \frac{\sin i}{\sin r}$. Combined with Eqs. (3.1), (3.3), and (3.4), the following is obtained:

$$\delta = n_1 \times \frac{2d}{\cos r} - 2d \times \tan r \times \sin i = 2n_1 d \times \cos r \tag{3.5}$$

In general, if light is perpendicular to the surface of the film, then both i and r are zero; the above equation can be cylindrical as $\delta = 2n_1 d$. When d is 1/4 of the wavelength of the incident light in the film, $d = 1/4\lambda_1$:

$$\delta = \frac{n_1 \lambda_1}{2} = \lambda_0/2 \tag{3.6}$$

At this time, the reflected light is eliminated by a thousand waders on the film surface. The following describes the inversion process of the optimal refractive index of the single-layer permeability-enhancing film. The reflectance of the sample surface after the Yung film is recorded as R and then by Fresnel's law:

$$R = \frac{r_1^2 + r_2^2 + 2r_1 r_2 \cos \Delta}{1 + r_1^2 + r_2^2 + 2r_1 r_2 \cos \Delta} \tag{3.7}$$

where r_1, r_2 are the Fresnel reflection coefficients of the external medium (here air) film and film-substrate surfaces, respectively, and $\Delta$is the dislocation angle due to the thickness of the film layer. In addition,

$$r_1 = \frac{n_0 - n_1}{n_0 + n_1} \tag{3.8}$$

$$r_2 = \frac{n_1 - n_2}{n_1 + n_2} \tag{3.9}$$

$$\Delta = \frac{4\pi d}{\lambda_0} \tag{3.10}$$

When light is incident vertically and the film thickness $d = 1/4\ \lambda_1$, $d = \lambda_0/4n_1$, the derivation of the above equation can be rewritten as

$$R = \left(\frac{n_1^2 - n_0 \times n_2}{n_1^2 + n_0 \times n_2} \right)^2 \tag{3.11}$$

Obviously, when $n_1 = \sqrt{n_0 \times n_2}$, R takes the minimum value 0.

For the above reasoning, a single-layer homogeneous transparency-enhancing film with optimal reflection reduction effect when light is directed should have both of the following characteristics: (i) the thickness of the film should be adjusted to one-fourth of the wavelength of the incident light in the film; (ii) the refractive index of the film should be adjusted to the square root of the product of the refractive index of the external medium (generally air) and the substrate medium.

The thickness of the transparency-enhancing film can be precisely controlled by certain means, but the single-layer film can only achieve excellent reflection reduction for a specific wavelength of light and cannot achieve broad-spectrum transparency. In addition, for the commonly used light-transmitting material glass (refractive index of 1.5 or so), the refractive index of its surface transparency film should be 1.22, while MgF_2 (refractive index of 1.38), as the inorganic material with the lowest refractive index in nature, still fails to meet requirement. Therefore, to improve the transparency performance of the film, researchers mainly adopt the following two ideas: first, the fruit of multilayer reflectance reduction film formed by superimposing multiple layers of uniformly permeable films with different refractive indices in the order of gradually increasing refractive indices; second, the introduction of nanopores in the film by certain physical or chemical means to produce a nonuniform reflectance reduction film with gradient refractive index.

3.2 Ultraviolet Light-Trapping Structures Derived from *Parnassius* Butterfly Wings

3.2.1 UV-ARS Mechanism of Original Butterfly Wings

Niu and coworkers [12] took butterfly as a bionic prototype and confirmed the prominent UV-selective antireflection effect of *Parnassius* butterfly. An accurate SiO_2 inverse replica of this nano-optical functional surface from butterfly wings was fabricated in multiscale using a synthesis method combining a sol–gel process and subsequent selective etching. It was found that the original ultraviolet-antireflection structures (UV-ARSs) of bio-template are well inherited by the replica including spacing, width, distribution, shape, formation, and so forth.

3.2.1.1 Reflective Spectra of Original Butterfly Wings

The surprisingly low reflectivity of butterfly in UV spectral region was confirmed by scientist, which reveals the secret that how butterfly can struggle against such an extremely bad environment. The digital photos of three species of original *Parnassius* butterfly are shown in Figure 3.3a–c. The widths of the forewing-span are 3.4, 3.0, and 2.3 cm, respectively. Figure 3.3d–f clearly shows the optical microscope images of the scales, including dimension, shape, and distribution of entire scales. It is found that the scales cover the wings with a shape of three leaves, which arrange regularly from the front to the end of the wings in the same sequence just like the tiles on the roof. The scales overlap and have the same angle (approximately 30°) to the wing plane. Under the optical microscope, each single scale is like a tiny minority shield with 80–100 μm in width and 140–200 μm in length. It is clear that these three species of butterflies have the same shape, similar size, and analogous distribution of scales. Figure 3.3g obviously shows that the reflectivity of both visible and infrared light range is always higher than 60% and that of UV range is all lower than 20%. Furthermore, the reflectivity of the range from 200 to 340 nm is lower than 10%. Therefore, this phenomenon is in accordance with our original expectations. All three of these butterflies have an outstanding antireflective effect for the UV light.

3.2.1.2 3D Visible Parameterized Models of Butterfly Feature Structures

The morphologies and microstructures of butterfly wing scales were observed clearly through field emission scanning electron microscope (FESEM) and transmission electron microscope (TEM). Figure 3.4 shows scanning electron microscope

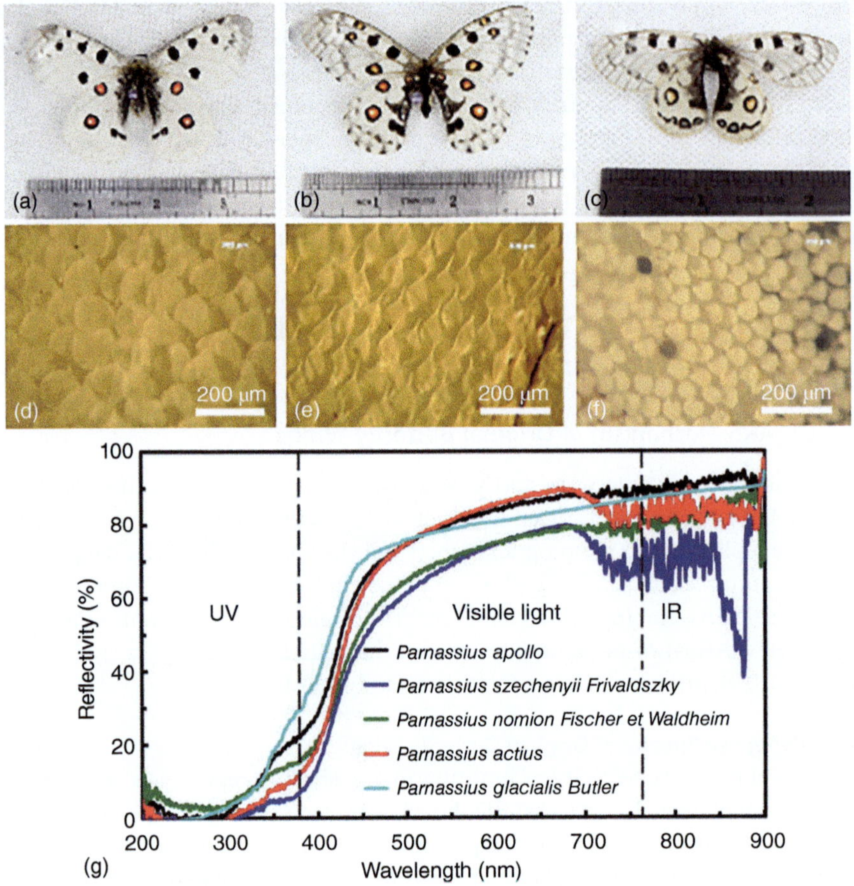

Figure 3.3 The appearance and the reflectance spectra of the original butterfly wings: (a) *Parnassius apollo*, (b) *Parnassius nomion*, and (c) *Parnassius szechenyii*. (d–f) The optical microscope images of the front surface of these three species of butterfly. (g) The reflectance spectra of five *Parnassius* butterfly, from which it is confirmed that all these butterflies have a lower reflectivity below 20% in the range of 200–350 nm (ultraviolet region) and a higher reflectivity in the visible light. So, this butterfly has excellent antireflective effect for the UV light. Source: Reprinted with permission from Han et al. [13]. Copyright 2013, American Institute of Physics.

(SEM) and TEM images of two kinds of butterflies (*Parnassius apollo* and *Parnassius nomion*) and a visible model. Figure 3.4a–c presents the images of the cover scales of *Parnassius apollo*. Figure 3.4d–f shows the images of the cover scales of *Parnassius nomion*. The average typical dimension of the scales is ~200 μm long and ~120 μm wide as seen in Figure 3.4a,d. The ridge separation distances in the cover scale are approximately 2 μm (Figure 3.4b,e), which are also consistent with that of back side scales of the respective butterflies. Figure 3.4c,f presents the enlarged top-view images of the comb teeth structures on the ridges. There is an inclination angle of each comb tooth relative to the surface of the base substrate. For description convenience, we define the direction from up to downside of the inclined shelf as

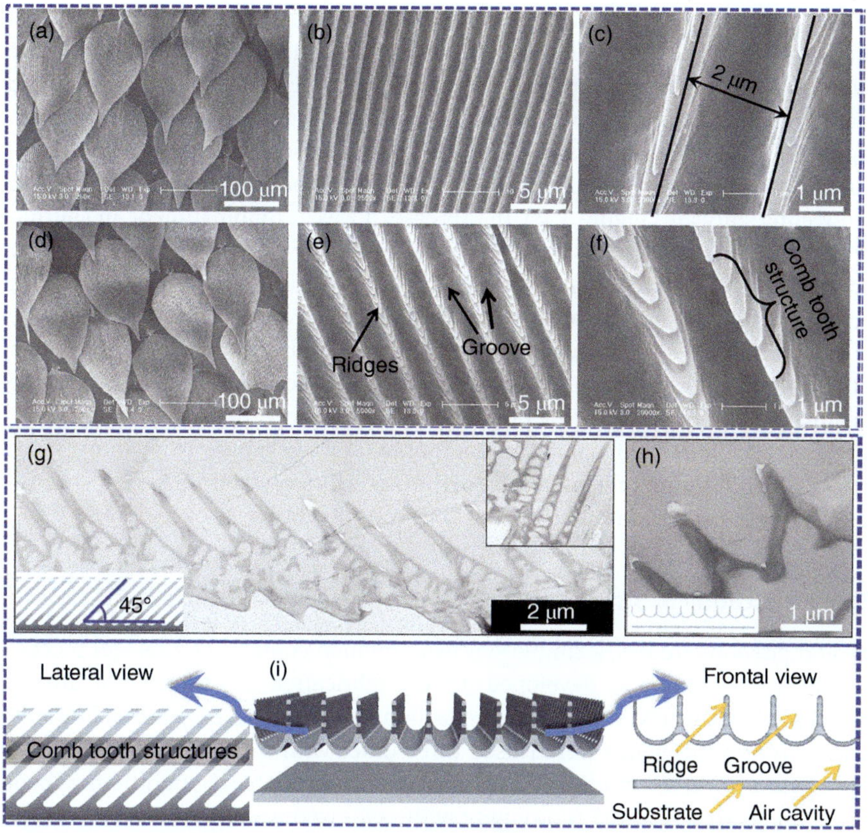

Figure 3.4 SEM and TEM images of two kinds of butterflies (*Parnassius apollo* and *Parnassius nomion*) and a visible 3D model. (a–c) Images of the front scales of butterfly *Parnassius apollo*. (d–f) Images of the front scales of butterfly *Parnassius nomion*. (g) Cross-sectional images of the cover scales viewing along their ridges. The inclination angle of each comb teeth relative to their flat base plane is about 45°. (h) Ultrastructure of the cross section perpendicular to the ridges. Obviously, it looks like a U-type structure, and its depth is about 2 µm. (i) 3D models built from feature size of the scales in *Parnassius* butterfly wings. Source: Reprinted with permission from Han et al. [13]. Copyright 2013, American Institute of Physics.

the tail to head. The distance between adjacent heads in the directions parallel to the ridge is about 1 µm. The shape of comb teeth gradually becomes more flat and narrow from the tail to head.

To explore the 3D structure of the scales, cross-sectional samples were prepared and investigated by TEM as shown in Figure 3.4g,h. The morphologies along with their ridges (the slicing is parallel to the ridge) are shown in Figure 3.4g, from which it is evident that the inclination angle of each comb tooth relative to their flat base plane is about 45° (the illustration in the lower-left corner of Figure 3.4g). More novel, by carefully observing the enlarged image of one comb teeth structure (the illustrations in the upper-left corner of Figure 3.4g), some oval air bags can be seen embedded in the comb teeth structures . The slender main comb teeth are ~8 µm in length and ~500 nm in average thickness with ~700 nm of adjacent spacing. Two adjacent ribs are filled with air. On a slightly larger scale, Figure 3.4h reveals the architectures of section perpendicular to the ridges. Obviously, it looks like a U-type structure and its depth is about 2 μm. Then, combining with the SEM and TEM images of the scales in *Parnassius* butterfly and its variety of feature size, the 3D optimized model was formed. After that, a visible 3D model was described by using 3D modeling software (Figure 3.4i). In this case, it was a combination of an alternative multilayer (chin and air) structures (viewing from the lateral side as left part in Figure 3.4i) and a diffraction grating (viewing from the frontal side as right part in Figure 3.4i). This design utilizes the beam redirecting property of the grating while still maintaining the selective characteristics of the 1D-PhC. In addition, this design cut through the layers of the 1D-PhC stack to generate the gratings and is hence able to seamlessly combine the diffraction and reflection properties of the gratings for optimal optical path length enhancement for the UV wavelength. This arrangement enhances the light-trapping capabilities of the butterfly scales within the UV spectral while allowing visible light wavelengths to be reflected back.

3.2.2 Fabrication of Structural Butterfly-Inspired UV-ARS Surfaces

Based on previous studies, a simple synthesis method combining a sol–gel process and selective etching was developed for the fabrication of an accurate inverse replica based on this UV-selective antireflection surface. First, the natural butterfly (*Parnassius nomion*) wings were rinsed three times with 0.65% NaCl solution and dipped into a solution of ether for 10 minutes under ambient conditions. Then, the specimens were washed thoroughly with deionized water and dried in air for 12 hours and subjected to a series of dehydration operations in 40%, 60%, 80%, 95%, and pure ethanol for 10 minutes for each phase and washed thoroughly with deionized water and then dried in air for 12 hours. Afterward, slices of butterfly wings (1 cm × 2 cm) were sandwiched between two glass slides (Figure 3.5a), and a suitable amount of the silica precursor, a volume of 5–8 μl, was added to the edge of the assembly (Figure 3.5b). Next, the entire assembly was heated at 130 °C for 25 minutes to evaporate residual solvent (Figure 3.5c). The whole assembly was dipped into a mixture of concentrated nitric acid and perchloric acid (1 : 1 in volume) while heating at 12 °C for 30 minutes (Figure 3.5d). The whole assembly was washed by ultrasonic oscillation for 10–20 minutes with deionized water. In fact, Figure 3.5e,f is the internal process between the original bio-template and the precursor corresponds to Figure 3.5d,c. The sol–gel precursor filled the space left between the ridges and was solidified; as a result, the notches become ridges and the comb teeth structures on ridges became oblique cone holes at the bottom of the notches.

3.2.3 Characterizations of Butterfly-Inspired UV-AR Surfaces

The morphology and structural features of the natural wings and the SiO_2 replica are presented in high magnification in Figures 3.3, 3.4, and 3.6. Lower magnification

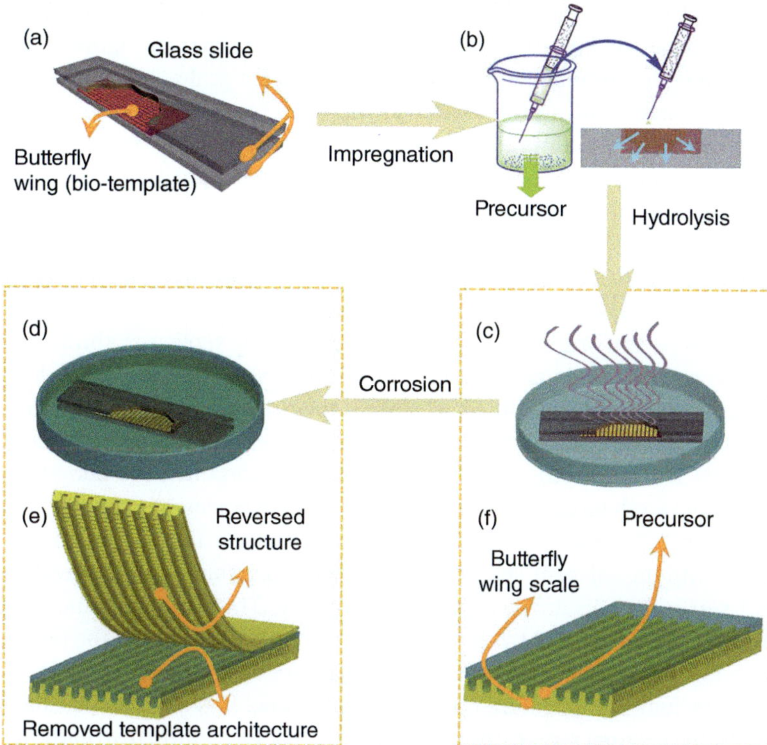

Figure 3.5 A schematic illustration for the fabrication process of the inverse replica based on the butterfly wing scales. (a) The clean slices of butterfly wings were sandwiched between two glass slides. (b) A suitable amount of the SiO_2 precursor was added to the edge of the assembly. (c) The entire assembly was heated at 120 °C for 20 minutes to evaporate residual solvent. (d) The butterfly wing slices were removed by acid-etching, the whole assembly in a mixture of concentrated nitric acid and perchloric acid (1 : 1 in volume) while heating at 150 °C for 20 minutes. (e and f) The internal process between the original bio-template and the precursor corresponds to (d) and (c), respectively. Source: Han et al. [12]/with permission of AIP Publishing LLC.

Figure 3.6 Images of the SiO_2 inverse replica of the UV-ARS wing scales. (a) The lower magnification image of the inverse replica. (b) The medium magnification image of the inverse replica. Ridges are lying in parallel, with an adjacent spacing of about 2 μm, agreeing with the width of the original separation distances analyzed previously. (c) Higher magnification of the inverse replica. The diameters of the inverse comb teeth structures were in the range 500–650 nm. Source: Reprinted with permission from Han et al. [13]. Copyright 2013, American Institute of Physics.

images of the SiO_2 inverse structure replica based on the UV-ARS are shown in Figure 3.6a. The images in Figure 3.3d–f are quite similar to that in Figure 3.6a, which indicates that the arrangement of the inverse replica scales is completely and accurately consistent with that of scales in the wing bio-templates. Scales of the replica are arranged in an ordered fashion in parallel, just like the scale of the UV-ARS wing shown in Figure 3.3d–f. From the upper-right illustration in Figure 3.6a, it is obvious that the width, length, and distribution of the replica of the scale are highly consistent with the original butterfly scales in Figure 3.4a,d. The slight difference between them may be caused because the original wing scales overlap one another. At a medium magnification, Figure 3.4e (5000×) shows the UV-ARS details of the original template, while the SiO_2 replica is shown in Figure 3.6b (5000×). It is obvious that the sol–gel precursor filled the space left between the ridges and became solidified; as a result the ridges became grooves and the grooves became ridges. In fact, the comb teeth structures on the ridges became oblique cone holes at the bottom of the grooves. More importantly, the period of the grating and comb teeth structures of the replica is highly consistent with that of the original structures. The medium magnification images show that the ridges are lying in parallel, with an adjacent spacing of about 2 μm, agreeing with the width of the original separation distances analyzed previously. This confirms that the original grooves are well inherited by the ridges of the inverse replica. On the other hand, the comb teeth structures on the original ridges are also well inherited by a complex shallow trough-like structure shown in Figure 3.6c. It is obvious that this replica resembles the original structures, including the spacing, width, distribution, shape, and formation. The diameters of the inverse comb teeth structures are in the range 500–650 nm. The sizes and shapes of these inverse structures match those of the original ones shown in Figure 3.4f, which confirms that these replicas are the inverse structures. The results shown above clearly indicate that the process can create a perfect inverse replica of this biological template. If the size shrinking effect is not considered, the inverse replica inherits the inverse structures of the template completely, including the nanometer-level venations, the micrometer-level ordered "ribs," and the ordered arranged organic scales.

3.3 Light-Trapping Surfaces for Visible Light Inspired by Butterfly Wings

3.3.1 Light-Trapping Mechanism of Original Butterfly Wings

Han et al. took butterfly as a bionic prototype and studied the relationship between surface exquisite structure and excellent light-trapping performance in detail. Butterflies are famous for their beautiful wing colors, which is the result of the interaction of microstructures and light on their wings. Also, the colors are more lasting and stable than chemical dye colors [14]. The structures on the wings of different butterflies are different, which results in variations in reflectivity. That is the reason why different wing colors can be observed. The relationship between the structures and functions was investigated from the aspects of theory and experiment.

3.3.1.1 Light-Trapping Performance of Original Butterfly Wings

To confirm that colors on butterfly wing scales is caused by structures, here, butterfly *Trogonoptera brookiana* was selected as the subject of the discoloration experiment. The appearance of the original wing in butterfly *T. brookiana* is quite different when we observe it from the front and back (Figure 3.7a,b). The scales on the front wing exhibit a uniform arrangement (Figure 3.7c). Reflection spectra in Figure 3.7d confirm a weaker signal in blue than that in the green region. During the discoloration experiment, it was found that the scale color changed to green when the alcohol was added. However, with the volatilization of alcohol, the color of the wing scales gradually recovered to their original blue. The whole discoloration process is shown in Figure 3.7e. In the control group, butterfly wings were soaked in ethanol for five minutes, and the extracted pigment was not detected. Besides, there was no residual color on the petri dish in the next experiment. Based on this, we experimentally verified the inexistence of the pigment on the butterfly wings, and the scale color is structure based.

This finding has also been verified in butterfly *Papilio peranthus* (Figure 3.8). It is concluded from the discoloration experiment that the scale color variation is caused by the change induced by ethanol rather than the loss of the pigment. This result is also consistent with our previous anticipated results. In fact, the principle of the ethanol discoloration experiment is that the refractive index of the multilayered filling medium on the butterfly wing scale is changed by ethanol solution. Then, the color changes immediately. After the alcohol is partially evaporated, the filling medium becomes a mixture of air and alcohol, so another color appears on the wings. When the alcohol is completely evaporated, the filling medium is restored to

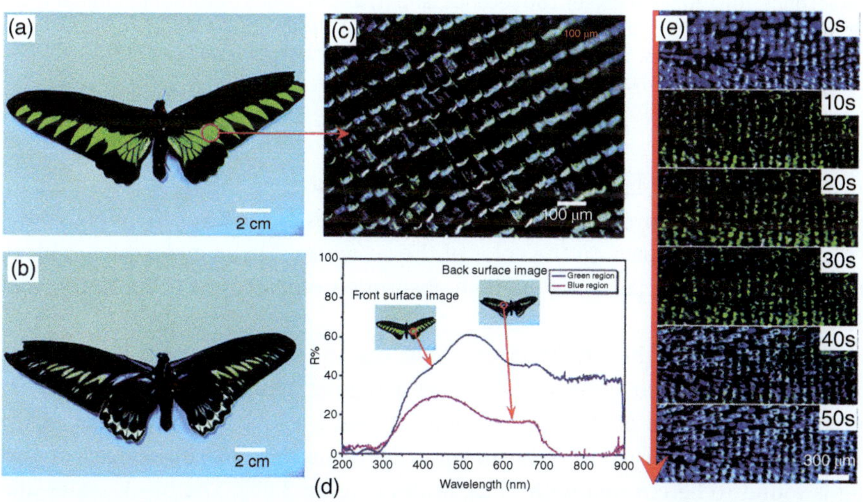

Figure 3.7 The appearance of the front (a), back (b), and optical microscope image (c) of the original butterfly (*Trogonoptera brookiana*) wing. (d) The reflectance spectra of different regions. (e) The discoloration experiments. Source: Reprinted with permission from Han et al. [5]. Copyright 2012, The Royal Society of Chemistry.

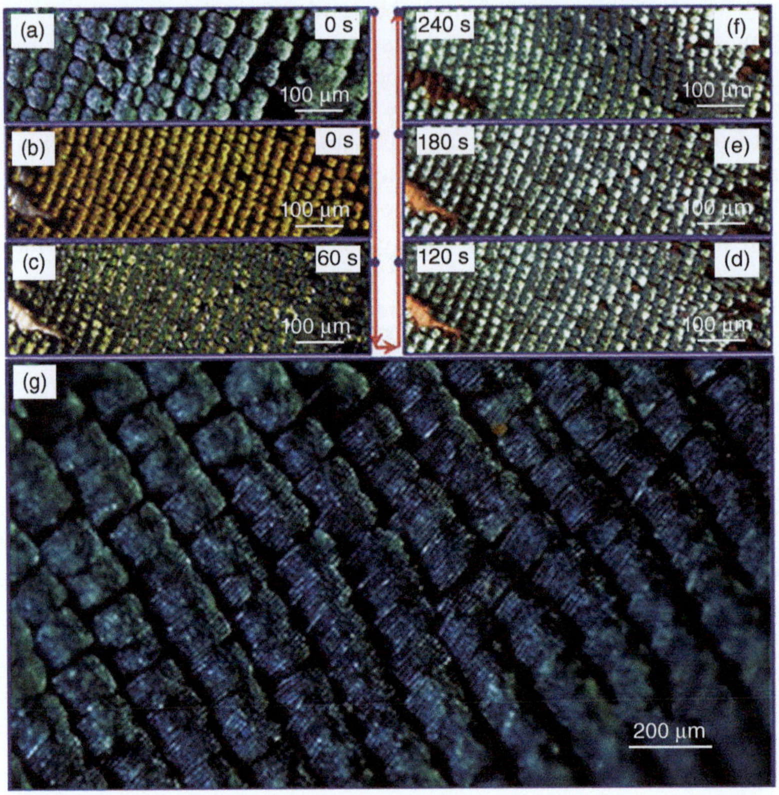

Figure 3.8 The four-minute discoloration process after (a) 0 seconds, (b) 0 seconds, (c) 60 seconds, (d) 120 seconds, (e) 180 seconds, and (f) 240 seconds (returning to initial blue color) indicates that the scale color is structure based. (g) Optical microscope image of butterfly *Papilio peranthus*. Source: Reprinted with permission from Han et al. [15]. Copyright 2013, Springer Nature.

its original state (multi-layer film of air and chitin), and then the color is restored to a bright green. The morphologies and microstructures of the butterfly wing scales have an important impact on the color of the structure. The complex hierarchical structure is also a key factor in the high absorption efficiency of other non-green visible light bands.

In fact, except for the bright green scale region, the black part of butterfly *Trogonoptera brookiana* is also covered with regular wing scales. As shown in Figure 3.9a, the wings of the original butterfly have a long, smooth, black strip. Through further observation with an optical microscope, the black part of the wing is found to be covered with black scales (Figure 3.9b). The black scales are alternately arranged in rows, and the scales overlap each other. Reflection spectra in the range of 400–900 nm in Figure 3.9c confirm that the black scales of butterflies perform better in the field of light-trapping area. Therefore, the black region is suitable as an experimental subject of broadband light trapping.

Figure 3.9 The macroscopic morphology of the butterfly wings and the reflectance spectroscopy of the black and green region of the butterfly wings. (a) Photograph of butterfly *Trogonoptera brookiana*. (b) Optical microscope image of the black butterfly wing scales. (c) The novel light-trapping property of the black wing scales was confirmed in the entire wave range. Source: Reprinted with permission from Han et al. [16]. Copyright 2015, Springer Nature.

3.3.1.2 3D Visible Parameterized Models of Butterfly Feature Structures

The morphologies and microstructures of butterfly wing scales were observed clearly through FESEM and TEM. The results in Figure 3.10 verify that the morphologies of butterfly scales are different, and the wing scales can be divided into two types. Type I scale is located in the upper layer and type II scale in the lower layer. Both scales have longitudinal ridges that run through the scales. However, there is a great difference between them. As shown in Figure 3.10a–c, the adjacent ridges on the scales are parallel, with an average distance of about 0.8 mm. Besides, some irregular holes are randomly distributed at the bottom of the ridges. In contrast, the distance between the two adjacent ridges in type II scale is about 1.5 mm (Figure 3.10d–f). The surface of the bottom scale consists of a set of protruding longitudinal quasi-parallel surfaces,

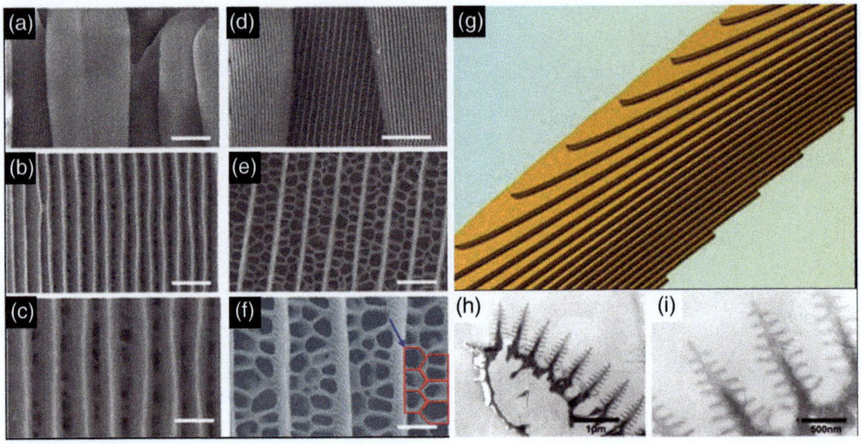

Figure 3.10 Field emission scanning electron microscope (FESEM) images of the butterfly in the blue region with different magnification: Type I scales (a–c) on the superstratum surface absorb light energy in a large range of wavelength, while Type II scales (d–f) under superstratum absorb the transmitted light, ensuring that most of the light energy is absorbed by the butterfly. Source: Reprinted with permission from Han et al. [5]. Copyright 2012, The Royal Society of Chemistry. (g) The model of stacking structure on the ridge. Source: Reprinted with permission from Han et al. [17] Copyright 2017, American Chemical Society. (h–i) High-magnification TEM images of a cross-sectional microstructure of the butterfly *Trogonoptera brookiana*. Scale bars: (a) 20 μm, (b) 2 μm, (c) 1 μm, (d) 10 μm, (e) 2 μm, (f) 1 μm, (g) 2 μm, (h) 1 μm, (i) 600 nm. Source: Reprinted with permission from Han et al. [5]. Copyright 2012, The Royal Society of Chemistry.

which is called a quasi-honeycomb-like structure (Figure 3.10f) [13]. The period of quasi-cellular structure is the same as certain wavelengths, which leads to the incident light being diffracted into the scale and absorbed by the substrate. The model in Figure 3.10g shows the microstructures on the ridge. A multilayer structure with a height of 1 mm and a width of 0.3 mm is found. It is clear that the structure of the ridge is like a bookshelf in the library [18]. According to Figure 3.10h–i, the distance between adjacent bookshelves is about 0.5 mm. The trunk width is 0.05–0.12 mm. Each shelf is composed of chitin/air overlapping multiple layers. Each ridge with the thickness about 50 and 180 mm has 7–9 chitin/air layers. Such a structure system can cause interference between adjacent ridges.

The models of the structures on the green scale and black scale are established in Figure 3.11. The structure parameter of the model came from the FESEM images of the original prototype. It is obvious that the structures in Figure 3.11a,b are quite different. The green scale looks like a shelf structure, while the black scale is close to the honeycomb structure. Han et al. experimentally verified the relationship between microstructures and light absorption. And the mechanism is mentioned in the next part.

3.3.1.3 Classic Optical Theory for Light-Trapping Performance

The phenomenon that the light traveling from one medium to another alters its route on the interface and continues spreading in the original medium is defined

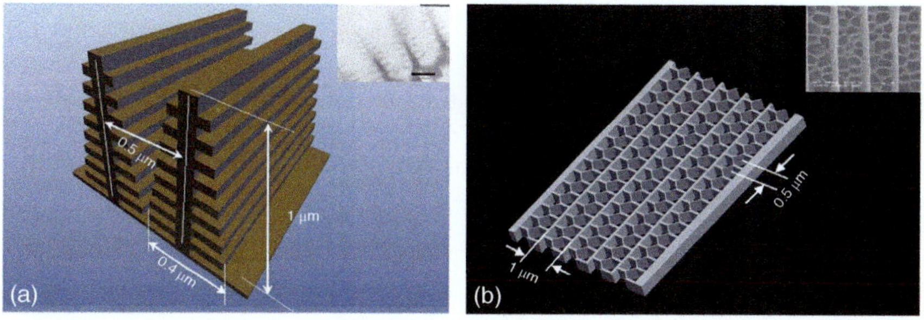

Figure 3.11 3D single-scale models of (a) shelf structure and (b) quasi-honeycomb-like structure. Scale bar (inset), (a) 500 nm; (b) 1 μm. Source: Reprinted with permission from Han et al. [5]. Copyright 2012, The Royal Society of Chemistry.

as reflection. Change of medium is also attributed to reflection. In fact, the standard for "change" depends on the parameter called refractive index (n). Light trapping can be performed if the difference of refractive index is reduced substantially.

3.3.1.3.1 Interference of Light Single Layer

First, a single-layer film was taken as an example to show the mechanism of interference of light applied for light trapping. The incident light shooting on the upper surface caused the first reflection with reflected light and refracted light. Then the refracted light shooting on the downward surface repeated the aforementioned process. Actually, reflected light from both the upper surface and downward surface are on the same side of the film. There is an offset if their phase difference of 180°, and it's a huge advantage for fabricating light-trapping materials. Theoretically, the characteristic matrix of a particular film can be calculated as Eq. (3.12):

$$\begin{bmatrix} B \\ C \end{bmatrix} = \begin{bmatrix} \cos\delta & \frac{i}{n}\sin\delta \\ in\sin\delta & \cos\delta \end{bmatrix} \begin{bmatrix} 1 \\ n_s \end{bmatrix} \tag{3.12}$$

B and C are the components of the electric and magnetic fields parallel to the interface between the film and the incident medium. Here, the input optical admittance is defined as $Y = C/B$. Besides, reflectance of that particular interface can be expressed as $R = |(n_0 - Y)/(n_0 + Y)|^2$, where n_0 is the refractive index (RI) of the incident medium. The film phase thickness is defined as $\delta = \left(2\pi d \cos\frac{\theta}{\lambda}\right)$, representing the change in the phase of the light wave when it travels through the film, and θ is the angle of incidence. Light trapping is achieved as the reflected waves, from the two optical interfaces "a" (air-film) and "b" (film substrate), are out of phase and consequently cancel each other by destructive interference, assuming their amplitudes are identical. There are some requirements leading to light trapping: (i) The two beams of light leaving the film must be 180° out of phase from low- to high-index medium. (ii) The optical thickness of the film must be an odd integer multiple of quarter wavelengths ($\lambda/4$). At normal incidence ($\theta = 0$), the characteristic matrix would be

$$Y = \frac{n^2}{n_s} \tag{3.13}$$

$$R = \left| \frac{(n_0 n_s - n^2)}{(n_0 n_s + n^2)} \right|^2 \tag{3.14}$$

As Eqs. (3.13) and (3.14) show, the light-trapping phenomenon of a single film ($R = 0$) must occur at a particular λ. For a double-layer thin film made up of two films processing a thickness of quarter wavelength, at normal incidence, reflectivity becomes

$$R = \left| \frac{(n_2^2 n_0 - n_s n_1^2)}{(n_2^2 n_0 + n_s n_1^2)} \right|^2 \tag{3.15}$$

where n_1 and n_2 are the refractive indices of the two layers in Eq. (3.4). Light trapping will happen when $(n_2/n_1)^2 = n_s/n_0$. The effective optical admittance is expressed by $\eta_s = Y \cos\theta$, $\eta_p = \frac{Y}{\cos\theta}$. The single film is a transition between the air and the substrate, while it is limited by the RI of materials. The refractive index of materials used for film is between n_0 and n_s, and the n_0 is the refractive index of air while n_s is that of the substrate. It indicates the film with a thickness of quarter wavelength must require $n \sim n_0$; however, this kind of material is rare.

3.3.1.3.2 Double Layer and Multilayer

Because of the limitations of single layers, double-layer even multilayer films appeal more attention. A double-layer coating is composed of a high-index layer (n_1) and a low-index layer (n_2). It has been proved that the effective reflectance (R_{eff}) of this double-layer structure decreases significantly around a wavelength of 500 nm when $n_0 < n_2 < n_s < n_1$. However, there is also a disadvantage that the band of wavelength is not wide enough. It is often termed as V-coating owing to its reflectance curve. Based on the shortage of bandwidths, a three-layer structure is designed with a thicker layer (M, $\lambda_t/2$ thickness) and two thinner layers (H, L, $\lambda_t/4$). The three-layer device can be arranged as in Figure 3.12. H layer has a high index (n_1) while L has a low index (n_2), and the RI of the middle layer satisfies $n_m > (n_1^2/n_s)$; hence, a broadband light trapping can be achieved.

3.3.1.4 Light-Trapping Mechanism of Butterfly Wings with Hierarchical Structures

The complex optical functions of butterfly wings are achieved by microstructures. The ridge structures existing on the surface of the butterfly scale (Figure 3.13a)

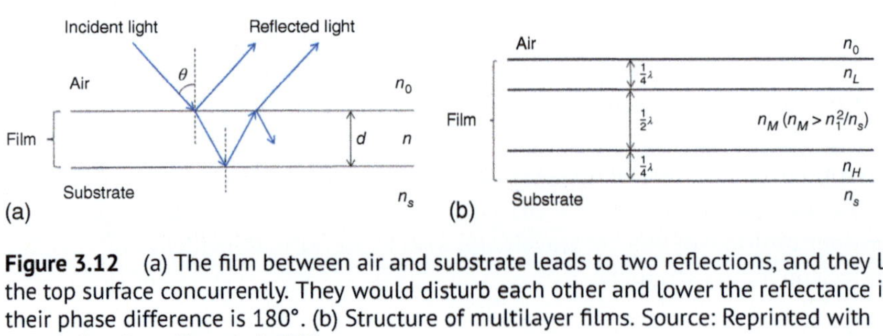

Figure 3.12 (a) The film between air and substrate leads to two reflections, and they leave the top surface concurrently. They would disturb each other and lower the reflectance if their phase difference is 180°. (b) Structure of multilayer films. Source: Reprinted with permission from Han et al. [19].

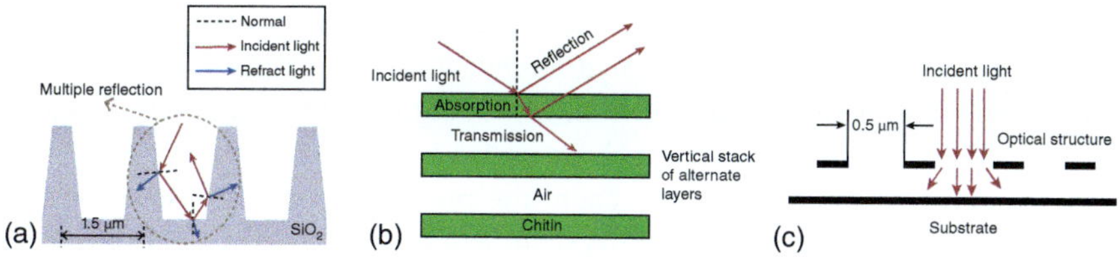

Figure 3.13 (a) The diagrammatic sketch of multiple reflections. Source: Reprinted with permission from Han et al. [16]/Springer Nature/CC BY-4. (b) The optical behaviors of chitin–air multilayered structure (green region). (c) The optical behaviors of the grating structure (black region). Source: Reprinted with permission from Han et al. [5]. Copyright 2012, The Royal Society of Chemistry.

play an important role in prolonging the light path, and multiple reflections occur between the ridges. Multiple optical behaviors occur at the interface between air and substrate when incident light shines on the material surface. The light-trapping structures on butterfly scales lead to multiple reflections and refractions. Most solar energy is absorbed, and a small part of light returns to air. The hump of the ridge top and the fold stripes on both sides of the ridge enhance the scattering of incident light on the structure and reduce the solar energy loss caused by reflectivity. According to the microstructures of green scale (type I) and black scale (type II), it can be found that the structural color is caused by the alternating arrangement of chitin and air in type I scales (Figure 3.13b). The distance between ribs and thickness of ribs determine which band of light will be reflected at a specific angle [20]. In this case, the arrangement of the chitin ribs allows most of the band solar energy to be absorbed after multiple reflections between the ridges, only a certain band of light reflecting into the air. As for type II scales, quasi-honeycomb structures are regarded as diffraction grating (Figure 3.13c). Since the size of the periodic parallel slits or grooves is similar to visible wavelengths, the incident light is induced inside the scale and will be absorbed after multiple internal reflection and scattering. When light enters the grating, it will be reflected back into the substrate so that the light will be reflected for multiple times. Thus, scales with nanostructures like this exhibit a significant effect of broadband antireflection.

3.3.2 Fabrication of Structural Butterfly-Inspired Light-Trapping Surfaces

Since the relationship between the structures and the optical behaviors of butterfly wings is investigated, bioinspired light-trapping films need to be prepared. There are many artificial methods for fabrication, such as nanometer imprinting, electron-beam lithography, self-assembly, chemical etching, electro-deposition. These technologies are relatively mature; however, it is difficult to accurately replicate the surface of wing scales with complex micro–nano classification structure. Therefore, it seems to be a better choice to use the prototype as a template. In this way, the original prototype can be used directly to ensure the perfect inheritance of structural parameters; meanwhile, the biological template method is not limited to biological species only, and it can be applied to most of the organisms.

Here, the original butterfly wing was used as a bionic prototype to transfer the light-trapping nanostructures to the polydimethylsiloxane (PDMS) substrate. The diagram of the biomimetic preparation is shown in Figure 3.14. First, the specimen was pretreated with acetone and ethyl alcohol for 10 minutes, respectively, to remove the impurities on the surface. Then, the precursor solution was obtained by the hydrolytic reactions of tetraethoxysilane (TEOS) in an acid environment. Next, the specimen was sandwiched between two glass slides with clips, and they could be defined as an assembly. The precursor solution was added at the edge of the assembly to ensure the specimen was soaked completely. Subsequently, the assembly was heated at 120 °C for 30 minutes to ensure the cross-linked curing process. Finally, the assembly was soaked into a mixture of hydrochloric acid and nitric acid (1 : 1 in volume) at 130 °C for 30 minutes, followed by a deionized water

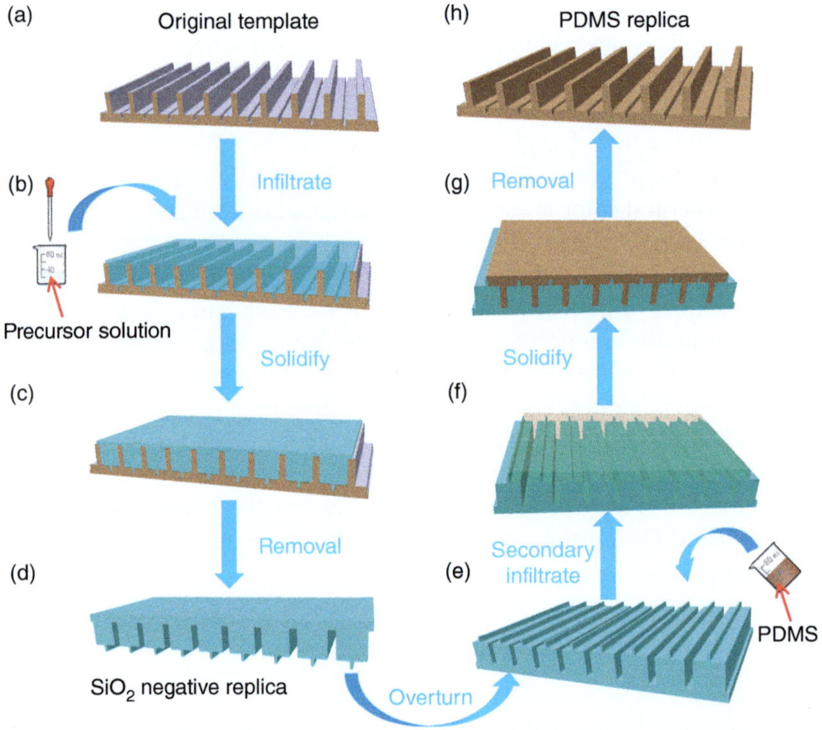

Figure 3.14 Fabrication process from the original nanostructured template of butterfly wings to the PDMS positive structures. (a) 3D nanostructured model of the original butterfly wings. (b) The precursor solution filled the space left between the micro-ridges with a micropipettor. (c) The precursor solution became solidified through heating to form ridges and notches. (d) The solid negative replica was obtained after the original bio-template material was etched away. (e) The prepared PDMS was poured onto the SiO_2 negative replica and got the PDMS/SiO_2 negative replica assembly. (f and g) The PDMS/SiO_2 negative replica assembly was placed in a vacuum oven to evacuate the air bubbles and heated to complete heating solidify process. (h) The PDMS positive replica was fabricated after cooling at room temperature. Source: Reprinted with permission from Han et al. [21]. Copyright 2015, Elsevier B.V.

washing process. In this process, the biological organism was completely corroded, leaving a SiO_2 negative replica on the glass sheet (Figure 3.14a–d). Later, the SiO_2 negative replica was used as a template to transfer the structures to the PDMS substrate. The PDMS and curing agent were mixed in a 10 : 1 volume ratio. When the mixture was stirred vigorously, it was added to the SiO_2 negative replica and heated at 80 °C for 2.5 hours. At last, the PDMS positive replica was obtained after peeling the film from the glass slide (Figure 3.14e–h).

3.3.3 Characterizations of Butterfly-Inspired Light-trapping Surfaces

3.3.3.1 Morphologies and Elemental Analysis of SiO_2 Negative Replica

As described earlier, the structures on butterfly wings were transferred to the glass substrate, and a SiO_2 negative replica was obtained. After the second replication, the bionic light-trapping film could be prepared. Here, the morphology of SiO_2 negative replica, which was derived from the green scales, was studied by the scanning electron microscope. As shown in Figure 3.15a, the scales are still lined up; however, they no longer overlap each other. The magnifying rectangular scale of the

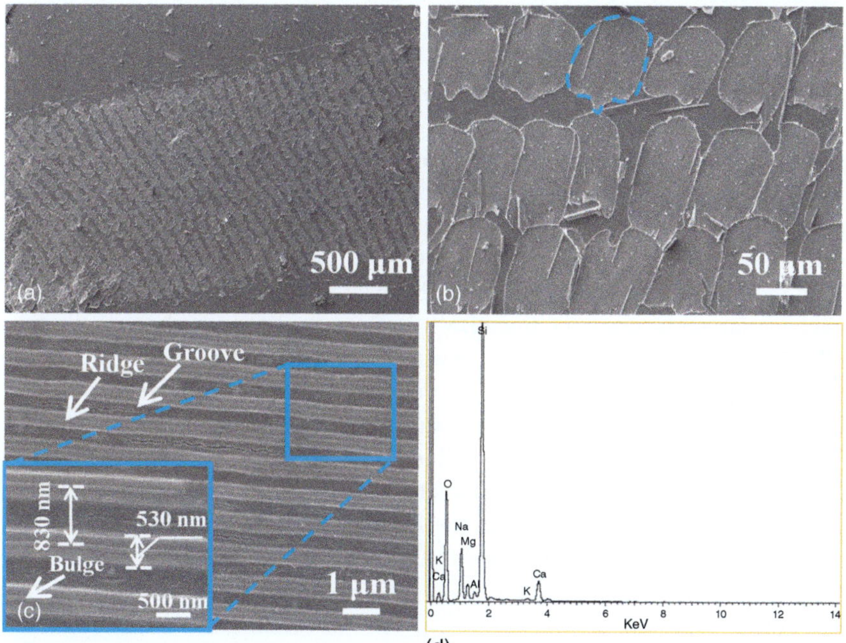

Figure 3.15 FESEM images of the SiO_2 negative replica. (a) Lower magnification image. It can be found that the scales are still arranged in rows. However, they no longer overlap each other. (b) Medium magnification image of the negative replica. It shows the rectangular scaly morphology of the negative replica of bio-templates scales, indicating the scales were completely inherited from the green scales. (c) High magnification images of the replica surface. The sizes and shapes of the negative groove-hump units matched with those of the ridge-groove units. (d) The EDS spectra obtained from the SiO_2 negative replica. Source: Reprinted with permission from Han et al. [21]. Copyright 2015, Elsevier B.V.

SiO_2 negative replica is shown in Figure 3.15b, the shape of which is close to that of the biological prototype. The dimensions of the scale on the negative replica are 100 nm in length and 60 nm in width. Each structural parameter is similar to that of the scale in the original template. The details of the single-scale SiO_2 negative replica are given in Figure 3.15c. The negative groove is formed by the ridges on the surface of the prototype. The distance between the ridges of the original scales is filled with sol–gel precursors, and the sol–gel precursors are solidified, so that the space of the original ridges becomes a negative groove, and the space between the ridges of the original scales becomes a hump. It is worth noting that a narrow protruding tissue was observed in the middle of each hump, which was present as a narrow dent in the original template. It was found that the pitch of the adjacent protruding and that of the ridge was 830 and 530 nm, respectively. It can be confirmed from the size and shape that the original template is inherited by the SiO_2 negative replica.

The element type and content of the surface of the SiO_2 negative replica were characterized by energy dispersive spectrometer (EDS). The analysis results (Figure 3.15d) show that the SiO_2 replica is mainly composed of silicon and oxygen. The atomic percentages of the two elements are 25.04% and 62.32%, respectively. There is no carbon peak in the spectrum, indicating that the original template is completely acid-etched. It can be confirmed that there will be no interference in the next step of the experiment by the original organism.

3.3.3.2 Morphologies and Elemental Analysis of PDMS Positive Replica

Figure 3.16a shows a digital photo of the positive structure of the bionic replica. The prepared PDMS positive replica is bright green, and it is proved that the optical microstructures of the butterfly wing scale have been copied successfully. The details of the PDMS positive replica were observed with FESEM (Figure 3.16b–d). The rectangular scale shape of the PDMS positive replica was shown at a low magnification rate, and its shape was close to that shown in SiO_2 negative replica. Figure 3.16c shows the integral ridges and grooves of the PDMS positive replica. The grooves are formed by the humps on the surface of the SiO_2 negative replica, and the space between the inverted ridges is filled with PDMS. The PDMS was heated for 120 minutes at 80 °C to ensure complete curing, making the microstructures in SiO_2 negative replica transferred to the PDMS substrate. It is obvious that the spacing of the adjacent raised ridges and the width of the grooves are about 810 and 500 nm, respectively, which are consistent with that of the biological prototypes (Figure 3.16d).

Fourier transform infrared spectroscopy (FTIR) spectroscopy is an important tool in the investigation of polymer, and FTIR spectrum of the PDMS positive replica surface is shown in Figure 3.17a. The peak at 1289 cm^{-1} was attributed to Si–O–Si stretching vibration bond. The bands at around 1263 and 803 cm^{-1} were assigned to the Si–C groups. Other characteristic peaks in the spectrum were assigned to –CH–, $-CH_2-$, and $-CH_3$ groups of the polymer backbone (2964, 2919, and 1415 cm^{-1}, respectively). It could be observed that the FTIR spectrum of the PDMS positive replica was consistent with the general pattern of the FTIR spectrum of the PDMS. The FTIR spectrum verified that the composition of the bionic PDMS positive

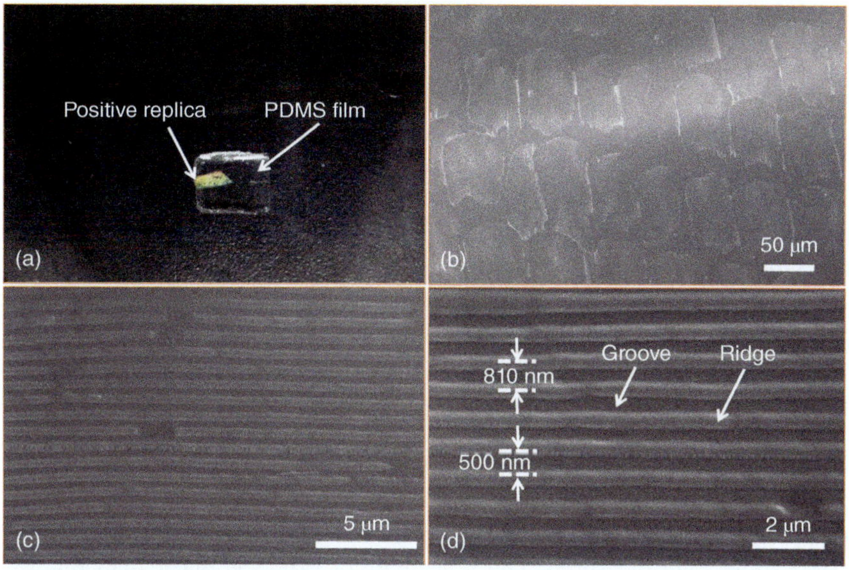

Figure 3.16 The macroscopic appearance and FESEM images of the PDMS positive replica. (a) The PDMS positive replica demonstrates brilliant green. Nevertheless, the PDMS film without optical nanostructures is transparent. (b) Lower magnification images of the positive replica surface. (c) Medium magnification images of the positive replica surface. It can be found that the PDMS positive replica maintains the ridges and grooves of the original scales. (d) High magnification images of the replica surface. The overall dimension of the ridge-groove units in the PDMS positive replica is very similar to those in the original scales and the hump-groove units in the SiO_2 negative replica. Source: Reprinted with permission from Han et al. [21]. Copyright 2015, Elsevier B.V.

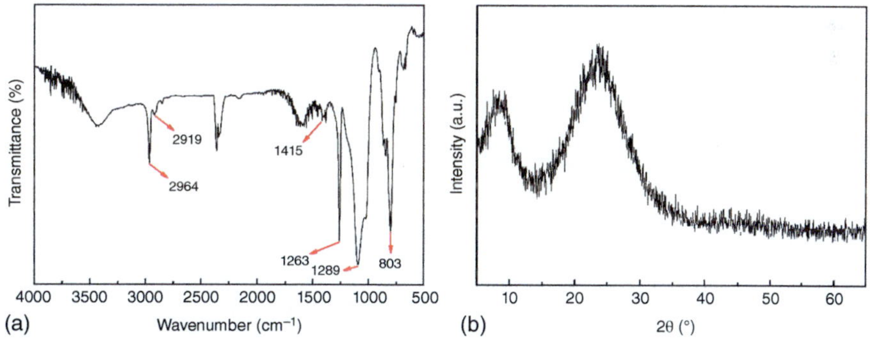

Figure 3.17 (a) FTIR spectrum of the PDMS positive replica surface. (b) XRD pattern of the SiO_2 negative replica. Source: Reprinted with permission from Han et al. [21]. Copyright 2015, Elsevier B.V.

replica was not changed after a series of treatment. The crystalline structure and morphologies of the prepared PDMS positive replica were characterized with X-ray powder diffraction (XRD). The XRD pattern of the SiO_2 negative replica is shown in Figure 3.17b. There were two broad peaks, which resulted from the amorphous silica [22]. It could be obtained that the SiO_2 negative replicas existed in the amorphous state.

3.3.4 Light-Trapping Performance of Butterfly-Inspired Structural Surfaces

To verify the effect of the light-capture ability of the bionic light-trapping structures on the substrate surface, reflectance spectra of the flat plate and structural PDMS are shown in Figure 3.18. Applying the negative quasi-honeycomb-like structure on the flat plate plays a key role in producing the super light-trapping property. And the average reflectance of the negative replica is 1/4 of the reflection of the flat plate. It is obvious that the negative SiO_2 replica displays a certain light trapping.

Similarly, the correlation optical performances of the PDMS positive replica are also characterized in Figure 3.19. The average reflectivity of the bionic PDMS positive structure is about 15%, and it is very similar to that of the biological prototype. The average reflectivity of the PDMS positive replica is only 3/7 of that of the flat PDMS film. Considering that flat PDMS and bioinspired poly(methylmethacrylate (PMMA) replicas are in the same light intensity environment and the material is uniform, it is obvious that the bionic light-trapping film possesses remarkable light-trapping performance. In addition, the reflection curves of the flat chitin and flat PDMS films were measured. The reflectivity of PDMS is lower than that of chitin, which indicates that the material also affects the optical properties of the surface. However, comparing the reflectivity curves of chitin, flat PDMS film, and bionic PDMS replica shows that the bioinspired microstructure of PDMS surface is very important in the generation of light-trapping performance. In brief, the bionic PDMS positive replica not only inherits the light-trapping structure of the biological prototype but also the optical characteristics.

In conclusion, both the composition and structures covering on the substrate have influence on the optical properties of the surface. The design and optimization of the

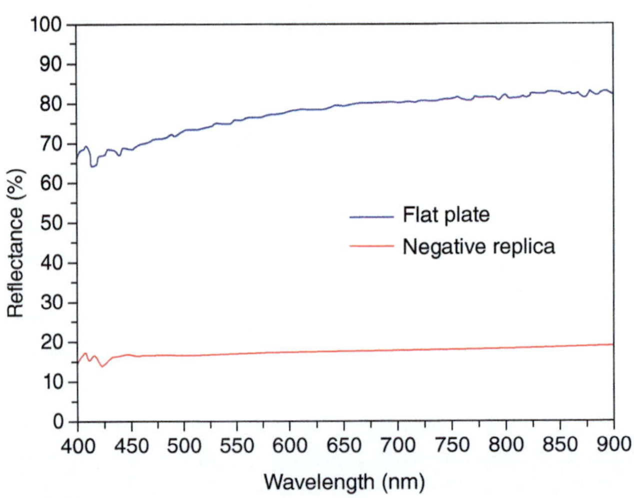

Figure 3.18 The reflectance curves of the flat plate and negative replica.
Source: Reprinted with permission from Han et al. [16]/Springer Nature/CC BY-4.0.

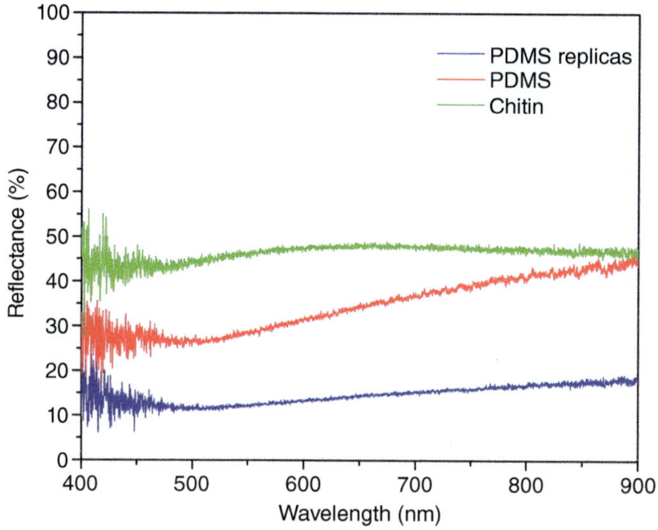

Figure 3.19 Reflectance spectra of the chitin, flat PDMS film, and PDMS positive replica. Source: Reprinted with permission from Han et al. [21]. Copyright 2015, Elsevier B.V.

bionic light-trapping surface could be promoted by selecting appropriate materials and constructing reasonable bionic microstructures.

3.3.5 Light-Trapping Mechanism of Butterfly-Inspired Structural Surfaces

3.3.5.1 3D Visible Parameterized Models of Biomimetic Feature Structures

To further explain the mechanism of light trapping, a 3D model of scale in green area was established. The 3D model of the original prototype and the prepared SiO_2 negative replicas should perfectly match with the biological prototype, so the prepared PDMS positive replica could completely inherit the original shape of the organism. In fact, the morphology of prepared PDMS positive replica was similar to the prototype, with a slight size difference. It could be explained as insufficient infiltrates of PDMS during replication. On the other hand, due to the possibility of volume shrinkage of PDMS materials during curing, the final structure on PDMS positive replica is slightly different from the prototype in detail. Although the difference still exists, it is within an acceptable range, as shown in Figure 3.20.

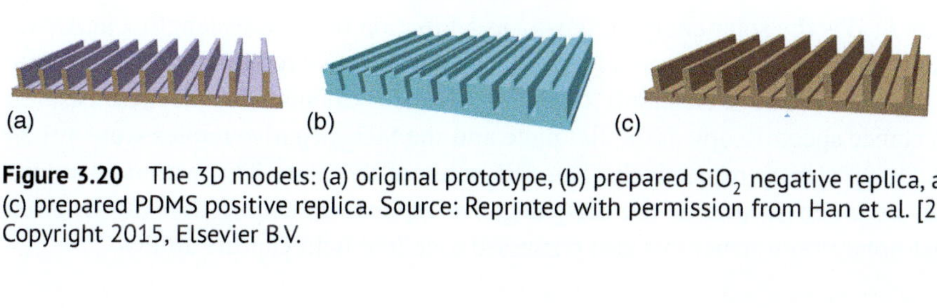

Figure 3.20 The 3D models: (a) original prototype, (b) prepared SiO_2 negative replica, and (c) prepared PDMS positive replica. Source: Reprinted with permission from Han et al. [21]. Copyright 2015, Elsevier B.V.

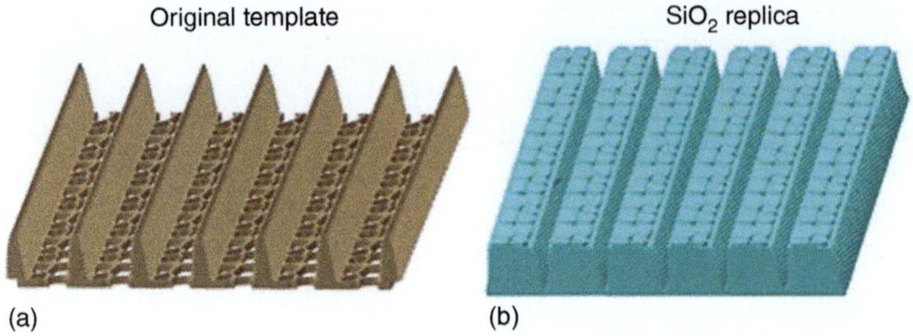

Figure 3.21 The 3D models: (a) The quasi-honeycomb structures distributed on the black scale. (b) The inverted quasi-honeycomb structures fabricated by a sol–gel process. Source: Reprinted with permission from Ha et al. [16]/Springer Nature/CC BY-4.0.

A 3D model of scale in black area is established in Figure 3.21. After the first step of replication, the inverted quasi-honeycomb structures were obtained from quasi-honeycomb structures distributed on the black scale, and there were extremely small grooves between the protruding structures. Limited by its fluxility and viscosity, PDMS is often accompanied by structural distortion when used as a structural transfer material. Therefore, PDMS is not suitable for the replication of a finer structure, especially for the grid grating structure on the black scale. Thus, the bionic preparation of black scale and the preparation of positive replica of black scale need to be further studied. However, the optical experimental results show that the reverse structure of the black scale also exhibit good optical properties, which is worthy of detailed investigation.

3.3.5.2 Light-Trapping Mechanism

The microstructures on the surface of the organisms are of great significance to light trapping. The related mechanism has been discussed in the biological part. Since the SiO_2 negative replica also exhibited outstanding trapping performance, in this part, we focus on explaining the light-trapping mechanism of the negative replica. Compared with plate SiO_2 surface, the reflectivity of the inverted quasi-honeycomb structures covering the SiO_2 negative replica is significantly lower. Here, light trapping of the inverted quasi-honeycomb structures is quantitatively characterized, and the loss of the solar energy caused by reflection is defined as solar energy loss (*SEL*):

$$SEL = \int_{400}^{900} I(\lambda)\alpha(\lambda)d\lambda \tag{3.16}$$

where $I(\lambda)$ is the solar energy intensity as a function of the wavelength λ at AM1.5 and $\alpha(\lambda)$ is the measured reflectance of the SiO_2 negative replica, and the flat plate is a function of the wavelength λ [23–26]. The *SEL* calculated from the measured reflectance spectroscopy of the flat plate and the SiO_2 negative replica were 491.34 and 110.98 W m^{-2}, respectively. The *SEL* value of the plate SiO_2 was at least four times as much as that of the SiO_2 negative replica, indicating that the inverted quasi-honeycomb structures also possessed excellent light capture ability.

3.3.5.3 Synergetic Multiple Light-Trapping Mechanism

Most of the butterfly scales have a double-layer arrangement, and the micron ridge structure on the top scale can promote multiple reflections, achieving the function of extending the light path. At the same time, the structure of chitin–air alternating arrangement formed by nanorib plate structures and the air can realize the selective reflection of the light wave. Since the reflected wavelength is relevant to the size of structure, the other wave bands are absorbed to form a bright and gorgeous structure color. In this method, light trapping was realized in partial band. For the bottom scale, the quasi-honeycomb structure was prevalent, and it was used as a micro-grating to finish the light trapping and absorption after multiple reflections. As a result of the absorption over the visible band, the scale usually appeared black. These micro/nanoscale hierarchical structures distributed on butterfly wing scales provided important design inspiration for light-trapping surfaces, and it was remarkable in the fields of solar energy, photo-thermal conversion, and so on.

References

1 Zhu, J., Yu, Z., Burkhard, G.F. et al. (2009). Optical absorption enhancement in amorphous silicon nanowire and nanocone arrays. *Nano Letters* 9 (1): 279–282.

2 Dennler, G., Forberich, K., Scharber, M.C. et al. (2007). Angle dependence of external and internal quantum efficiencies in bulk-heterojunction organic solar cells. *Journal of Applied Physics* 102 (5): 054516.

3 Fan, T., Chow, S., and Zhang, D. (2009). Biomorphic mineralization: from biology to materials. *Progress in Materials Science* 54 (5): 542–659.

4 Li, Y., Zhang, J., Zhu, S. et al. (2009). Biomimetic surfaces for high-performance optics. *Advanced Materials* 21 (46): 4731.

5 Han, Z., Niu, S., Shang, C. et al. (2012). Light trapping structures in wing scales of butterfly *Trogonoptera brookiana*. *Nanoscale* 4 (9): 2879–2883.

6 Wang, H., Sun, K., Tao, F. et al. (2013). 3D honeycomb-like structured graphene and its high efficiency as a counter-electrode catalyst for dye-sensitized solar cells. *Angewandte Chemie International Edition* 125 (35): 9380–9384.

7 Čampa, A., Krč, J., and Topič, M. (2009). Analysis and optimisation of microcrystalline silicon solar cells with periodic sinusoidal textured interfaces by two-dimensional optical simulations. *Journal of Applied Physics* 105 (8): 083107.

8 Sai, H. and Kondo, M. (2009). Effect of self-orderly textured back reflectors on light trapping in thin-film microcrystalline silicon solar cells. *Journal of Applied Physics* 105 (9): 094511.

9 Sai, H., Kanamori, Y., Arafune, K. et al. (2007). Light trapping effect of submicron surface textures in crystalline Si solar cells. *Progress in Photovoltaics: Research and Applications* 15 (5): 415–423.

10 Llopis, F. and Tobías, I. (2006). Texture profile and aspect ratio influence on the front reflectance of solar cells. *Journal of Applied Physics* 100 (12): 134504.

11 Lee, Y.C., Huang, C.F., Chang, J.Y., and Wu, M.L. (2008). Enhanced light trapping based on guided mode resonance effect for thin-film silicon solar cells with two filling-factor gratings. *Optics Express* 16 (11): 7969–7975.

12 Han, Z., Niu, S., Li, W., and Ren, L. (2013). Preparation of bionic nanostructures from butterfly wings and their low reflectivity of ultraviolet. *Applied Physics Letters* 102 (23): 233702.

13 Han, Z., Niu, S., Yang, M. et al. (2013). An ingenious replica templated from the light trapping structure in butterfly wing scales. *Nanoscale* 5 (18): 8500–8506.

14 Fu, F., Shang, L., Chen, Z. et al. (2018). Bioinspired living structural color hydrogels. *Science Robotics* 3 (16): eaar8580.

15 Han, Z., Niu, S., Zhang, L. et al. (2013). Light trapping effect in wing scales of butterfly *Papilio peranthus* and its simulations. *Journal of Bionic Engineering* 10 (2): 162–169.

16 Han, Z., Li, B., Mu, Z. et al. (2015). An ingenious super light trapping surface templated from butterfly wing scales. *Nanoscale Research Letters* 10 (1): 1052.

17 Han, Z., Mu, Z., Li, B. et al. (2017). Bioinspired omnidirectional self-stable reflectors with multiscale hierarchical structures. *ACS Applied Materials & Interfaces* 9 (34): 29285–29294.

18 Stratakis, E., Bonse, J., Heitz, J. et al. (2020). Laser engineering of biomimetic surfaces. *Materials Science & Engineering R: Reports* 141: 100562.

19 Han, Z., Wang, Z., Feng, X. et al. (2016). Antireflective surface inspired from biology: a review. *Biosurface and Biotribology* 2 (4): 137–150.

20 Ingram, A.L. and Parker, A.R. (2008). A review of the diversity and evolution of photonic structures in butterflies, incorporating the work of John Huxley (The Natural History Museum, London from 1961 to 1990). *Philosophical Transactions of the Royal Society of London. Series B, Biological Sciences* 363 (1502): 2465–2480.

21 Han, Z., Li, B., Mu, Z. et al. (2015). Fabrication of the replica templated from butterfly wing scales with complex light trapping structures. *Applied Surface Science* 355: 290–297.

22 Yu, X., Wang, M., and Li, H. (2000). Study on the nitrobenzene hydrogenation over a Pd-B/SiO_2 amorphous catalyst. *Applied Catalysis A: General* 202 (1): 17–22.

23 Wang, Z., Zhang, R., Wang, S. et al. (2015). Broadband optical absorption by tunable Mie resonances in silicon nanocone arrays. *Scientific Reports* 5: 7810.

24 Cihan, A.F., Curto, A.G., Raza, S. et al. (2018). Silicon Mie resonators for highly directional light emission from monolayer MoS_2. *Nature Photonics* 12 (5): 284–290.

25 Xi, J., Schubert, M., Kim, J. et al. (2007). Optical thin-film materials with low refractive index for broadband elimination of Fresnel reflection. *Nature Photonics* 1 (3): 176–179.

26 Wu, Z., Wang, Z., Wang, S., and Zhong, Z. (2014). Substantial influence on solar energy harnessing ability by geometries of ordered Si nanowire array. *Nanoscale Research Letters* 9 (1): 495.

4

Transparent Antireflective (AR) Surfaces Inspired by Cicada Wings

Light is one of the most important ways to obtain information. At the same time, there is inestimable energy stored in light, which may be a huge treasure for further energy reform. It is closely related to daily life, exact measurement, optoelectronic field, and photothermal field. Light is of great significance to both human beings and the natural environment. Generally speaking, 90% of the messages we get come from the sense of sight. It means that a good observation surface is quite important for people. On the one hand, strong specular reflection can usually destroy the exactitude of the information we get with horrible effect on our eyes. On the other hand, the new energy, represented by solar energy, has attracted a plenty of researchers because of the growing demand for energy and severity of the environment.

However, unwanted reflection is always confusing, leading to some negative influences such as dazzling light, interference during measuring, and low efficiency of optoelectronic conversion as well as photothermal conversion. Therefore, antireflective coatings (ARCs) or materials are highly appealing for optical fields [1, 2]. Based on this, effective light management – antireflection, light trapping, light scattering, and light diffraction – as one of the seductive strategies, has drawn considerable attention to enhance the optical performance for widespread applications in solar cells [3], light-emitting diodes (LEDs) [4], and organic light-emitting diodes (OLEDs) [5].

The Fresnel reflection, the main source, occurs as light impinges on the air/polymer interface due to an abrupt change in the refractive index (RI) between air and substrates. Antireflection, a way to depress optical reflection, can be achieved by reducing mirror reflection with a rough surface, and antireflective materials have been investigated widely using a reflection-reducing coating or by forming a light-trapping structure.

As an excellent antireflection prototype, the hierarchical structure on the butterfly wings focuses on light absorption. However, antireflection surfaces on some display devices require not only lower reflectivity but also a satisfactory degree of transparency, so antireflection films with high transparency and low reflectivity are very important for display surface.

Until now, all kinds of electronic products have been widely used in daily life. However, the readability of mobile phones, computers, dashboards, and displays is often disturbed in the sun or strong light, which has drawn the attention of

Nature-Inspired Structured Functional Surfaces: Design, Fabrication, Characterization, and Applications, First Edition. Zhiwu Han.

researchers. Generally speaking, "clarity" refers to the contrast between light and shadow. For most transmission-based displays, light received by the eye consists of ambient light and transmitted light from the display. Therefore, when an image is exposed to strong light or external light, the shadow is weakened, which means the contrast is reduced. Consequently, the quality of image readability is degraded. The quality of image readability is quantitatively defined by the ambient contrast ratio (ACR) as follows:

$$\mathrm{ACR} = \frac{L_{\mathrm{ON}} + R_{\mathrm{L}} \cdot L_{\mathrm{ambient}}}{L_{\mathrm{OFF}} + R_{\mathrm{L}} \cdot L_{\mathrm{ambient}}} = 1 + \frac{L_{\mathrm{ON}} - L_{\mathrm{OFF}}}{L_{\mathrm{OFF}} + R_{\mathrm{L}} \cdot L_{\mathrm{ambient}}} \tag{4.1}$$

$$R_{\mathrm{L}} = \frac{\int_{\lambda_1}^{\lambda_2} V(\lambda)R(\lambda)S(\lambda)\mathrm{d}\lambda}{\int_{\lambda_1}^{\lambda_2} V(\lambda)S(\lambda)\mathrm{d}\lambda} \tag{4.2}$$

where L_{ON} and L_{OFF} are luminance of on and off pixels, respectively, excluding glare. L_{ambient} is the ambient luminance. R_{L} is the luminous reflectance of the display [6]. The luminous reflectance is defined as the normalized integrated product of $V(\lambda)\cdot R(\lambda)\cdot S(\lambda)$, where $V(\lambda)$ is the standard photonic curve, $R(\lambda)$ is the total reflectance of the system, and $S(\lambda)$ is the source spectral power distribution. $\alpha R(\lambda)S(\lambda)$ represents the reflected spectral radiance due to ambient light, where a geometric factor α converts radiant exitance into radiance.

It has been verified that higher ACR values lead to clearer vision [6, 7]. According to Eqs. (4.1) and (4.2), there are two ways to solve the problem. The first one improves the readability of images by increasing the intensity of L_{ON}, and it seems easier to achieve. However, screens with high luminance consume more electric energy, and it will cause serious damage to vision. Another way is to decrease $R(\lambda)$, and it means improving screen brightness or minimizing surface reflection. Consequently, the latter one is easier to achieve.

Therefore, a biological surface with high transparency and low reflectivity is suitable for this study. Han et al. from Jilin University studied the cicada wings in detail to investigate the relationship between nanostructures and excellent antireflection properties. As a typical antireflective prototype, cicada wings are thin and transparent, possessing outstanding multifunctional features such as high transparency, low reflectivity, and self-cleaning. The composite function not only provides a guarantee for its survival but also a natural blueprint for the study of bionic antireflective surfaces.

4.1 High Transparent Antireflective (AR) Surfaces of Original Cicada Wings

4.1.1 Optical Properties of Cicada Wing Surfaces

To explore the excellent surface properties of cicada wings, a static water contact angle (CA) measurement instrument (OCA 20, Dataphysics, Germany) was used

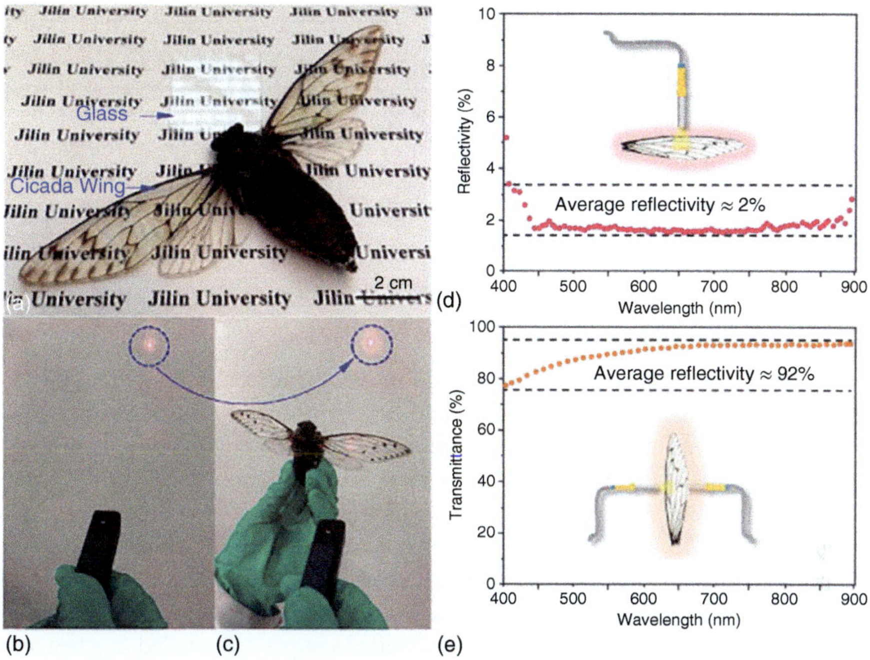

Figure 4.1 High reflection-reducing surface of cicada wings and their self-cleaning performance. (a–c) High reflection reduction characteristics of natural cicada wings. (d–e) Cicada reflectance and transmission spectra. (f–h) Superhydrophobic characteristics of transparent zone of cicada wings. Source: Reprinted with permission from Wang et al. [8]. Copyright 2020, Springer Nature.

and an optical fiber spectrometer (Ocean Optics USB 4000) to quantitatively characterize the reflection and transmission spectra. The wing samples taken in this study were all cicadas that died naturally, with a body length of about 4–5 cm and a wingspan of about 9–13 cm. As shown in Figure 4.1, when the cicada specimen was placed on a background with the word "Jilin University," the text behind the cicada could be clearly observed in the transparent area of the cicada wing, which clearly indicates that the cicada wing has very high light transmittability. By comparing with the transparent glass slide under the same light source, the excellent anti-glare effect of the cicada wing surface is clearly reflected, which shows that it has outstanding antireflection performance. At the same time, the concentrated light path dispersion occurs when the laser beam passes through the surface of the cicada, indicating that the cicada also has a certain scattering effect. Based on the above preliminary test results, it is reasonable to choose this cicada as the bionic prototype.

Further observation showed that the wings of cicada are divided into forewings and hindwings. In appearance, they are thin and transparent and can be roughly divided into transparent area and ridgel-vein area. The grid ridges distributed on the surface of cicada wings belong to the support structure, which can improve the overall strength of the wings and protect the cicada wings from irreversible deformation or rupture due to airflow impact in the movement process. The optical properties of cicada wing surface were further quantitatively characterized by the spectrometer. The average reflectance of the cicada wings in the visible light band was about 2%, much lower than the reflectance of 8% on the glass surface, and the average transmittance was as high as 92%. The high antireflection characteristics of the cicada wings in a wide band provide an excellent natural blueprint for the study of bionic antireflection surfaces in this paper.

4.1.2 Microstructure and Composition of High Reflection Reduction Surface of Cicada Wing

The excellent performance of the biological functional surface is attributed to the joint action of its surface microstructure and composition. To further explore the relationship between the excellent antireflection performance of cicadas and the microstructure of their surface distribution, field emission scanning electron microscope (FESEM) (JSM-6700F, JEOL) and atomic force microscope (AFM) (Bruker DIMENSION ICON) were used to observe the structure of cicadas comprehensively. To avoid natural cicada surface adhesion of dust, grease, protein, and other impurities and not to affect the accuracy of the experiment, the cicada sample should be preprocessed as follows: (i) the biological sample was soaked in a beaker containing the required amount of acetone, and ultrasonic washing took place for 15 minutes after sealing, to remove impurities such as surface adhesion of protein and fat; (ii) then, the biological sample was taken out, soaked in a beaker containing a large amount of deionized water, sealed, and washed by ultrasonic cleaning for 15 minutes; then the sample was taken out and dried in the dark for use.

The cleaned cicada wings were cut to an appropriate size (about $4\,mm \times 4\,mm$) and pasted on the conductive adhesive. Due to the poor conductivity of the organism itself, it needed to be sprayed with gold to prepare for subsequent observation. As shown in Figure 4.2, for the supporting structure of the black ridge veins of cicada wing, there are micron-sized conical hair-like structures distributed on it, with a depth ratio of about 5–20. At the same time, after observing the tissue of the transparent area of the cicada wing, it was found that the section thickness of the transparent area of the cicada wing is only 8–10 μm, and the structure is completely different from that of the ridge veins. The transparent area of the cicada wing is divided into upper and lower layers, and the subwavelength-level domed cone array structure is uniformly distributed, the superstructure area corresponds to the front of cicada and the lower structure area to the ventral surface of cicada. The DC array structures maintain a high degree of consistency in the transparent area. Each independent structural unit presents a streamlined structure similar to a dome-shaped cone. The diameter of the base circle is about 180 ± 10 nm, the diameter of the top circle is about 90 ± 10 nm, and the overall height is about 325 ± 15 nm. The distance between adjacent structural units, the structural period, is about 200 ± 10 nm. The overall arrangement of the array structure presents a very regular hexagonal tight arrangement.

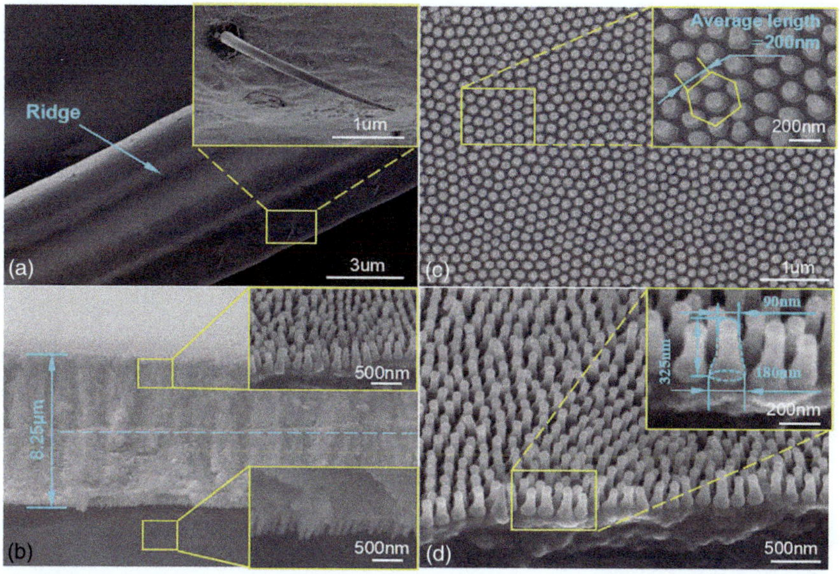

Figure 4.2 SEM images of surface microstructures of cicada (*Megapomponia intermedia*) wing. (a) The ridge. (b) The cross section of transparent part on the wing. (c)–(d) Arrays on the wing. Source: Reprinted with permission from Han et al. [9]. Copyright 2019, American Chemical Society.

To observe the morphology of the dome-shaped cone array structure on the surface of cicadas more stereoscopic, AFM with a pointed probe was used to describe the surface of cicadas deeply and finely. The scanning area was 5 μm × 5 μm in the frontal and ventral transparent areas of cicada wings. Figure 4.3a,c shows the three-dimensional images of the front of cicada wings, respectively. Protrusions in some positions shown in the figure may be due to the uneven surface of cicada wings or errors caused by the disturbance in the force of the structural unit when the probe contacts the structural surface. Taking three parallel sections arbitrarily on the front of cicada wings, the obtained section curve data is shown in Figure 4.3b. Taking the average height of DC array as the reference, the section can reflect the change in normal height of the individual structural unit involved, denoted Δy. The height variations of structural units on sections 1, 2, and 3 are all less than 30 nm, and most of the section curves overlap. On the ventral surface of cicada wings, we also obtained similar results after changing the interception method in different directions, indicating that the ventral surface of cicada wings also has relatively stable morphological distribution characteristics. These results are consistent with the scanning electron microscope (SEM) observation results, which indicate that there are regular DC array structures on both sides of the ventral transparent area of cicada wings, and these structures display uniform, stable, and isotropic distribution characteristics.

With the detailed characterization of the surface morphology of cicadas, we further studied the main components of cicadas and quantitatively described the

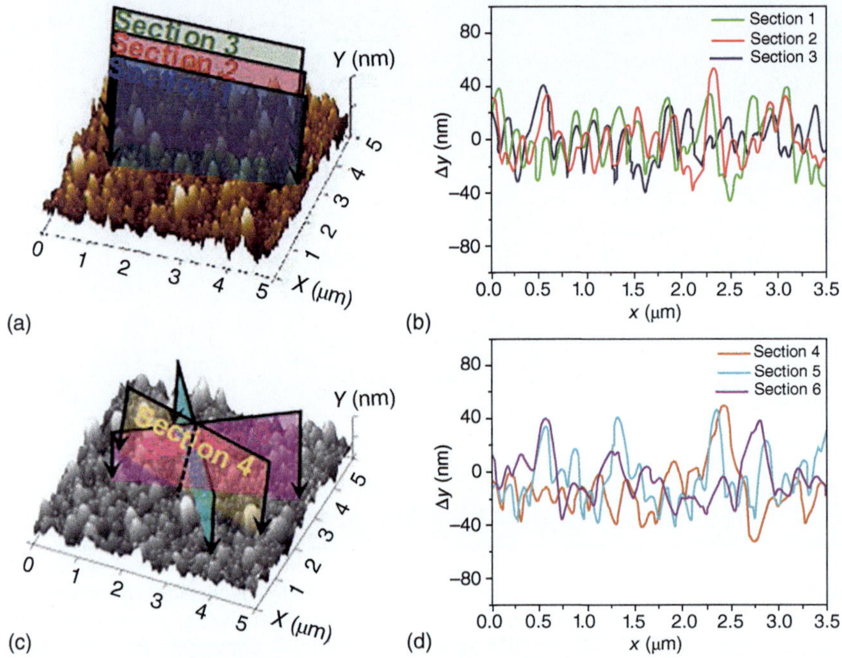

Figure 4.3 AFM observation and analysis of morphology distribution of cicada (*Megapomponia intermedia*) on the anterior ventral surface. (a)–(b) The structures from section 1, 2, 3. (c)–(d) The structures from section 4, 5, 6. Source: Reprinted with permission from Wang et al. [8]. Copyright 2020, Springer Nature. Reprinted with permission from Han et al. [9]. Copyright 2019, American Chemical Society.

proportion of the main elements with energy chromatography (energy-dispersive X-ray spectroscope [EDS], OXFORD X-Max^{N} 150). Then Fourier transform infrared spectrometer (FTIR, Bruker EQUINOX 55) was used to determine the main components of cicada wings. EDS analysis results are shown in Figure 4.4. The main components of the surface structure of cicadas include carbon (C), nitrogen (N), and oxygen (O). The vast majority of biological materials contain organic components, and carbon, nitrogen, and oxygen are important elements. These three elements account for the largest proportion in the total amount of elements, 64.31%, 14.02%, and 21.27%, respectively. At the same time, there is an obvious characteristic peak of gold (Au) among the strength peaks of many elements. This is because cicada wings are not conductive, so gold-spraying treatment can improve their electrical conductivity for the analysis of corresponding morphology and element composition. The surface composition of cicada also contains very little chlorine (Cl) and a negligible amount of potassium (K).

After the basic elements determination, combining with the determination of functional groups helps to further analyze the main components of biological sample materials. Therefore, functional groups on the surface of cicada were determined by FTIR spectroscopy. Since most components of insect wings are chitin,

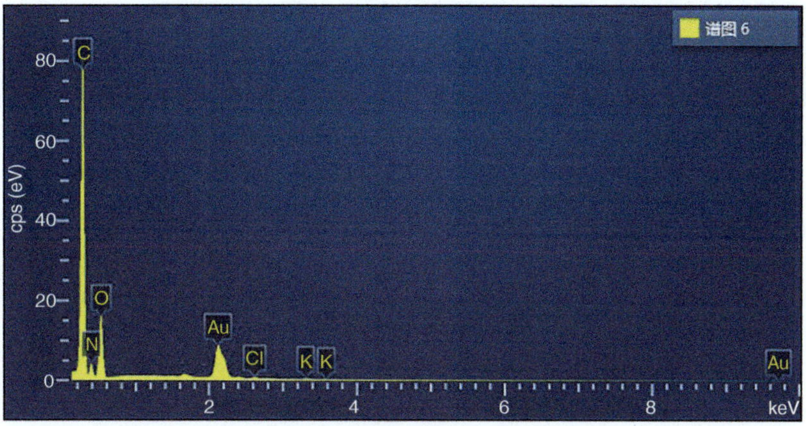

Figure 4.4 EDS analysis of cicada (*Megapomponia intermedia*) wings.

Figure 4.5 FTIR spectra of cicada (*Megapomponia intermedia*) wings and chitin.

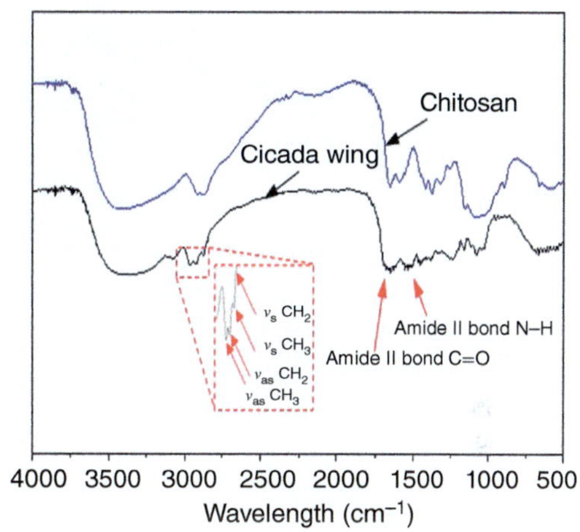

the FTIR spectrum curve of chitin is specifically selected as a reference to compare the actual measured components of cicada wings with the chitin components. The results are shown in Figure 4.5. Compared with the single chitin component, the characteristic peaks of cicada wings are different and their components are more complex. The characteristic peaks at 1550 and 1680 cm^{-1} are the amide band II and the amide band I of the amide functional group. The former is caused by the N–H bending coupled C=O stretching vibration and the latter by the N–H bending coupled C–N stretching vibration. These two characteristic peaks indicate that cicada wings contain chitin and protein components [10, 11]. The characteristic peaks from 2848 to 2968 cm^{-1} are the symmetric stretching peaks of CH_2 and CH_3 and their antisymmetric stretching peaks. Based on the above results, it can be concluded that there is a waxy layer on the surface of cicada wings [12].

4.1.3 Mechanism of High Reflection Reduction on Cicada's Surface

4.1.3.1 Reflective and Transmittance Spectra of Original Cicada Wings

The typical cicada wing was investigated in terms of theory and experimental preparation to reveal the relationship between the structure and function. Here, the cicada wing (*Megapomponia intermedia*), which exists all over the world in tropical climates, was selected as a bionic prototype. There are some advantages with cicada wings: the typical "moth-eye structure" on compound eyes is absent. Since the cicada wing is flat and large in area, it is very suitable as a bionic prototype. As shown in Figure 4.6, when the laser passes through the transparent area of the

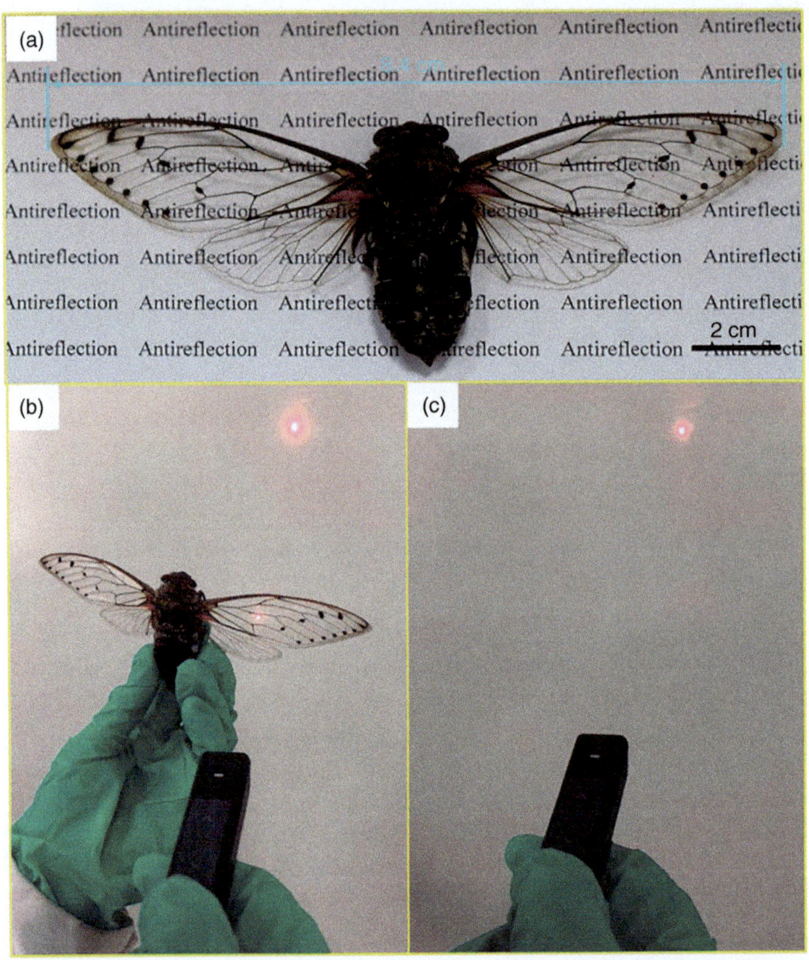

Figure 4.6 The macroscopic morphology of cicada wings: (a) A digital photograph of a cicada. (b and c) The comparison of spot sizes on background when the laser beams passing through cicada wings and not passing through cicada wings. Source: Reprinted with permission from Han et al. [9]. Copyright 2019, American Chemical Society.

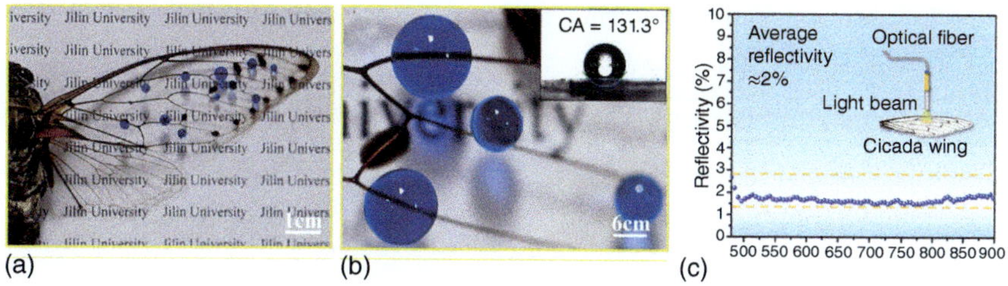

Figure 4.7 Cicada *Megapomponia intermedia* wing. (a) Digital photograph of cicada *Megapomponia intermedia* wing placed on a paper sheet. (b) Droplets on the cicada wing and the static water contact angle is 131.3°. (c) The reflectivity and transmittance spectra of cicada wings. Source: Reprinted with permission from Han et al. [9]. Copyright 2019, American Chemical Society.

wing, the strong beam is dispersed and weakened, leaving an enlarged spot on the background.

Figure 4.7a shows the fiber network structures used for supporting, which gives the wing a certain strength. Besides, the words on the paper could be seen clearly through the wings. Also, the 131.3° water CA confirms that the cicada wings possess remarkable hydrophobicity (Figure 4.7b). The optical property of the natural cicada was characterized by the marine optical spectrometer. The results showed that the cicada wing exhibited low reflectivity in most visible bands, and the average reflectance was less than 4%. Meanwhile, the cicada wing also maintained high transmittance, which indicates the multifunctions of natural cicada wings (Figure 4.7c).

4.1.3.2 Another Classic Optical Theory for Antireflective Performance

The antireflection phenomenon is often accompanied by a rough structure of a surface, which is defined as the microgeometry of a surface with small spacing and peak valleys. The macroscopic rough surface depends on scattering to suppress reflection. However, the morphology of microscopic rough surfaces tend to be irregular; meanwhile, the characteristic dimensions are smaller than the optical wavelength. Hence, it is meaningful to propose a model to analyze the complex interactions between light and these rough surfaces with effective medium theory or effective medium approximation [13–15]. Random microscopic rough surfaces can be used as a series of multilayer effective media. Its effective refractive index usually depends on the volume fraction (f) of each rough layer and the estimated value based on the development of ellipsometric techniques [16]. Typical expressions for the effective refractive index are demonstrated in Eq. (4.3):

$$\frac{(n^2 - n_1^2)}{(n^2 + 2n_1^2)} = (1 - f_1)\frac{(n_2^2 - n_1^2)}{(n_2^2 + 2n_1^2)} \tag{4.3}$$

where n_1 and n_2 are the effective refractive indexes of two constituent layers, f_1 and f_2 $(=1 - f_1)$ are the corresponding volume fractions. In this model, the relation holds only if the layer (n_2) is assumed to be surrounded by the other layer (n_1). On the other hand, the formula can also be deformed as Eq. (4.4):

$$f_1 \frac{(n_1^2 - n^2)}{(n_1^2 + 2n^2)} + f_2 \frac{(n_2^2 - n^2)}{(n_2^2 + 2n^2)} = 0 \tag{4.4}$$

The term n is assumed to be a homogeneous mixture of two constituent layers and can be extended to multiple numbers of constituent layers with Eq. (4.5):

$$\sum_{i=1}^{n} f_i \frac{(n_i^2 - n^2)}{(n_i^2 + 2n^2)} = 0 \tag{4.5}$$

Gradient-RI can be termed as a single nonuniform layer with a changed refractive index from the upper surface to lower surface, obscuring the interface between the air and the substrate [17, 18]. An optical thin film with gradient-RI was first analyzed and fabricated in 1960 [19]. There are several different profiles such as linear, cubic, quantic. Besides, this film used for broadband and omnidirectional AR possesses great potential in applications. Typical expressions for the continuous gradient-RI with linear, cubic, and quantic profiles [20] are given as follows:

Linear index profile :

$$n = n_0 + (n_s - n_0)t, 0 \leq t \leq 1 \tag{4.6}$$

Cubic index profile :

$$n = n_0 + (n_s - n_0)(3t^2 - 2t^2) \tag{4.7}$$

Quantic index profile:

$$n = n_0 + (n_s - n_0)t(10t^3 - 15t^4 + 6t^5) \tag{4.8}$$

The antireflection mechanism of the cicada wing was investigated with finite-difference time-domain (FDTD) simulation. The model of AR nanostructures is displayed in Figure 4.8a, and the structural parameters were obtained according to the SEM and AFM images. Here, two-dimensional spatial simulations were conducted. The boundary conditions in the x-direction were periodic, and perfectly matched layer (PML) absorbing boundary conditions were applied to the boundary surrounding the computational domain in the y-direction. The normal incident plane was set with a 500 nm height above the top of the nanostructures. The refractive index of air and nanostructures were $n_{air} = 1$ and $n_{chitin} = 1.7$.

Optical spectra of the cicada wing were analyzed with FDTD simulation. The results in Figure 4.8b confirmed its high transmittance of >80% and low reflection of <4% at wavelengths of 350–800 nm; meanwhile, almost all plane light waves transmitted across the surface. Besides, as shown in Figure 4.8c, FDTD results indicated that intensive electric field enhancement occurs around the nanocones, and the maximum amplitude of electric field is 1.4. Although this enhancement occurs only in a small area, it has also been verified that the nanostructures on cicada wings play an important role in changing the electric field distribution around the surface.

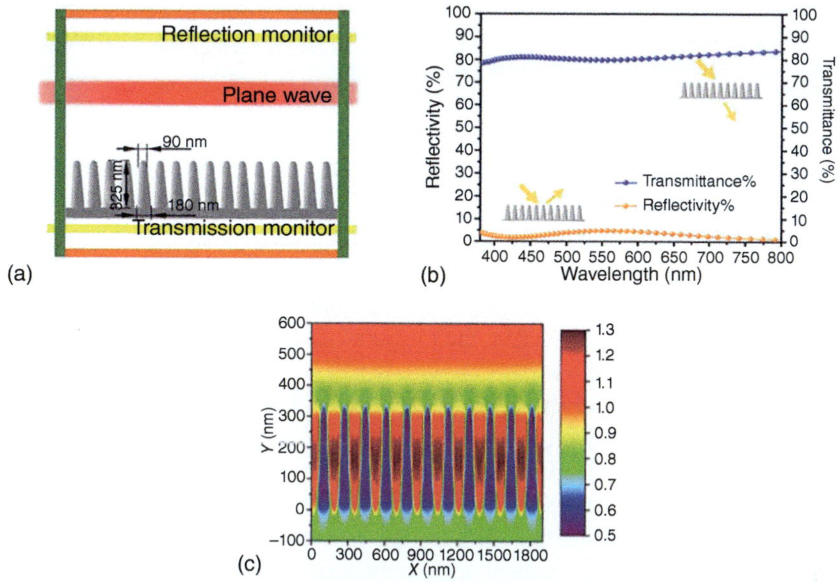

Figure 4.8 The FDTD simulation results of the cicada wings and their antireflective mechanism. (a) The antireflective model of the cicada wing surfaces and its parameters for simulation. (b) The simulation results of the cicada optical wings. (c) Electric field (*E*) profile at the $z = 0$ plane for the bump structure when plane wave light was injected from top to bottom. Source: Reprinted with permission from Han et al. [9]. Copyright 2019, American Chemical Society.

4.2 Accurate Fabrication of Cicada-Inspired AR Surfaces

4.2.1 Preparation Technology

Subwavelength structures have been investigated widely because their tiny enough pitch can suppress diffraction effects. Artificial media with optical properties synthesized by deliberate structuring have been used for over 50 years in diffractive free-space optics [21]. Subsequently, structural media were studied in detail with the Rayleigh theory [22]. However, transmission of electromagnetic waves in a subwavelength structural medium was explored in the 1940s. A comprehensive theoretical study of subwavelength structures was developed by Rytov. In the 1950s, Rytov showed that these structures can behave as homogeneous media with an equivalent refractive index that depends on their geometry and the polarization state of the incident wave. The subwavelength structures in the cornea of moth eyes were found by Bernhard and Miller in the 1960s [23]. Owing to outstanding antireflection of the structural moth-eye cornea, the subwavelength structures were used as a camouflage mechanism for protecting moths from predators at night. Inspired by the moth eyes, antireflective structures (ARSs) were prepared by Clapham and Hutley in the 1970s [24]. Since then, subwavelength structures are widely applied in free-space optics, including birefringence optics and polarizers and antireflection surfaces [25]. With the development of photonic and high-resolution lithography, subwavelength structures are used in integrated optical devices. Recently, these structures with simple geometries have been patterned to obtain variable refractive index [26–31]. Research on optimizing the structural parameters and applying them to the integrated photon field or the antireflective functional surface is still underway.

In fact, the line between ARCs and ARSs is becoming blurred. The coating mentioned here refers to the surface based on the principle of 1/4 wavelength interference cancelation or the antireflection surface with rough structures constructed by spraying. However, the ARSs refer to the antireflective surface with relatively regular structure arrays. Although the preparation technology of ARCs is more mature and the cost is lower than that of ARSs, the mechanical strength of the ARCs is less desirable [32].

Due to the outstanding advantages of ARSs in surface quality and overall mechanical strength, it has been a research hotspot in the field of antireflection. For ARSs, their structural parameters and the interaction between structures and incident light are very important to realize final antireflection performance. However, selecting and optimizing structural parameters has been a challenge in current research. Complex functions of organisms in nature are often realized with simplest structures. The periodic nanostructures on moth eyes were first discovered as ARSs by Bernhard in 1967. Based on the previous studies, it has been investigated that there are micro-/nanostructures on the surface of organisms, which play a decisive role in the realization of special functions of biological surfaces. Therefore, it is of great significance to study bionic antireflection structures. In recent decades, bionic antireflective research has attracted significant attention. Numerous biological surfaces, such as moth eyes, cicada wings, butterfly wings, have shown excellent antireflection performance. Figure 4.9 displays a series of research on bioinspired AR surfaces. The fabrication of nano-/microstructures is the key to further investigation. Although the replication of AR structures has been accomplished, the process is complex. Besides, high cost also limits its practical applications.

Recently, reversible AR surfaces fabricated with shape memory polymer (SMP) are also appealing[35]. Inspired by dragonfly wing, Yang and coworkers prepared exquisite, inclined structures with reversible AR performance [36]. Reflectivity of self-erasable AR surface coating with nanocone arrays can be switched from 4.5% to 0.6%, averaged over the visible spectral range by controlling the temperature below or above 80 °C (Figure 4.10) [37]. The coupling between the material and structure have greater potential in applications such as smart windows, smart materials.

In addition, some ARSs that are common in organisms are listed in Table 4.1 [38]. In general, ARSs are varied and can be divided into two categories. The first way to suppress reflection is to extend the light path by trapping light, and it is common in the hierarchical structures on the surface of butterfly scales and rose petals. Glare could also be suppressed by increasing light transmission and reducing Fresnel reflection, and it is realized by subwavelength structures on the surface of most moth eyes and butterfly eyes or cicada wings.

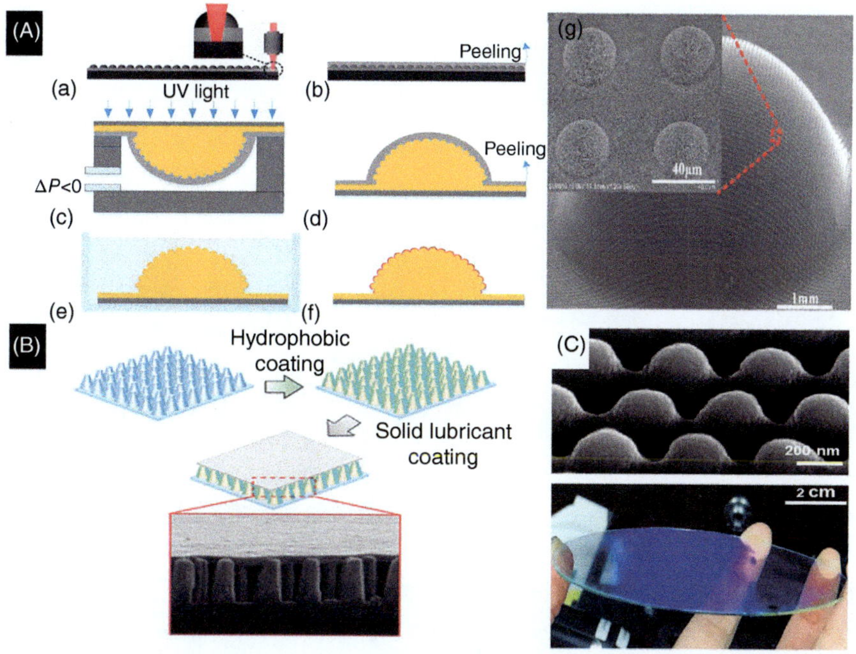

Figure 4.9 (A) Artificial compound eyes. (a–f) Fabrication process. (g) SEM image of fabricated compound eyes. The inset shows partial magnification. Source: Reprinted with the permission from Li et al. [33]. Copyright 2019, American Chemical Society. (B) Fabrication process of the moth-eye structural surface and the SEM image of fabricated AR material. Source: Reprinted with the permission from Han et al. [34]. Copyright 2020, American Chemical Society. (C) Large-area AR surface covering with nanoarrays. Source: Reprinted with the permission from Bochet-Modaresialam et al. [18]. Copyright 2020, American Chemical Society.

4.2.2 Structure and Composition Analysis

4.2.2.1 3D Visible Parameterized Models of Cicada Feature Structures

To clearly describe the nanostructures distributed on the surface of cicada wings, field emission scanning electron microscope and atomic force microscope are used to observe the transparent region of cicada wings. The exquisite nanostructures observed by FESEM confirmed their complete hexagonal compact arrangement, and the distance between adjacent vertices was about 200 nm (Figure 4.11a). As observed in Figure 4.11b, the substrate diameter of the nanostructure was 180 nm, and the height of a single nanohump was about 325 nm. The results showed that the shape of the nanostructures is not a simple cone or cylinder, but a parabola with a large aspect ratio (ARO). Their shape can be rather described as streamlines. Figure 4.11c shows that the cicada wing is so thin with a thickness of about 8.25 μm. Besides, the front and back of the cicada wing have the same structures. In addition, the outline of the nanostructure arrays can be recognized more intuitively in Figure 4.11d.

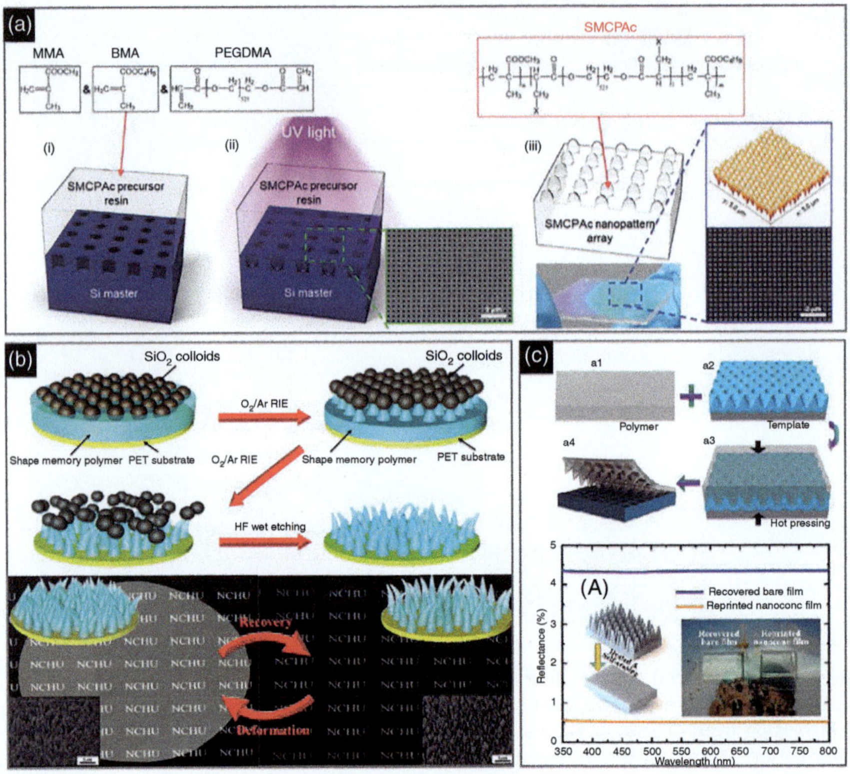

Figure 4.10 (a) Reversible AR surfaces. Source: Reprinted with the permission from Park et al. [35]. Copyright 2017, Royal Society of Chemistry. (b) The fabrication process for creating dragonfly wing-inspired antireflection coatings. Source: Reprinted with the permission from Tseng et al. [36]. Copyright 2020, American Chemical Society. (c) Large-area AR surface covering with nanoarrays. Source: Reprinted with the permission from Li et al. [37]. Copyright 1996–2021 MDPI (Basel, Switzerland).

The 3D model of the nanostructures on cicada wings was established with data obtained from the FESEM and SEM images (Figure 4.12a). The nanocones on the cicada wings were on a hexagonal tight arrangement. The array period was around hundreds of nanometers, leading to interactions with incident light. Therefore, this kind of nanoarray plays a very effective role in the optical behaviors of the surface over the visible light range. Besides, the single nanostructure looks like a cone with a dome, and its structure parameters have been marked in the section view in Figure 4.12b.

4.2.2.2 Fabrication of Cicada-Inspired AR Surfaces

With the development of bioinspired AR surfaces, exquisite structures on creatures have drawn huge attention owing to their various applications such as flexible displays, photovoltaic industry, medical endoscopes, smart windows, and other fields. At present, many technologies have been used to duplicate the exquisite and ordered nanostructures, such as electroless plating and electroplating [39, 40],

Table 4.1 Typical biological surfaces with unique AR structures and intriguing functions.

Biological surfaces	AR structures	AR mechanism	Functions
Moth eye	Nanonipple structures	Continuous gradient refractive index ARC	Antireflection, antifogging
Butterfly eye			Antireflection
Moth wing	Nanopillar structures		Antireflection
Butterfly wing	Hierarchical structures (parallel ridges, inclined ridge-lamellae structures, quasi-honeycomb structures, parallel-laminae structures, concave multilayer structures), nanopillar structures, and nanohole structures	Destructive interference, multiple refraction, and continuous gradient refractive index ARC	Antireflection, light-trapping, antifogging, structural color, self-cleaning, superhydrophobicity, directional adhesion, chemical sensing capability
Fly eye	Nanonipple structures	Continuous gradient refractive index ARC	Antireflection, antifogging
Mosquito eye			Antireflection, Antifogging, superhydrophobicity
Cicada wing			Antireflection, superhydrophobicity, antibiosis
Owlfly eye			Antireflection
Body surfaces of hyperiids	Nanoprotuberance structures and monolayers of spheres	Continuous gradient refractive index ARC and quarter-wavelength ARC	Antireflection
Bird of paradise feather	Modified barbule array structures	Multiple scattering of light	Antireflection, structural absorption
Beetle eye	Maze-like nanostructures	Continuous gradient refractive Index ARC	Antireflection
Rose petal	Hierarchical structures (nanostructures on microstructures)	Continuous gradient refractive index ARC, increasing light path length, and multiple internal reflections	Antireflection, Light trapping, structural color, high adhesion, superhydrophobicity
Lotus leave	Hierarchical structures (nanostructures on microstructures)	Continuous gradient refractive index ARC, increasing light path length, and multiple internal reflections	Antireflection, light trapping, self-cleaning, low adhesion, superhydrophobicity

Source: Reprinted with permission from Han et al. [38]. Copyright 2019, Elsevier Ltd.

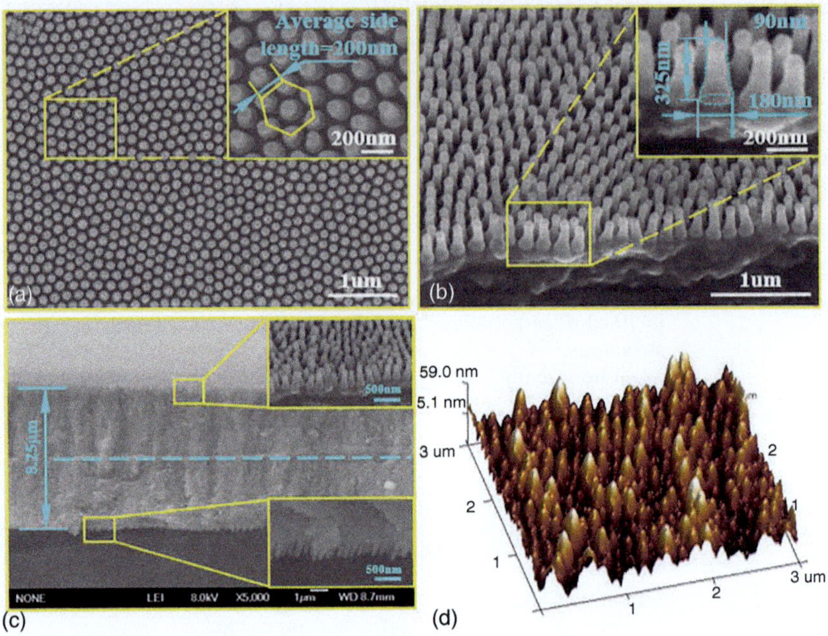

Figure 4.11 The morphology of the original prototype. (a) The image and enlarged view of cicada wings from top view. (b) The image of the cross section of cicada wings observed from side view. (c) The structure of cicada wings in both positive and negative directions. (d) The AFM image of cicada wings. Source: Reprinted with permission from Han et al. [9]. Copyright 2019, American Chemical Society.

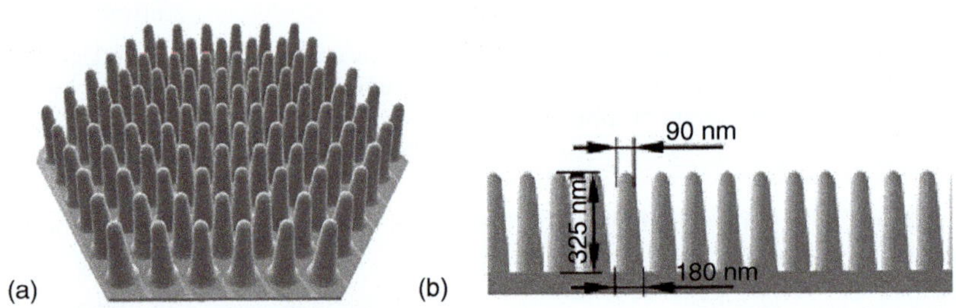

Figure 4.12 The 3D model of the nanostructures on cicada wings. (a) Their array. (b) The structural parameters of the arrays.

self-assembled technology [41], template method [8, 42, 43], and so on. These methods can solve the problems in part, but there is still a long way to go. There are deviations between original organism structures and that of biomimetic replica, and it might lead to a discount on functions. For such excellent surface with high transmission and low antireflection, it is believed to inherit completely from the original prototype, so the bionic preparation is carried out by using the organisms as a biological template directly. This part provided a simple way to fabricate biomimetic nanocones with a modified sol–gel progress.

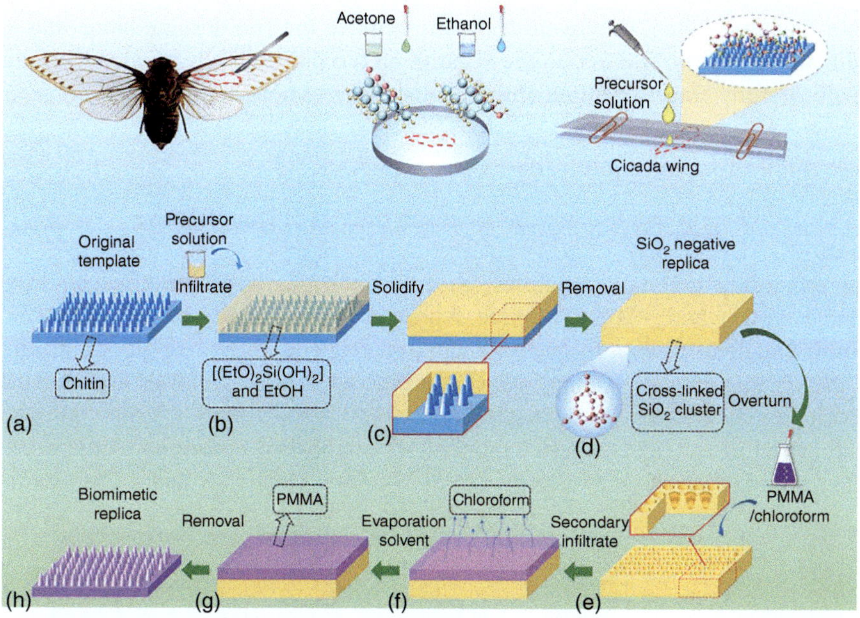

Figure 4.13 Schematic diagram of rebuilding biomimetic replica, showing cicada-wing original template, fabrication of SiO_2 negative replica, and transfer of the pattern onto the PMMA positive replica. (a) Original template obtained from cicada wings. (b) The precursor solution added on the original template. (c) Heating curing process. (d) The SiO_2 negative replica. (e) The mixture of PMMA/ chloroform was added on the surface of SiO_2 negative replica. (f) Evaporation solvent process. (g) Solidified PMMA film. (h) The PMMA positive replica. Source: Reprinted with permission from Han et al. [9]. Copyright 2019, American Chemical Society.

The cicada *Megapomponia intermedia* (Distant, 1905) was selected as the biological prototype. In this part, the cicada wing was used as the original template to guarantee the perfect inheritance of biological structures. The preparation process of the SiO_2 negative replica is shown in Figure 4.13a–d, and it has been expatiated in detail in our previous work [44]. Firstly, the cicada wing samples were soaked in ether and alcohol to remove the impurities. Then, the biological specimen was sandwiched between two pieces of glass and the precursor reacted with tetraethylortho-silicate and attenuated hydrochloric acid was added to the edge of the glass slide. Subsequently, they were moved into a vacuum-drying oven to solidify the precursor. Finally, the biological specimen was removed in acid and the SiO_2 negative replica was left on the glass slide. Since the diameter of the hole was so tiny, the fluid with higher surface energy could not fill into the nanopores; hence, it is necessary to choose a material that could be soaked into the holes distributed on the SiO_2 negative replica. In this work, the polymethyl methacrylate (PMMA) with a smaller relative molecular weight was used to manufacture the positive replica (Figure 4.13e–h). First, the PMMA was pretreated with a vacuum drying oven to remove moisture (due to its hygroscopicity). Then, the PMMA powder was dissolved in chloroform, followed by coating on the SiO_2 negative replica with a rubber machine. Then, they were heated to remove the trichloromethane, and a solidified PMMA film was left on the negative SiO_2 replica. Finally, peeling the thin film off carefully, the bioinspired antireflective film was obtained.

4.2.3 AR Performance of Cicada-Inspired Surfaces

4.2.3.1 Morphology and Composition Characterizations of SiO_2 Negative Replica

After rebuilding the nanostructures, the morphologies and structures of SiO_2 negative replica and PMMA positive replica were characterized by AFM and FESEM. The detailed data for replica including single nanocone height, diameter, and the spacing between nanocones were expressed, and the analysis of elements and components was carried out subsequently.

The model of the SiO_2 negative replica was established according to the morphologies and structural parameters in FESEM images (Figure 4.14a,b). The results

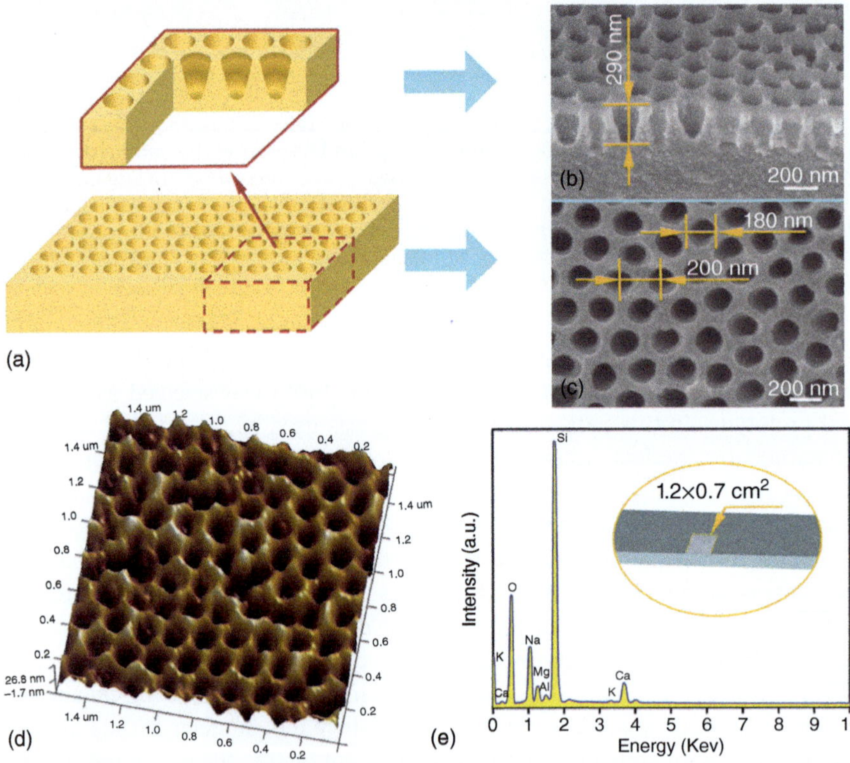

Figure 4.14 The characteristic results of SiO_2 negative replica. (a) The model of the SiO_2 negative replica. (b) The FESEM images of the cross section of SiO_2 negative replica. (c) The FESEM image of SiO_2 negative replica observed from the top view. (d) The AFM image of SiO_2 negative replica. (e) The energy dispersive spectroscopy spectrum of SiO_2 negative replica. Source: Reprinted with permission from Han et al. [9]. Copyright 2019, American Chemical Society.

of negative structures observed from the top view and side view are expressed in Figure 4.14b,c. The shape of the nanoholes on negative SiO_2 replica looks like the inverted cones, and the distribution is identical to the original arrays on the cicada wing (Figure 4.14d). The diameter of the tapered bottom is about 180 nm while the distance between the inverted cones is about 200 nm, and they are consistent with the biological prototype. However, the height of the inverted cones is slightly shorter than that of the arrays on the prototype. It could be attributed to incomplete infiltration of precursor solution, and it is within acceptable levels.

The EDS spectrum of the SiO_2 negative replica is used to analyze the element types. The peaks of silicon and oxygenium are obvious (Figure 4.14e). The SiO_2 negative replica is mainly composed of 24.28% silicon and 62% oxygenium. Since the cicada wing is mainly composed of chitin, there is no carbon peak in the spectrum. Hence, it illustrates that the original template has been removed completely and the SiO_2 negative replica is ready for the next step.

4.2.3.2 Characterizations of PMMA Positive Replica

The morphologies and arrays on positive PMMA replica are characterized in Figure 4.15. This AR film displays satisfactory flexibility, and its thickness is about 180–190 μm. Besides, the nanostructures on the positive PMMA replica are observed with FESEM in the top view and the side view (45° inclination). The ordered array on positive PMMA replica is consistent with the hexagonal arrangement distributed on the cicada wing. Its morphologies and structural parameters are also almost the same with the SiO_2 negative replica and the original template. Consequently, it can be confirmed that the morphologies and structures of the positive PMMA replica inherit the original template well. The materials analysis of the positive PMMA replica with the FTIR spectrum is depicted in Figure 4.15e. The peaks at 2995 and 2951 cm^{-1} were caused by –CH and –CH_2– stretching vibration bonds. The peak at 1730 cm^{-1} represents C=O stretching vibration bond. Other peaks from 1100 to 1300 cm^{-1} were attributed to C—O—C stretching vibration bond. The FTIR spectrum of the PMMA positive replica matches the FTIR spectrum of PMMA [45].

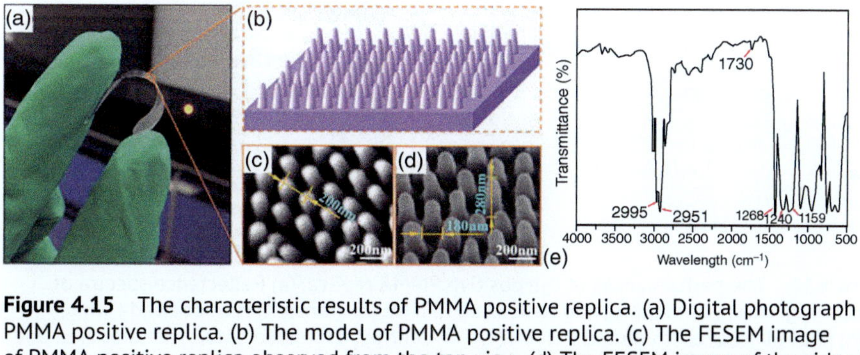

Figure 4.15 The characteristic results of PMMA positive replica. (a) Digital photograph of PMMA positive replica. (b) The model of PMMA positive replica. (c) The FESEM image of PMMA positive replica observed from the top view. (d) The FESEM image of the side view of PMMA positive replica. (e) The FTIR spectra of PMMA positive replica. Source: Reprinted with permission from Han et al. [9]. Copyright 2019, American Chemical Society.

Consequently, it can be confirmed that the component of the positive replica is indeed PMMA.

4.2.3.3 Characterizations of PMMA Positive Replica

A structured PMMA thin film was successfully prepared, which was inherited from the biological prototype by an improved biological template method, and the performance characterization was carried out. Figure 4.16a shows the reflectance spectra of a planar chitin substrate, a flat PMMA film, a cicada wing, and a bioinspired structural PMMA film. It is known that the main component of the cicada wing is chitin [46]; however reflectivity of a planar chitin substrate is much higher than that of the cicada wing. This trend is also found in flat PMMA films and structural PMMA films. It has been experimentally verified that special nanostructures truly promote AR performance. Besides, the bioinspired AR film exhibited an angle-dependent AR performance, and the reflectivity spectra measured with different incident angles (30°, 40°, 50°, 60°) confirmed this. The reflectance takes on a raising trend with increasing incident angle (Figure 4.17). Moreover, Figure 4.16b shows the transmittance spectra of flat PMMA, glass slide, bioinspired structural PMMA, glass slide covered with structural PMMA, and glass slide covered with flat PMMA. Although

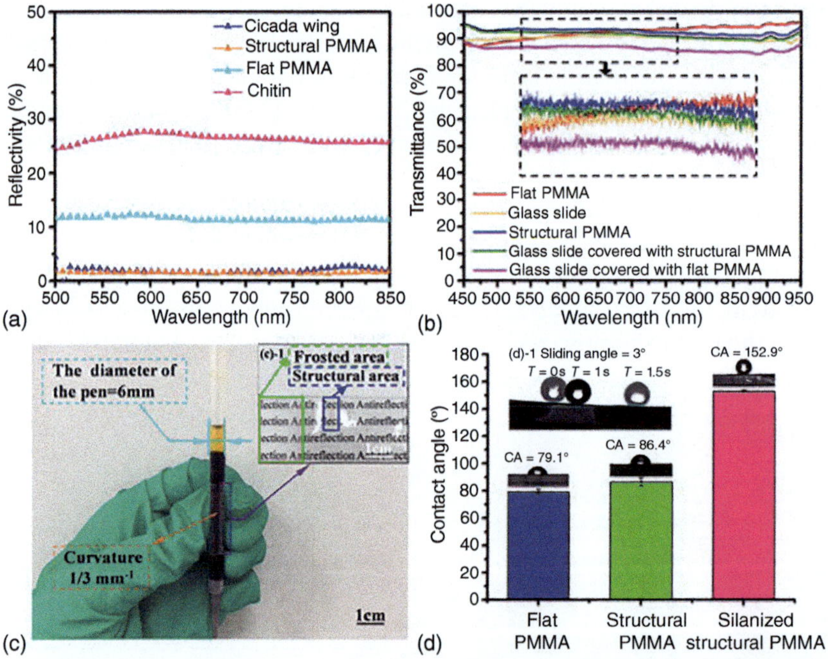

Figure 4.16 The performances of the positive PMMA replica. (a) Reflectance spectra of cicada wing, chitin, flat PMMA, and structured PMMA. (b) Transmittance of cicada wing, chitin, flat PMMA, and structured PMMA. (c) The contrast of frosted area, structural area and flat area under an illuminant. (d) The connect angle of flat PMMA, structural PMMA, and silylanized structural PMMA. Source: Reprinted with permission from Han et al. [9]. Copyright 2019, American Chemical Society.

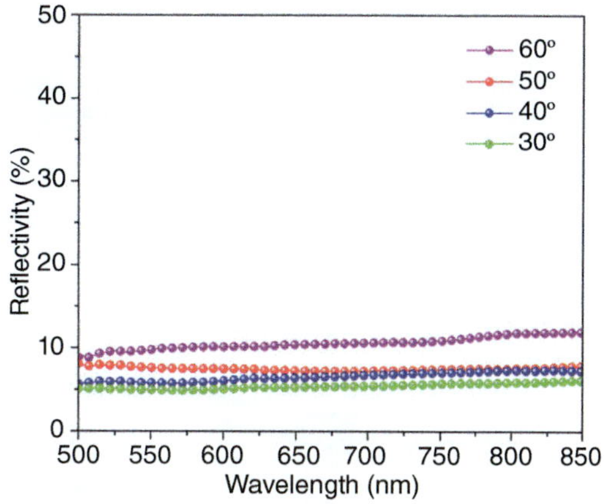

Figure 4.17 The reflectivity curves of the antireflective film with different incident angles.

transmittance of the glass slide covered with structured PMMA film is relatively lower than that of a single structured PMMA film, averaging 92.5% for 450–950 nm, it improves the transmittance of the glass slide. In contrast, the glass slide covered with flat PMMA film is also considered. Compared with a single glass side, the transmittance of the glass side covered with flat PMMA film is lower, averaging 87.2%. The drop in transmittance is attributed to an abrupt change in the refractive index of the interface between air and flat PMMA. It is obvious that the nanocone arrays have an influence on the optical properties of the surface, increasing light transmission and reducing light reflection. In brief, the replica with mimetic nanocone arrays has inherited excellent optical properties from the biological specimen. At present, flexibility of the AR film has become an important indicator owing to the development of flexible electronic displays and thin-film solar cells [47, 48]. Therefore, a pen tube with a diameter of 6 mm is selected to characterize the flexibility of positive AR replica. Meanwhile, the inset in Figure 4.16c displays the differences in appearance between structural PMMA area (blue frame), frosted PMMA area (green frame), and flat PMMA area (the remaining part). It could be seen that the flat PMMA glares severely, while the frosted PMMA area looks much better but declines picture resolution. The nanostructures on surface not only affect the optical properties but also change the surface wettability greatly. Here, the surface wettability is evaluated with water CA. The CAs of flat PMMA, structural PMMA, and silanized structural PMMA are shown in Figure 4.16d. The CA of structural PMMA film is 86.4°, higher than that of flat PMMA film. The surface of structural PMMA film is silanized to achieve a superhydrophobic state [42]. Besides, a water slide angle (SA) test is also carried out and recorded. The SA of the surface is about 3°, and the duration of the process is less than 10 seconds. These results meet the requirement of a superhydrophobic state well and confirm its satisfactory self-cleaning performance.

Although biomimetic self-cleaning AR materials with excellent performance can be obtained with cicada wings as the original template, the ridges on cicada wings limited the structural area on AR film. Based on this, the preparation process has been improved and the whole cicada wing can be replicated with polydimethylsiloxane (PDMS) [8]. Figure 4.18a,b displays the PDMS AR surface under strong

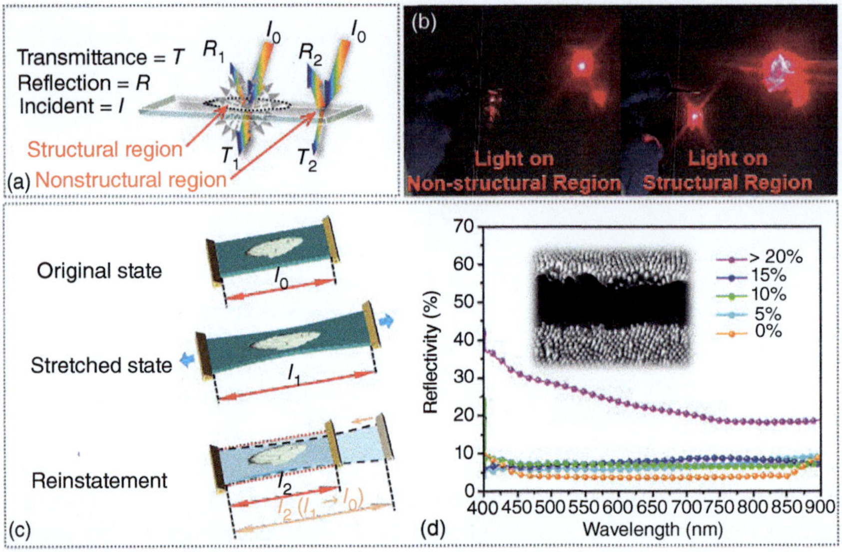

Figure 4.18 (a) Schematic diagram of the light pathway through the different regions of the replica. (b) The replica under strong light. (c) Schematic diagram of the stretching test. (d) The reflectivity curves of the PDMS replica after stretch-recovery process.
Source: Reprinted with permission from Wang et al. [8]. Copyright 2020, Springer Nature Switzerland AG. Part of Springer Nature.

light. Compared with the nonstructural area, the incident light was scattered when it passed through the structural region. It was obvious that the nanoarrays on the AR surface dispersed the concentrated light path. At the same time, the PDMS AR surface was stretched with different extension ratios to evaluate its elasticity (Figure 4.18c,d). The morphologies of the PDMS AR surface will not be destroyed when the tensile rate does not exceed 20%, indicating satisfactory elasticity of the PDMS replica. It is very important for further application of antireflective surfaces in curved surfaces and flexible screens.

According to the microstructures, SEM picture on the structured PMMA thin film prepared in the prior art, the detailed structural parameters of the positive structure are obtained. The actual parameters of the structure are then modeled, and the established model is shown in Figure 4.19.

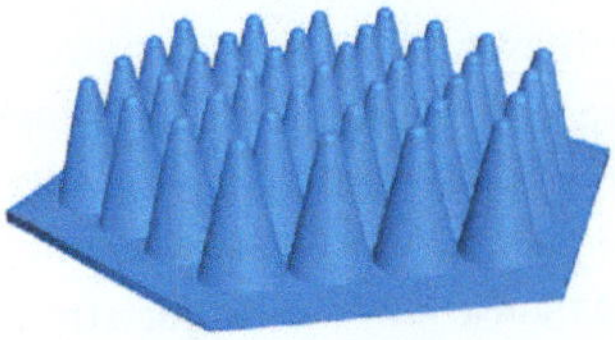

Figure 4.19 The model of the nanostructures on PMMA positive replica.

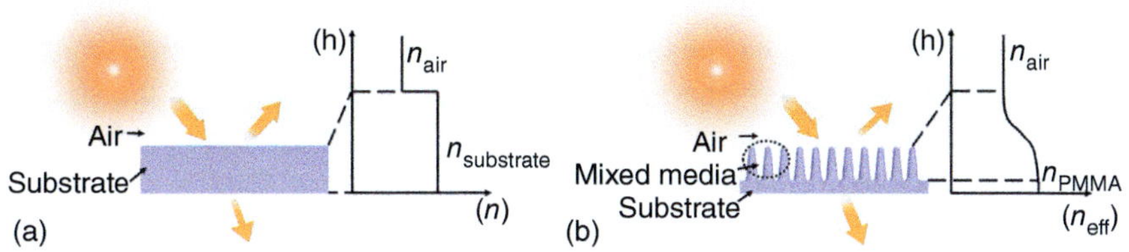

Figure 4.20 Schematics of mechanism for antireflection of the cicada wing.
Source: Reprinted with permission from Han et al. [9]. Copyright 2019, American Chemical Society.

Given that the Fresnel reflection is caused by the abrupt change in index, reflection will be suppressed if the interface between different substances is blurred. Fortunately, the nanocones covering on cicada wings have exhibited near-perfect AR performance, which provides some innovative ideas for the development of bioinspired AR materials. Besides, the interface between the air and the nanoarrays on cicada wings could be defined as a mixed media layer, and the refractive index for normal incidence can be expressed according to the effective medium theory [49]:

$$n_{\mathrm{eff}}^2 = n_{\mathrm{a}}^2 f + n_{\mathrm{s}}^2(1-f) \tag{4.9}$$

where n_{eff} is the effective refractive index of air–substrate mixed media, n_{a} is the refractive index of air, and n_{s} is the refractive index of substrate. The f in this formula is the volume percentage of the nanocones arraying on the cicada wing surface.

In general, air is one of the most common media, and the refractive index of most solids is much larger than that of air. Therefore, the abrupt refractive index change is inevitable in the gas–solid interface, leading to strong reflection at the interface (Figure 4.20a). In contrast, the nanocones on the substrate are surrounded by air, and they have formed mixed media, which can achieve smooth transition from the air to substrate. The n_{eff} curve in Figure 4.20b is calculated using Eq. (4.9). It is easy to see that the effective refractive index of mixed media changes smoothly along the nanocone height, indicating that the influence of nanostructures on optical behaviors should not be ignored.

4.3 Large-Area Preparation of Cicada-Inspired AR Surfaces

Traditional technologies for biomimetic fabrication include self-assembly approach [50], chemical vapor deposition [51], etching [52], lithography techniques [25], laser fabrication [53], and template assistance [54]. Among these methods, bio-template fabrication could guarantee high fidelity of pattern replication. However, due to the size limitation of biological organisms, the AR membrane prepared by biological template is too small to be applied in practical applications. Therefore, it is urgent to develop a method that can complete large-scale bioinspired AR surface preparation with minimum resource consumption. In addition, the size of the structure also directly affects the performance of the surface. For example, diffraction grating structures are very sensitive to changes in structural parameters. In other words, when the structural parameters change greatly, the original function of the structural plane may be lost. In the preparation process, it is difficult to ensure that the design parameters are completely consistent with the actual parameters.

A simple and conventional method for fabricating large-area bionic augmented reality arrays was developed using recyclable porous templates. It not only proposes the design concept of an extremely simplified structure to simulate the main functional characteristics of biological AR array but also innovatively realizes the scale insensitivity of bioinspired AR surface, which is very friendly for further applications.

4.3.1 Preparation Technology

4.3.1.1 Test Materials and Reagents

Porous anodic aluminum oxide (AAO) template was used as the original structural template, and the materials and chemical reagents used are given in Table 4.2.

4.3.1.2 Bioinspired Large-Area Preparation Template and Its Pretreatment

Porous anodic alumina materials have drawn extensive attention because of their highly regular surface structure, controllable preparation process, and stability of the obtained morphology. At present, its preparation process has been relatively mature, using high purity (99.999%) aluminum plate as anode, graphite plate as cathode, at 4 °C and 21 V voltage in the mixed solution of perchloric acid and ethanol (volume ratio of 1 : 4) for electrochemical polishing time of nine minutes, to reduce the surface roughness. To prepare for subsequent anodic oxidation, the polished aluminum plate is then washed with deionized water and allowed to dry in the shade.

The aluminum plate fixed as the anode and graphite as the cathode, about 2 cm apart, are immersed in an electrolyte for electrolysis. To obtain a better surface shape, a mild anodizing process (double anodizing) is widely used. In general,

Table 4.2 Test materials and chemical reagents used.

Material	Molecular formula	Specifications	Manufacturer
Acetone	C_3H_6O	A.R.	Changtai Weiner Chemical (Shandong) Co.
Anhydrous ethanol	CH_3CH_2OH	A.R.	Sinopharm Chemical Reagent Co.
NOA61	—	—	Cheng Jianda Photoelectric (Shenzhen) Co.
Deionized water	H_2O	Level 1	Sangon Biotech (Shanghai) Co.
V-type taper hole AAO plate	Al_2O_3	—	Osborne (Shenzhen) Co.

initial anodization results in an unsatisfactory structure, where the oxide layer is usually dissolved by a mixture of chromic acid and phosphoric acid, forming pits on the aluminum surface. These pits provide the direction for secondary anodic oxidation, and the structure after secondary oxidation is more orderly and the surface morphology is better. In this process, the composition and concentration of the electrolyte solution, the voltage in the electrolytic process, and the temperature of the electrolytic environment can all affect the surface morphology of the product; that is, the morphology of the final anodic oxidation structure can be controlled by changing the conditions of the electrolytic reaction. In addition, the AAO template is removed by chemical corroding process with 8 wt% $CuSO_4$ solution and 10 wt% NaOH solution soaking for one and two hours, respectively. It increases the material cost, thereby increasing the preparation cycle.

The structured AAO template will be pretreated initially. The purpose of pretreatment is to enhance the overall stiffness of the template. Since demolding by mechanical stripping may cause deformation of the AAO template, the AB component epoxy resin is mixed at a 3 : 1 volume ratio, and the AAO template is then bonded and fixed on the glass surface with a fully stirred epoxy resin. After heating and curing, the overall stiffness of the template can be strengthened so that the template can be reused for many times.

4.3.1.3 Structure Design and Surface Microstructure of AAO Template

Porous AAO template is suitable for the preparation of micro-/nanostructures due to its adjustable parameters in the anodizing process. As mentioned, porous templates are used to provide structure blueprints for bioinspired preparation, rather than cicadas, to achieve large-area AR membranes. It is worth noting that bioinspired AR materials based on AAO template are not exactly the same as the original AR arrays on cicada wings in terms of structure morphology and characteristic size. This method brings some changes to the AR array prepared, especially the structural parameters such as height, diameter, and period. Meanwhile, the variation of structural parameters may affect the performance of the prepared surface. Based on this, further experiments are carried out to reveal the relationship between the interface optical performance and the structural parameters of the arrays. It is obvious that reflectivity of the arrays is affected by multiple factors, and a larger ARO promotes the AR property when the arrays are in the same period. However, the influence of shape must be considered in the case of arrays of the same height. Here, six types of AAO templates are selected to prepare the AR arrays, and optical behaviors of the arrays with different morphologies and different heights are considered (Figure 4.21).

Six kinds of porous structure templates were selected to consider optical behaviors of the arrays with different morphologies and scales. The reflectance spectra in Figure 4.21b show that for porous arrays, both the structural parameters and the size of holes have influence on the AR properties. For near-wavelength structures with the same period (IV, V, VI) with similar morphologies, the reflectance of their surfaces decreases significantly with the increase in the hole depth. This is because light was reflected for multiple times in the near-wavelength holes. The increase

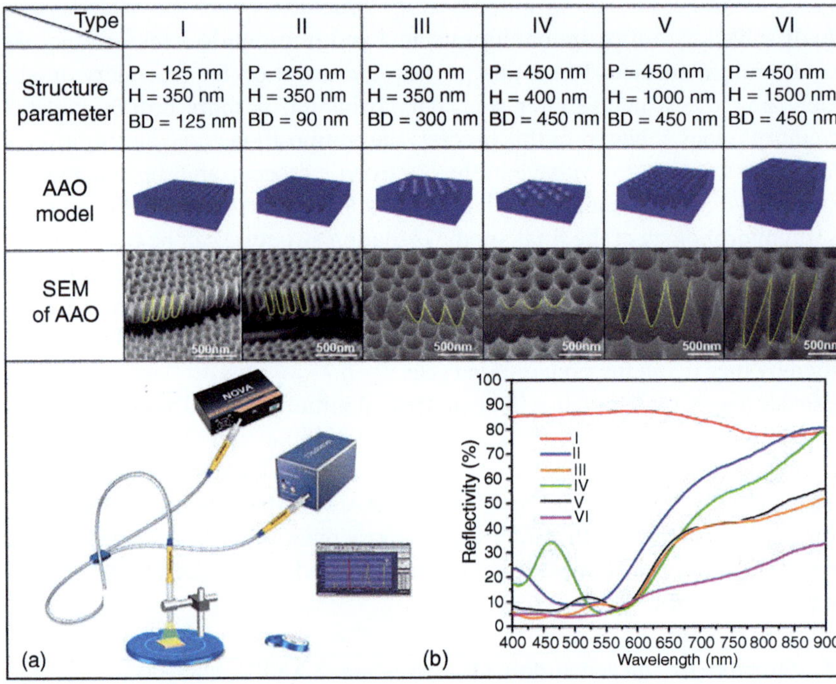

Figure 4.21 Six porous AAO template structures. (a and b) AAO template reflectance spectroscopy determination schematic diagram and results, P = period, the structure period, H = height, the structure height, BD = base circle diameter, the structure base circle diameter. Source: Reprinted with the permission from Wang et al. [55]. Copyright 2021, American Chemical Society.

in the hole depth will lengthen the light path, which enhances the light-trapping performance and decreases the reflectivity of porous structures. However, the same scenario is not found in subwavelength porous structures with the same height (I, II, III) arrays. This indicates that reflectivity of arrays is affected by many factors. When the microstructure of an array is in the same period, a large ARO promotes the AR characteristics. Once an array of the same height is involved, the influence of morphology characteristics on the AR effect of the array structure must be considered.

4.3.1.4 Large-Area Fabrication Technology and Parameter Optimization of Bioinspired AR Array Structure

This part provides a simple and controllable preparation technology to fabricate large-scale bioinspired AR surface, and the process is shown in Figure 4.22. Parabolic pore structures on the surface of the AAO template are similar to that of cicada surface, a precondition of choosing templates.

Pretreatment for recycled AAO template: Template reuse is the key to cost-effective fabrication. Here, we provide a simple way to solve the problem. The AR film is peeled off from the AAO template directly. To guarantee perfect separation and recycling, the AAO should be flat and undeformed, and the material used for replication should be strong enough to avoid tearing or breaking during

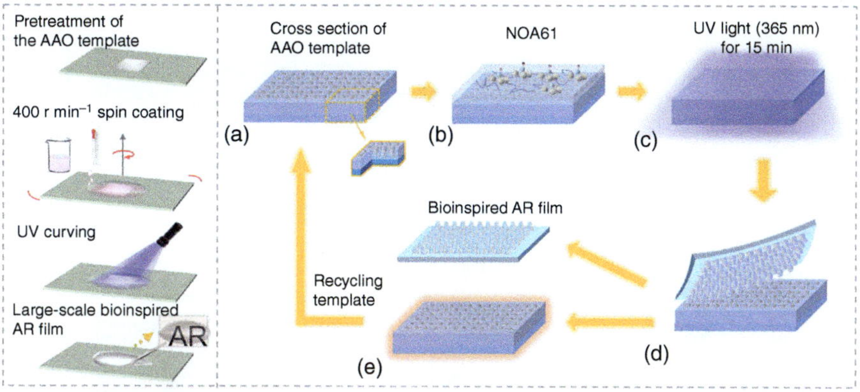

Figure 4.22 Process flowchart of large-area controllable preparation of bioinspired array structures. (a) The template. (b)–(c) The curing process. (d)–(e) The stripping process. Source: Reprinted with the permission from Wang et al. [55]. Copyright 2021, American Chemical Society.

the entire process. On the other hand, the replica should be thin enough to obtain a flexible film. Hence, epoxy resin (A : B = 1 : 3) is coated on the surface of glass sheet, and the porous AAO is applied on it. After heating for 20 minutes, the epoxy resin is curved, and the AAO template is fixed on the slide. Then NOA61 is coated on the surface of AAO, and subsequently exposed to ultraviolet (UV) light of 365 nm for 15 minutes. Finally, the film is peeled off, and the AAO template can be reused.

This method possesses the advantages of simple operation, low cost, good surface quality, strong structure controllability, good repeatability, green and environmental protection, and has outstanding advantages in large-area controllable preparation of bionic antireflection surfaces.

4.3.1.5 Durability Analysis of AAO Template

Figure 4.23a shows the AFM image of bioinspired AR thin film prepared by a reusable AAO template. The AR array structures prepared by the first copy, the fifth, and the tenth times have almost no difference in morphologies.

Figure 4.23b shows the morphology distribution of AAO template surface before and after recycling. The insets in Figure 4.23c record the peeling procedure of cured replica. It can be found that no residue of NOA61 attaches to the surface when peeling. The insets in Figure 4.23d clearly demonstrate that the nanopores on the repeatedly used template remain intact. To further investigate the cyclic stability of porous AAO templates, the surface elements of unused and 10-time recycled AAO templates were analyzed. The results in Figure 4.23c,d show that the main elemental components of the AAO template before and after the cycle are 66.75% aluminum and 33.25% oxygen. Because NOA61 contains a large amount of C element, there is no carbon peak in the AAO template after recycling, indicating that there is no NOA61 component on the surface of the template, which means that the AAO template is almost free from residue in the recycling process.

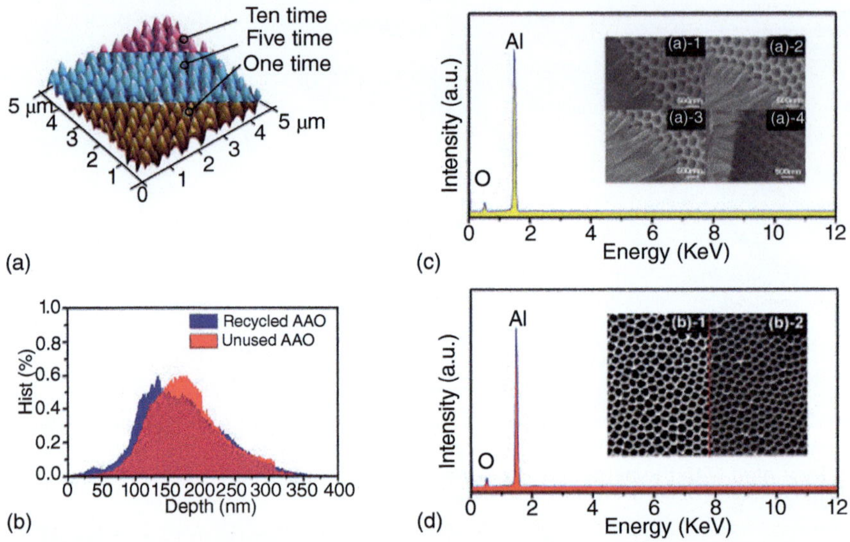

Figure 4.23 Cyclic stability of AAO template. (a) Comparison of surface morphology after repeated replication. (b) AAO template hole depth distribution before and after recycling. (c) Energy spectrum analysis results of unused AAO templates, illustrated with SEM images of the stripping process of cured Ar films. (d) Energy spectrum analysis results of AAO template after recycling. Source: Reprinted with the permission from Wang et al. [55]. Copyright 2021, American Chemical Society.

4.3.2 Structure and Composition Analysis

4.3.2.1 The Morphologies of Bioinspired AR Structures

The morphologies of bioinspired AR films based on AAO template depend on the structure characteristic parameters of the original template and the morphological inheritance ability of the material in the process of structure transfer. The former can realize structural change by choosing different electrochemical reaction conditions in the process of template preparation, while the latter can ensure the stability of the structure morphology in the transfer process of different substrate materials by choosing materials with low surface energy and strong fluidity. Six types of AAO templates were selected, and their morphologies were observed to evaluate the surface quality of the prepared large-area bioinspired AR materials.

Figure 4.24a shows the three-dimensional visualization model and microstructure of the surface structure of NOA61 bioinspired AR material. Type I, II, and III are subwavelength array structures, while type IV, V, and VI are near-wavelength array structures. These array structural parameters are consistent with the corresponding hole parameters of the AAO template, indicating that uncured NOA61 material has good fluidity and the shrinkage rate is small when cured under ultraviolet radiation. Figure 4.24b shows six kinds of bioinspired AR films. By comparison with glass sheets, it can be found that this film has extremely high transparency and high

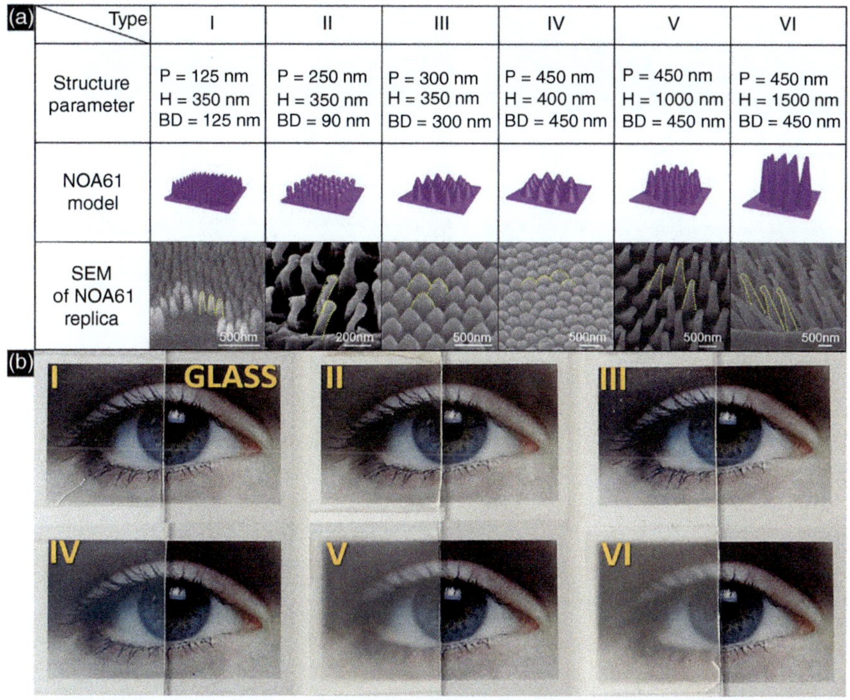

Figure 4.24 Six NOA61 biomimetic antireflection structures and their physical pictures. (a) Six NOA61 biomimetic AR array microstructures. (b) Definition testing of six NOA61 biomimetic AR materials. Source: Reprinted with the permission from Wang et al. [55]. Copyright 2021, American Chemical Society.

surface quality. Meanwhile, haze of the surface will be more obvious if the scale of structural parameters increases.

4.3.2.2 Chemical Composition of the Bioinspired AR Surface

Figure 4.25 shows the composition characterization results of the replica. The region shown in Figure 4.25a is the energy spectrum scanning range, and EDS results of this region are shown in Figure 4.25b,c. The main elements in the material are carbon and oxygen, which are consistent with their chemical composition, and their element distribution is very uniform. There is no abnormal change in element distribution due to the existence of structure. No aluminum element appears in the material, indicating that the template was not destroyed in the preparation and separation process. To further explore the composition of functional groups on the surface of the material, the material was further investigated with FTIR. The results are shown in Figure 4.25d. The peaks at 2960 and 2870 cm^{-1} are caused by the stretching vibration of $-CH_3$ group. The peaks from 1750 to 1680 cm^{-1} are caused by the stretching vibration of the carbonyl group, and the peak at 1380 cm^{-1} is caused by the bending vibration of $-CH_3$ plane. Since most of the functional groups on the material surface are hydrophilic groups, it is speculated that the surface has certain hydrophilicity.

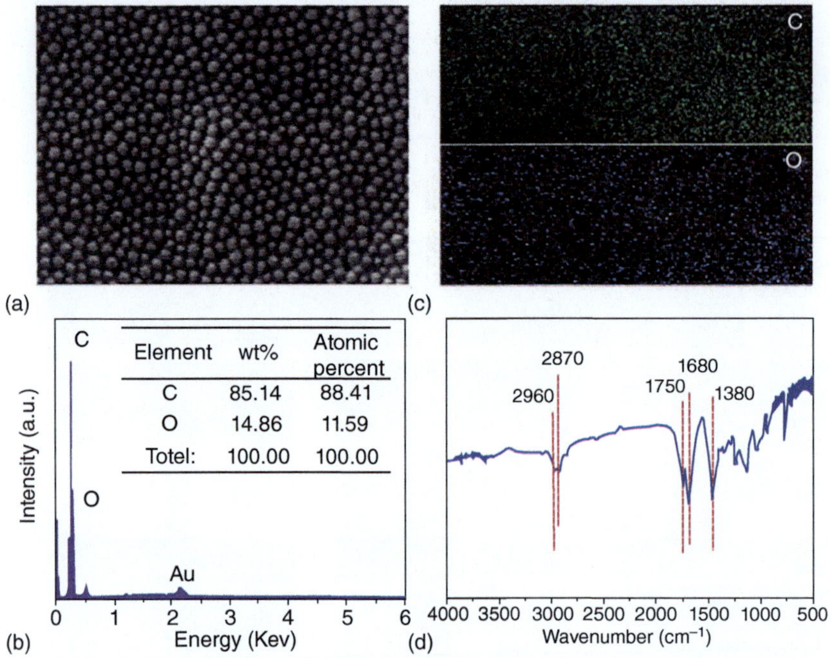

Figure 4.25 Surface composition of large-area bioinspired AR material based on AAO template. (a) Scanning area. (b) EDS spectrum. (c) Element distribution maps. (d) FTIR spectra.

4.3.3 AR Performance of Bioinspired AR Surfaces

4.3.3.1 The Effect of Structural Parameter Changes on Surface Optical Function

To quantitatively characterize optical properties, the optical spectra of the prepared AR films were measured for five times and the average value was obtained. The results are shown in Figure 4.26. During the measurement, the angle of incidence (AOI) was fixed at 0° for normal incident irradiation. The wavelength of the incidence ranged from 400 to 900 nm, covering most of the visible light band. As shown in Figure 4.26a, the reflectivity curves of type IV, V, and VI replicas are lower than that of type I, II, and III. Reflectivity is further reduced by the increase in the height of nanostructures with the same period. Besides, type I and III nanostructures display slightly better antireflective behavior than that of type II. It means that the hexagonal arrangement has a certain effect on decreasing reflection. Here, the ratio of height to period is defined as the ARO. It is obvious that reflectivity of the arrays is affected by multiple factors, and a larger ARO promotes the AR property when the arrays are in the same period. However, the influence of shape must be considered for arrays of the same height.

Figure 4.26b shows the mirror transmittance of the replicas with different arrays fabricated with different AAO templates. It is obvious that the arrays with subwavelength structures (type I, II, III) display satisfactory transmittance. Besides, the arrays with P = 450 nm (type IV, V, VI) show lower specular transmittance values

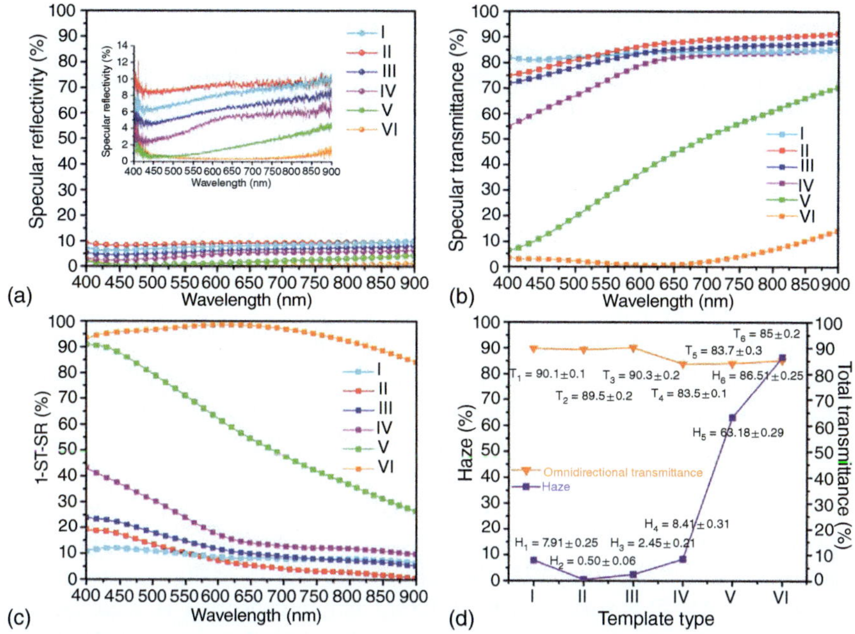

Figure 4.26 Optical performances of the replicas. (a) Specular reflectivity curves of different types of AR arrays and the enlarged drawing (inset) for a clearer comparison. (b) Specular transmittance curves of different types of AR arrays. (c) 1-SR-ST curves of different types of AR arrays. (d) The average haze and total transmittance values of different types of AR arrays.

in the visible wavelength range. Transmittance is further reduced by increasing the nanostructure height. The decrease in specular transmittance is mainly attributed to strong light scattering between the nanocones. When the period of the nanopillar arrays is equal to the wavelength of incident light, diffraction and scattering of the arrays cannot be ignored. Due to the existence of the nanostructures, multiple optical behaviors occur at the interface. Except for specular reflection and specular transmittance, the scattering between the nanostructures and light absorption also occur. This part is defined as S (S = 1-SR-ST). Since the AR films are thin enough to ignore light absorption of the films, the S curve mainly indicates scattering. Figure 4.26c shows the S curves of different AR films. The scattering curves of type I, II, and III arrays are similar. It is proved that the subwavelength structure is insensitive to light scattering. Meanwhile, scattering increases with increasing ARO if they possess the same period (type IV, V, VI). The haze and total transmittance of the AR films are also measured for further investigation. As shown in Figure 4.26d, the arrays with larger periods exhibit significantly higher hazes. All AR films display good transparency even for the arrays with larger periods. Some results can be found by comparing Figure 4.26c,d. The AR films with type IV, V, and VI arrays display lower specular transmittance while their total transmittance is equivalent to that of AR films with a smaller period. It is notable that the concentrated light path will be dispersed when it goes through larger periodic structures. In other words, there is huge luminous flux through the film while only a small part of the light remains in the original direction. In conclusion, all the prepared AR films are able to decrease reflection. In fact, the structural parameters of the microstructure arrays have great influence on the optical properties of the surface. There is a scale effect between these structures and the wavelength of incident light. For the subwavelength nanostructures, they display excellent mirror transmittance, little scattering, and little haze. So they are more suitable to be applied to the surface of displays and instrument panels. In near-wavelength microstructures, specular transmittance accounts for only a small part of total transmittance. There is much scattering between the nanostructures, leading to high haze of the film. So these near-wavelength arrays prepared with some suitable materials can be used as flexible solar cells, sensors, and so on. Since the AAO templates can be controlled by changing their preparation conditions, the structural parameter design can be carried out for different requirements. In this method, controllable fabrication of AR films can be achieved.

In actual scenes, AR materials are often used in curved surfaces, so excellent flexibility of materials and antireflection performance under bending state are also necessary indicators. Taking type IV structure surface as an example, reflectivity of a bioinspired AR film surface with different bending degrees is recorded in Figure 4.27. Whether the bending direction is upward or downward, reflectivity of the film surface decreases with the increase in bending degree, and reflectivity of the film surface in the bending state is less than that in the horizontal state. Reflectivity of the bending film is less than that of the film placed horizontally. When the film is curved, the direction of the reflection changes from vertical incidence to other directions. The beam is dispersed in this way, and that is why the curved film displays much less reflection. The greater the bending degree, the more the reflected light direction deviates from the vertical direction, so specular reflectivity decreases, indicating that the bioinspired AR material still maintains excellent AR performance when it is applied to the curved surface.

4.3.3.2 Angle-Dependent Optical Performances of the Bioinspired AR Surfaces

Based on the perspective of actual observation, the AR performance of surfaces is often angle dependent. Therefore, the optical performances of the surfaces are measured with various angles in Figure 4.28. The reflectivity trend of the six structures changing with angles is similar. When the light incident angle increases, the average reflectivity also tends to rise, indicating the AR performance drops slightly. However, they still have an obvious AR effect, and the non-close-packed AR arrays show stronger angle-dependent optical performance.

4.3.4 Scale-Insensitive AR Mechanism of Cicada-Inspired Structural Surfaces

4.3.4.1 Construction and Electric Field Simulation of Biomimetic AR Microarray 3D Model

Biology in nature provides a perfect strategy to reduce reflection. However, it is difficult to ensure that the dimensional accuracy in the processing is consistent with

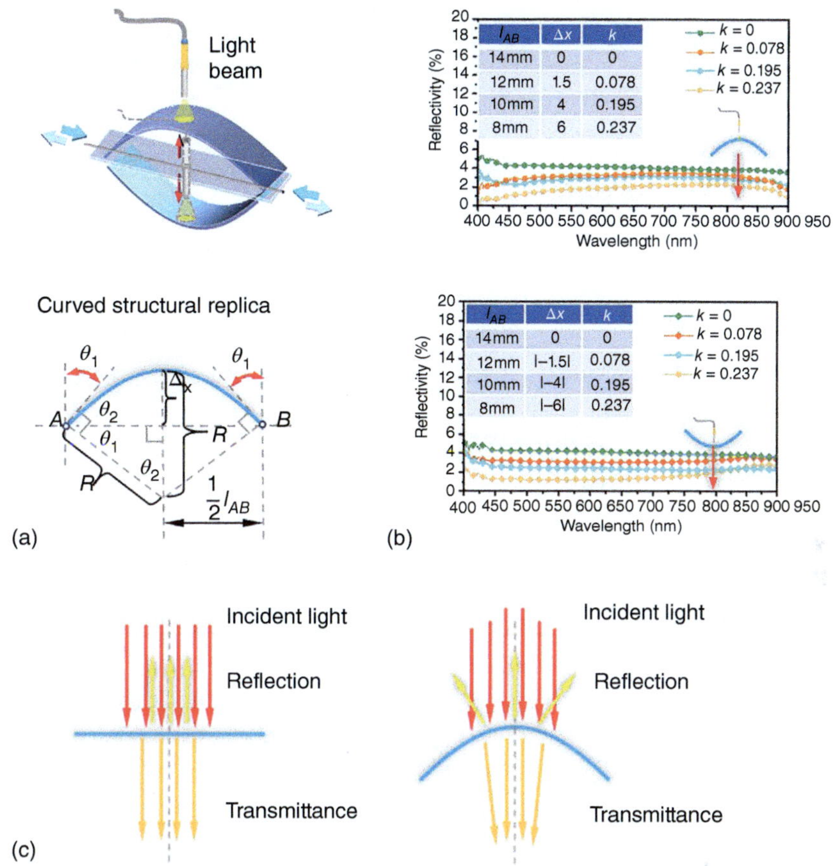

Figure 4.27 Antireflection performance of the curved AR surfaces. (a) Schematic diagram of the measurement of reflectance spectra. (b) Reflection spectra of the curved AR surfaces. (c) Various optical behaviors on flat and curved surfaces. Source: Reprinted with the permission from Wang et al. [55]. Copyright 2021, American Chemical Society.

the original biological structure. This will significantly affect the optical properties of the interface, especially when some diffraction grating structures are involved. Therefore, it is desirable to retain the function of the AR surface even when the structural parameters vary within a certain range, and large-scale manufacturing of scale-insensitive AR structures remains a major challenge.

A large-area biologically inspired flexible AR membrane with micro-/nano-cones was fabricated using a recoverable porous template. Based on six kinds of bioinspired AR array structures on the surface of the optical properties analysis result shows that the bionic structure for the interface optical effect plays an important role, from the subwavelength structure to close to the wavelength structure, the structure scale increases, the surface of optical behavior also changes, although in the process of this change, the material of scattering increases, fog degree enhancement. The AR film inspired by the microstructures of cicada wings still maintains good reflection

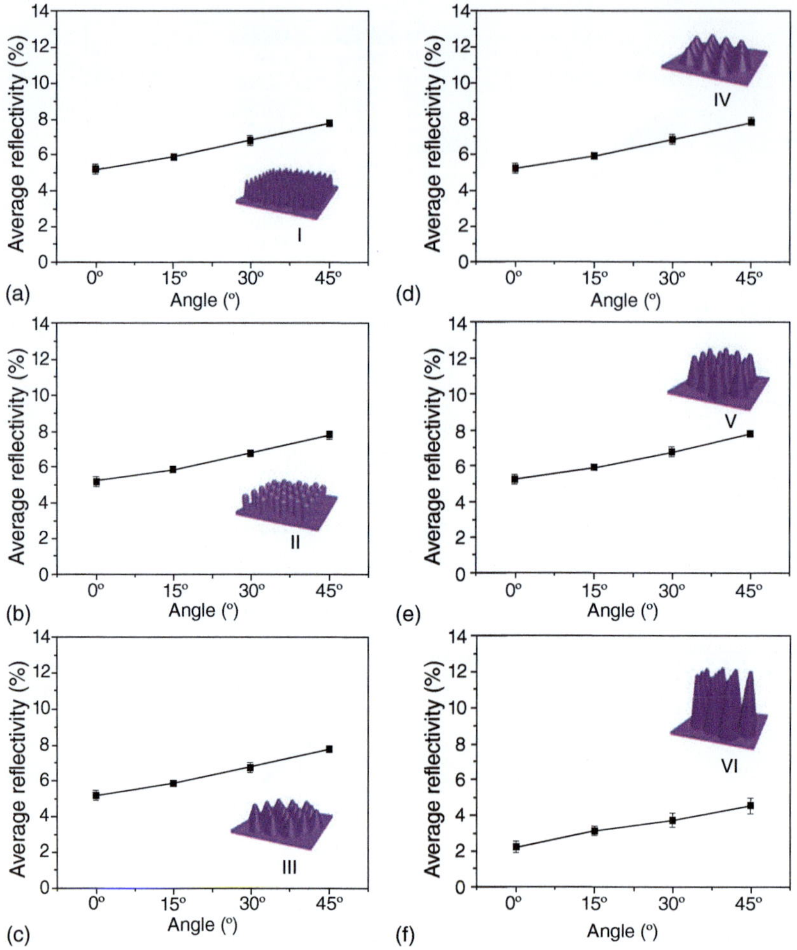

Figure 4.28 Angle-dependent antireflection properties of different structures. (a)–(e) The results of type-I, II, III, IV, V, VI structures, respectively.

inhibition effect, and the mechanism of this antireflection characteristic, which is insensitive to the change of structural scale, is worthy of further investigation.

Based on this, the models of replicas are built to further investigate the structural effects on the transmission of incident light. Here, FDTD simulations are used to exhibit the distribution of solar-spectrum-weighted (400–900 nm) electrical field intensity ($|E|^2$). As shown in Figure 4.29, the electric field between the structures is obviously enhanced, indicating strong multiple reflections between the arrays. Figure 4.29a–c displays the electrical field distribution of the interfaces covered with type I, II, and III arrays, respectively. Apparently, these three subwavelength structures are similar in the ability of electric field enhancement. However, the enhancement regions are closely related to the morphology of nanostructures. Meanwhile, the electrical field distribution of structures with a larger period indicates that more obvious electric field enhancement in arrays can be obtained if the height of conical

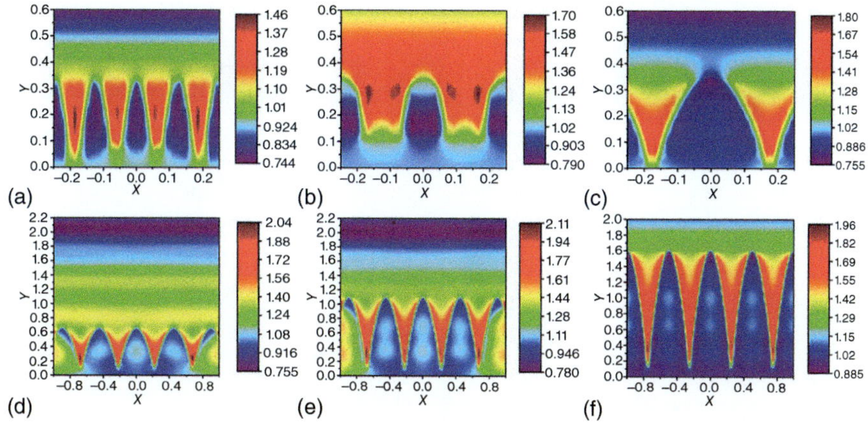

Figure 4.29 Surface electric field intensity distribution diagram of six bioinspired AR structures. Source: Reprinted with the permission from Wang et al. [55]. Copyright 2021, American Chemical Society.

structures increases. Moreover, the electric field intensity decreases quickly when it spreads in the air. We attribute it to energy dispersion when it transmits from the structures to the air. Among all the structures, the electric field enhancement region of type VI structure is the largest. The relative intensity is also higher. It proves that there are strong scattering and multiple reflections between type VI arrays. The above results indicate that there are strong scattering and multiple reflections between the near-wavelength bioinspired AR arrays, which are consistent with the scattering and hazing analysis results measured in the previous section. These arrays have achieved certain light-trapping performance.

4.3.4.2 Scale Insensitivity of Biomimetic AR Microarray and Its Antireflection Mechanism

Equivalent medium theory (EMT) and FDTD can be used to analyze the optical properties of microscopic arrays. For subwavelength nanostructures, EMT can easily quantify the structural effects. At the same time, FDTD simulation can intuitively show the distribution of electric field intensity near microarray structures. The EMT theory proposed by Maxwell–Garnett considers the structured interface as a uniform mixing medium of the substrate material and air, which makes the analysis of the optical properties of the structured interface more convenient. Morphology and dimension both affect the reflection performance of bionic structural surfaces. In this section, anti-reflection mechanism of subwavelength structure arrays (I, II, III) was explored by EMT theory and FDTD method.

Type I, II, and III structures are highly identical, but slightly different in terms of structural morphology. Figure 4.30a shows the longitudinal cross sections of the three structural forms. The morphologies of type I and III arrays are parabolic structures, while type II arrays are different because of their non-close-packed arrangements and the cylindrical structures. Therefore, the calculation of the effective refractive index (n_{eff}) for type II arrays can be simplified into two parts.

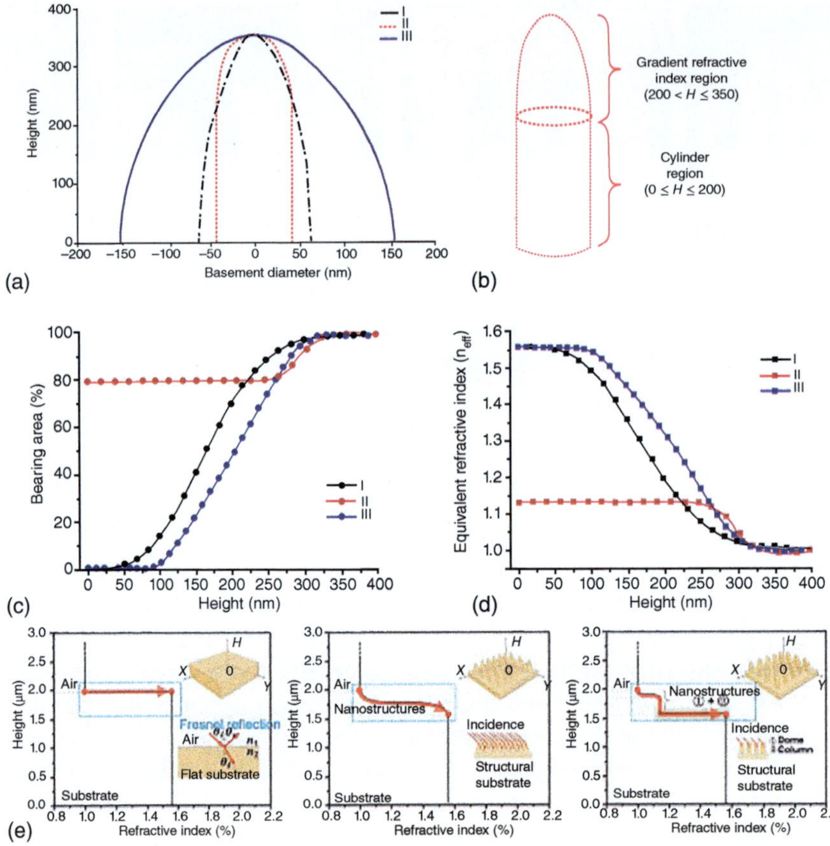

Figure 4.30 Three kinds of array structures and their antireflection mechanism. (a) Longitudinal section diagram of the array structural element. (b) Diagram of type II structural morphology division. (c) AFM measured the duty ratio of three kinds of array structures. (d) The equivalent refractive index function curves of the three kinds of arrays. (e) Schematic diagram of refraction index distribution at the interface of planar, non-close arranged parabolic structures and close-packed arranged parabolic structures. Source: Reprinted with the permission from Wang et al. [55]. Copyright 2021, American Chemical Society.

In the $0 \leq H \leq 200$ region, of the morphology of type II arrays can be regarded as a cylinder. The volume ratio of the substrate material in the mixed medium does not change with the increase in height. In the $200 < H \leq 350$ region, the morphology of type II arrays looks like a parabola. It can be calculated with Eq. (4.9). The bearing area of these three types of arrays is analyzed from the bottom ($H = 0$) to the top ($H = 350$) with the help of AFM. And the n_{eff} curves are calculated. As mentioned, type II arrays can be regarded as the combination of column (forming the constant refraction medium) and dome (forming the gradient refraction medium). Therefore, the calculated height-refractive index curves take on obvious two stages from air to substrate material. The trend of the first stage is similar to that of moth-eye structures, while the second stage is similar to that of the flat interface. In other words, the refractive index in the second stage still changes abruptly. It indicates the reason why the type II structures do not perform as well as the other types on AR property.

The results of the gradient refractive index for different surfaces are shown in Figure 4.30e. It is obvious that the nanostructures eliminate the abrupt refractive index change between the substrate and air. In this way, the Fresnel reflection can be suppressed effectively. It can explain the reason why the structured AR films reflect less than unstructured films do. Meanwhile, it is obvious that the non-close-packed arrays partly have achieved AR properties. In other words, type II structures can also achieve a partial gradient of refractive index to some extent, but there is still a sudden change in the refractive index (the second stage) on the whole, which is also the reason why the AR performance of type II structures is inferior to that of I and III structures, but still better than that of unstructured planes.

As for type IV, V, and VI arrays, their structural parameters are equal to the wavelength of incident light. Larger structural parameters bring higher-order diffraction. As shown in Figure 4.31, the spread of light at the interface is complex, and the zero-order diffraction component is reduced. In this way, these structures reduce specular reflection over a broad range of wavelength.

Based on this, scattering of type IV, V, and VI arrays must be considered. Since their morphologies are similar, the ARO becomes the main factor affecting the optical properties. As the ARO increases, light is reflected several times between the structures, extending the light path and forming the light-trapping effect. This explains the decrease in reflectivity for type IV, V, and VI arrays. These results indicate that the optical properties of the surface can be changed effectively by forming appropriate micro-arrays.

All the prepared AR films can achieve the AR performance, but the ways to inhibit reflection are different. In fact, the structural parameters of AR arrays have great influence on the optical properties of the surface, and there is a scale effect between these structures and the wavelength of the incident light. Subwavelength nanostructures exhibit excellent specular transmittance, almost no scattering and very little haze. Therefore, they are more suitable for applications such as the surface of displays or instrument panels. For near-wavelength microstructures, specular transmittance is only a small fraction of total transmittance. There is more scattering between near-wavelength arrays, resulting in higher haze. Therefore, these

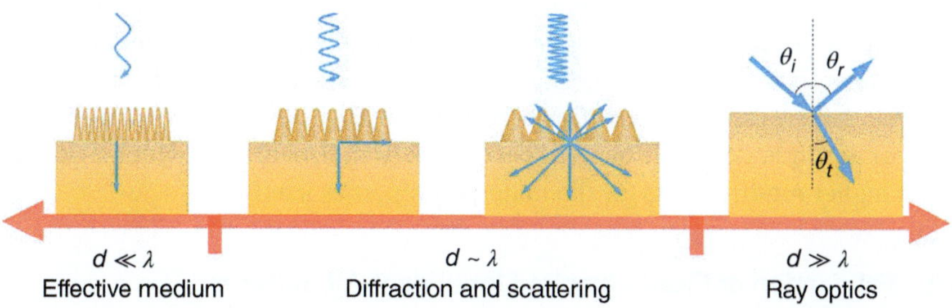

Figure 4.31 Multiple optical behaviors of structured surfaces of different scales. Source: Based on Ross et al. [56]. Copyright 2019, American Chemical Society.

near-wavelength array structures can be prepared with some suitable materials for application in flexible solar cells, sensors, etc.

The parabolic structures inspired by cicada wings show scale-insensitive AR properties. When the AR structures at the interface change from subwavelength to near-wavelength, reflection can still be reduced by the bioinspired surface with nano- or micron arrays, but the surface reflection reduction pathways of subwavelength nanocones and microncones are completely different. Compared with subwavelength nanocones, the increased size of the micron cone changes the original optical behaviors, reducing the proportion of transmitted light, while increasing scattering and absorption. In other words, part of the light will be converted from specular transmission to scattering or absorption when the structural parameters increase.

4.4 Intelligent AR Cicada-Inspired Structures

The morphologies of microstructures or nanostructures on AR surface are crucial to the overall performance of the interface. Both microstructures and nanostructures in natural creatures such as rose, cicada, moth exhibit good AR performance. Especially, subwavelength structures produce a mixture of air and substrate material to form the gradient refractive index layer on the interface. In this way, the gradient refractive index layer blurs the interface when the refractive index changes generally, reducing reflection and increasing transmitted light. The microstructures with near-wavelength structural parameters usually realize the enhancement of interface light capture ability by increasing scattering, leading to a reduction in reflection and increase in scattered light. These AR strategies rely on the subtle morphologies of micro-array structures. However, mechanical robustness of microstructures is generally not good, and even prone to irreversible damage by strong external forces. In practical applications, once the AR film is applied to the touch surface (such as electronic equipment, instrument panels, mobile electronic equipment touch screen), the surface structure will inevitably undergo force deformation. As a result, the optical behaviors at the interface will change accordingly, and the original excellent AR performance will also be diminished. In the future, flexible display devices will be the general trend of the development of display screens. To ensure that bioinspired AR materials can be widely used in such flexible display electronic devices, selected materials should meet the requirements of high transmission and outstanding flexibility. At the same time, for structured functional materials, the deformation recovery ability of microstructures when subject to multiple deformations has become very indispensable, which also puts forward higher requirements for the material properties itself.

4.4.1 Fabrication of Cicada-Inspired Intelligent AR Surfaces

SMP materials can solve the above problems because they have a good shape memory function [35]. SMP refers to a polymer with a certain initial shape that changes its initial shape and is fixed under certain conditions, and then passes through external stimulus conditions (e.g. heat, electricity, light, chemical induction). A polymer material that can be changed and restored to its original shape is called SMP. SMP materials belong to the category of smart materials, and there is most extensive research on temperature-controlled shape memory materials. The introduction of SMP materials solves the problem that the micro-array structure is easily deformed by force. However, most of the current SMPs are in the centimeter scale, and the memory shape materials on the micrometer or even nanometer scale are still to be explored [36]. Therefore, it is very important to study the shape recovery mechanism of structured SMP materials and their influence on the optical behaviors of microstructures at the interface. Here, a clean porous AAO template is used to provide the original structure prepared by bionics, and the structure is transferred through the ultraviolet curing technology to obtain a bionic antireflection structure with higher similarity to the biological prototype structure. Since the material used is a SMP, the prepared bionic antireflection material has the characteristics of shape self-recovery under a certain external force, ensuring that its AR performance will not be attenuated due to external force contact during actual use. The schematic diagram of its bionic preparation is shown in Figure 4.32.

First, the AAO template was cleaned with ethanol and deionized water for 10 minutes, respectively. Then it was dried in the shade for later use. A and B components of epoxy resin were mixed in a volume ratio of 3 : 1, respectively. After complete mixing of A and B components, the epoxy resin glue was spread on the glass slide, the AAO template was fixed on the glass slide coated with the colloid, and the assembly was heated at 60 °C for four hours to make the resin fully cure, and the AAO template pretreatment is completed. Then, precursor materials composed of monomers, a cross-linker, and a photoinitiator were mixed together using a vortex mixer. Two monomers, methyl methacrylate (MMA) and butyl methacrylate (BMA), were combined in a 1 : 1 (2 ml: 2 ml) volumetric ratio. The cross-linking agent, polyethyleneglycol dimethacrylate (PEGDMA, $M_w = 550$, Aladdin, USA), formed cross-linking points between pre-polymeric chains. The

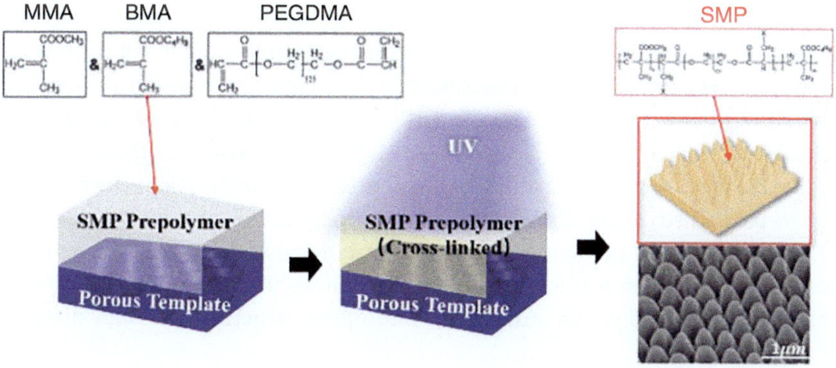

Figure 4.32 Preparation process of the intelligent AR surfaces.

photoinitiator, Phenylbis (2, 4, 6-trimethylbenzoyl) phosphine oxide (BAPOs), was also purchased from Sigma Aldrich. The volume ratio of the total monomers and the cross-linker was about 1 : 2 (4 ml: 8 ml), and the amount of the cross-linker was 0.2 wt% (0.018 g) of the total weight of the precursor. The synthesized SMP was denoted as shape memory copolyacrylate (SMCPAc) [56]. After that, another glass slide was covered on the bonded AAO template, and the assembly was fixed with clips to guarantee the relative position. Subsequently, the SMP precursor mixture was added to the AAO template from the gap between the upper and lower slides. Under the action of capillary force, the SMP precursor mixture could fill the nanoholes smoothly, and then the assembly was irradiated with 365-nm-wavelength ultraviolet light for 10 minutes to make sure the SMP precursor mixture was fully cross-linked and cured. Finally, the bioinspired AR material with shape memory capabilities was obtained after peeling it off from the AAO template.

4.4.2 Characterizations of Cicada-Inspired AR Surfaces

4.4.2.1 Morphologies of the Replica

To investigate whether this kind of shape memory material can completely replicate the structure morphology, the AAO template and the prepared bionic structures were observed. Figure 4.33a is the SEM image of the AAO template in sectional view. The morphology of the structure is in a parabolic shape with the close-packed arrangement. The top circle diameter is about 450 nm, and the depth of the structure is about 450 nm. After further investigation by the AFM microscope in Figure 4.33c–g, the distribution and three-dimensional structure characteristics of the AR arrays can be understood intuitively.

The parabolic inverted cone hole structures covering the surface of the AAO template are in close-packed arrangement. The inner wall of the structure is smooth and the surface quality is intact. The prepared SMP biomimetic reversible AR structures are highly consistent with the AAO structures in terms of shape and array distribution, and its cross-sectional shape is also perfectly matched with the AAO structures. After the analysis of the height of the prepared structures, it is found that their distribution characteristics are similar to nonstandard normal distribution. It indicates that the height of most microstructures is concentrated in 100–300 nm, and the reason for the difference in height distribution is that the microstructure of the AR surface is inherently different. In addition, the surface is not absolutely flat; uneven surface will also cause the deviation in the structure test height. Besides, because it is difficult for the AFM probe to reach the bottom of the close-packed structure arrays, the measured height will be relatively low. In general, this preparation method can realize high-efficiency structure transfer and has greater advantages in terms of structural morphology maintenance. The prepared biomimetic reversible AR surface has a complete structure and high surface quality.

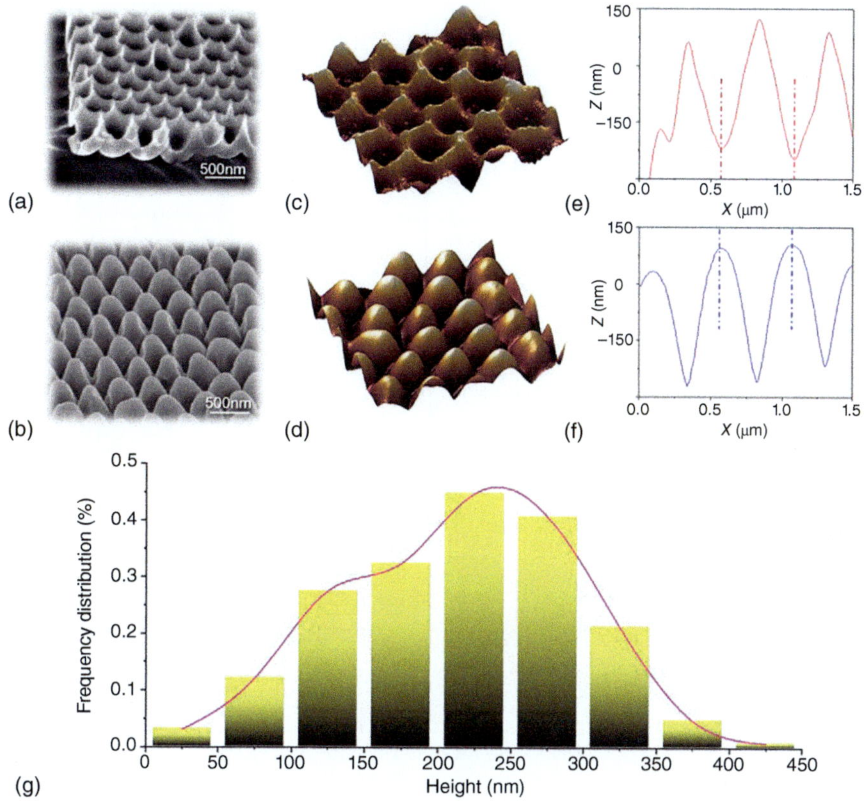

Figure 4.33 The morphologies of the AAO template and prepared AR arrays. (a and b) SEM image, (c and d) AFM image of the surface of the reverse structure on AAO template and the positive structure. (e and f) Cross-sectional shape of negative structure and positive structure. (g) The height distribution map of the positive structure of the bionic reversible antireflection surface.

4.4.2.2 Composition Characterizations of the Replica

Figure 4.34a is the Fourier infrared spectrum of the prepared SMP reversible AR material. The characteristic peaks located at 2955, 2870, 1720, 1140, and 1105 cm^{-1} are $-CH_3$ vibration, $-CH_2-$ vibration, carbonyl vibration, CO vibration of MMA/BMA, respectively, and CO vibration of cross-connected body cause. After the material is further measured by X-ray photoelectron p spectroscope (XPS), the position and area percentage of each element peak can be obtained. The first peak is at 284.50 eV (area percentage is about 35.36%), which is related to the binding energy of carbon to C–C and C–H, the second peak at 285.30 eV (19.75%), which is related to CH_2–CO_2, and the third peak at 285.30 eV (19.75%). 286.10 eV (27.71%), related to CO, the fourth peak at 286.90 eV (7.67%), related to O=C=O, the last peak at 288.61 eV (9.51%), related to O=CO, all peak positions and peak areas.

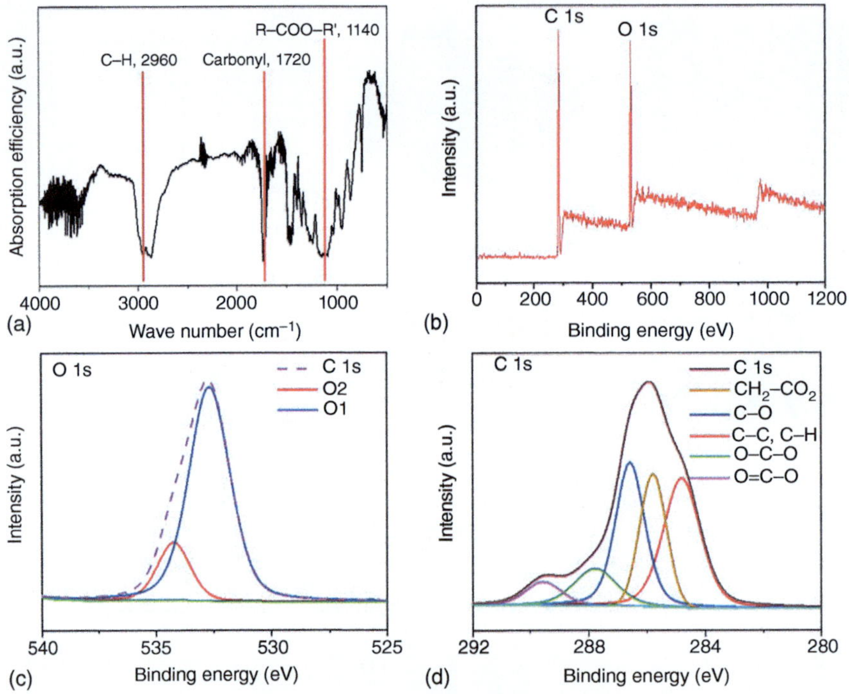

Figure 4.34 Bioinspired reversible AR surface composition determination. (a) Fourier infrared spectrum of bioinspired reversible AR materials. (b–d) XPS spectra of bioinspired reversible AR materials.

The percentages are consistent with the reference data [57]; the main component of the material is an acrylic copolymer.

To further determine the recovery temperature range of the SMP material, a differential scanning calorimeter (DSC) was used to measure the glass transition temperature (glass transition temperature, T_g) of the prepared bioinspired reversible AR material. The temperature is in the range of 20–550 °C during the test. The temperature rise rate was 10 °C min^{-1}, and the temperature drop rate was 5 °C min^{-1}, and the number of scans was set to 3. Figure 4.35 shows the DSC results. Analysis of the heat flow and thermogravimetric curve shows that T_g of the material is 42 °C and the melting temperature is 195 °C. It can be determined that the material begins to deform when the temperature is increased to 42 °C. When the temperature continues to rise to 195 °C, the material melts; the temperature continues to rise to 280 °C, and the material starts to decompose. Therefore, all subsequent testing temperatures should be less than 280 °C. When the temperature rises to about 450 °C, the material is completely decomposed and the material residual carbon rate is 0%.

4.4.2.3 Reversible AR Performance of Cicada-Inspired Structural Surfaces

Most functional surfaces may be subjected to a certain external force, which is a big challenge to the stability of the function of the surface. To explore whether

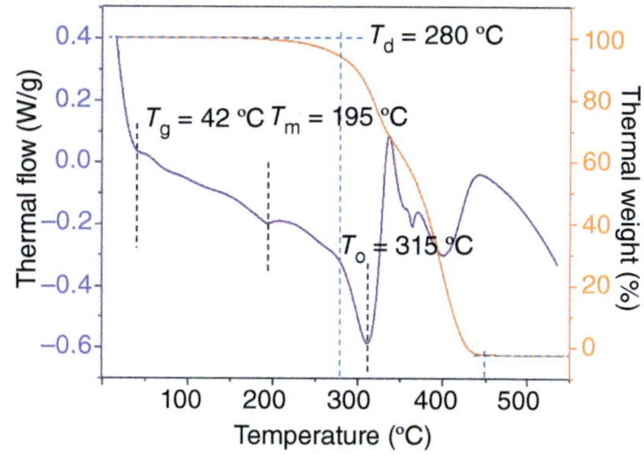

Figure 4.35 The DSC results of the SMP AR material.

the structural deformation will have a significant impact on the surface performance, the specular transmission spectra of the flat film, bioinspired AR film, and force-deformed bioinspired AR film were measured, respectively. The wavelength range of incident light is 400–900 nm, and the incidence angle changes from 15° to 50°. As shown in Figure 4.36, the increase in the angle causes a decrease in specular transmittance of the interface. The bioinspired surface has a more sensitive angle-dependent AR effect for light waves with a wavelength of less than 650 nm. When the light wavelength is large, the structure is equal to the light wavelength, or even smaller. Light waves are more difficult to identify the structure; therefore, the angle-dependent transmittance of the biomimetic structured surface in the long wavelength range is significantly reduced (Figure 4.36a,c). When the morphologies of the prepared AR structures are destroyed by force, their specular transmittance decreases obviously with the increase in the angle, and the changing trend is different from the results of the initial state of the bioinspired AR surface, indicating that its AR performance is changed when the morphologies of the microstructures are damaged.

Since the bioinspired AR surface displays significant angle-dependent optical properties, its imaging features observed at different angles under sunlight are photographed and recorded in Figure 4.36e. No structural color can be observed on the flat surface, while significant specular reflection can be found when facing a large incident angle; it' is one of the reasons for the decrease in specular transmittance when the incident angle increases, which is consistent with the results in Figure 4.36a. For bioinspired AR materials, complex diffraction, scattering, and other optical behaviors are caused by the interaction between the AR arrays on the surface and the incident light, leading to obvious angle-dependent optical properties, and it is beneficial for anti-counterfeiting materials. However, the film with deformed arrays displays a significantly weakened structural color, indicating that the microstructure morphology is one of the important factors affecting the inherent optical characteristics of the structured functional surface, and the ability of shape memory is crucial to the maintenance of the interface effect.

To further investigate the effect of shape memory ability on maintaining excellent optical performance of bionic reducing reflection materials, we characterized

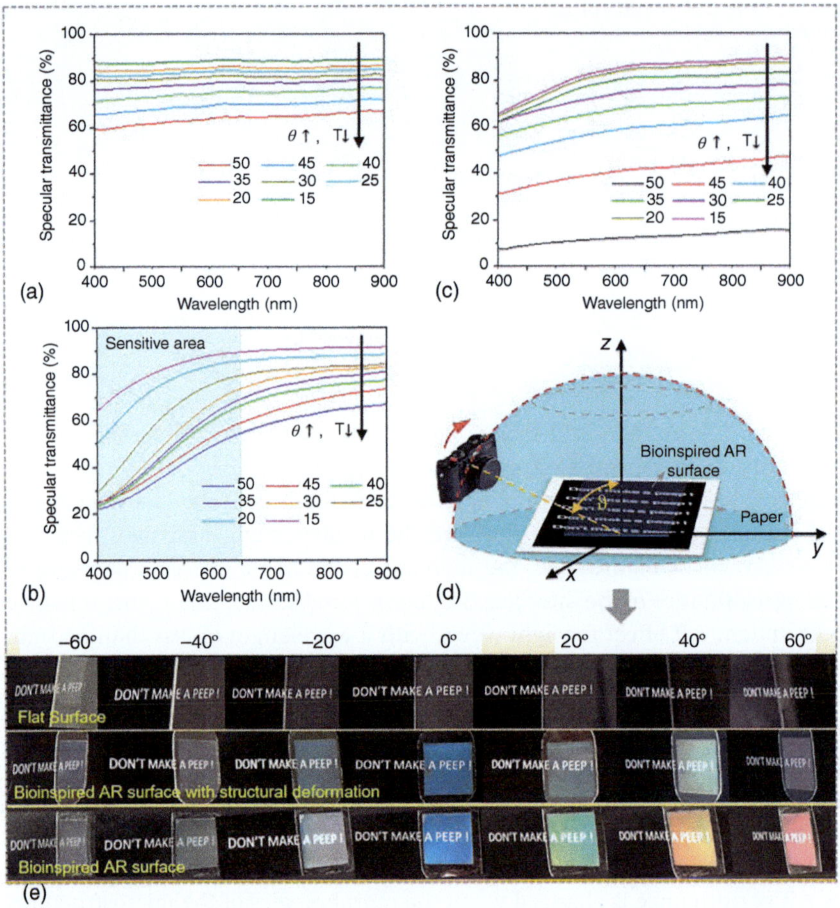

Figure 4.36 Specular transmission spectra and angle-dependence of the material in the visible optical band depend on visual discoloration process (a) specular transmittance of the unstructured surface at varying inclination; (b) specular transmittance of the bionic structured surface under changing inclination; (c) specular transmittance of the bionic structured surface after external force deformation under changing inclination; (d) schematic diagram of material image characteristics and its observation results; (e) angle-dependent discoloration process.

total transmittance and haze of the material in the visible light band with different incident angles (0°, 15°, 30°, 45°, and 60°), and the results in Figure 4.37 are the average values obtained after five measurements. In general, transmittance of the material decreases when the incidence angle increases. For flat surfaces, total transmittance does not decrease significantly, and their transmittance remains nearly 90% even at a large inclination of 60°. For SMP AR surfaces, when the incidence angle increases, total transmission decreases from 90% to 70%, due to the presence of a microscopic array structure at the interface that causes scattering and diffraction of light (Figure 4.37b). In Figure 4.37c, the arrays on SMP AR surface are deformed and

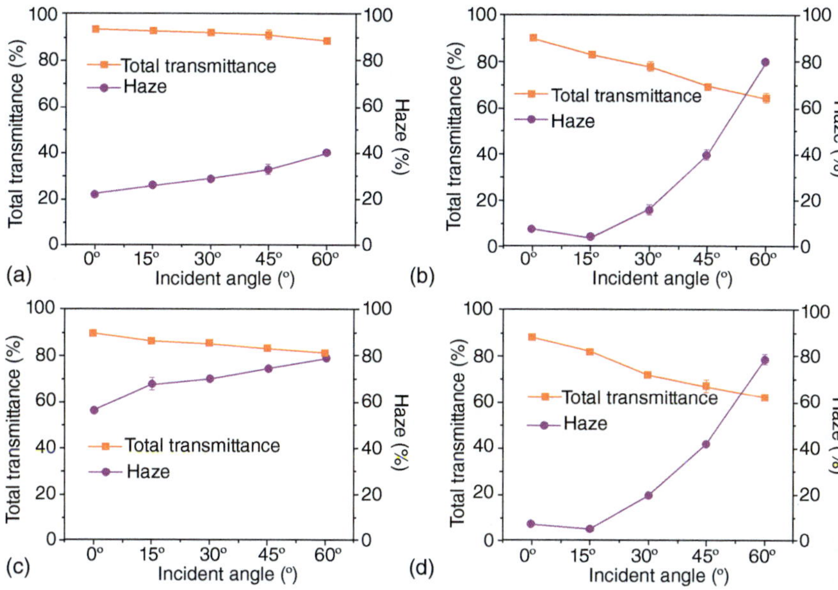

Figure 4.37 Total transmittance and haze of (a) flat surface, (b) bioinspired AR surface, (c) bioinspired AR surface with deformed arrays, and (d) recovered bioinspired AR surface.

its transmission trend changing with angle is similar to that of the SMP AR surface, but the changing amplitude is significantly reduced, indicating that the structure morphology change on AR arrays will affect its optical performance. Subsequently, the deformed SMP AR structures are heated at 50°C for 30 minutes to recover to the original state (Figure 4.37d), and it is found that its transmittance is nearly the same as that of the original state, indicating that the shape recovery function of the microstructure is very important for its omnidirectional AR performance.

After analyzing haze of the material, we have learned that the trend between haze of the material and total transmission is opposite. It is worth noting that during vertical incidence, the bioinspired AR surface has lower haze with an average value below 10%; however, surface haze increases substantially when the incident angle increases. Compared with the 20% change amplitude (20% from 0°, 40% of 60°), the average haze value of the structured surface reaches 80% at 60°, nearly seven times higher than the original haze value, indicating that the incident angle is more obvious to promote the scattering effect of the structured surface, with the application potential for angle visualization and anti-peep display. On the other hand, the surface with deformed arrays shows higher haze than the original structured surface at normal incidence. Even during vertical incidence, there will be a certain inclination between light and the structure, so it shows higher haze. After heating, until the deformed structure it returns to the original state, the trend of haze is consistent with that before deformation, indicating that the shape memory ability of the array structure also has an impact on overall haze of the material.

4.4.3 Modification Treatment and Application Exploration

4.4.3.1 Preparation of Bioinspired Self-cleaning AR Materials

Hydrophobic SiO_2 nanoparticles (diameter: 20–50 nm) are dispersed in an alcohol with an 8 wt% ratio. Then the film is soaked and covered with hydrophobic SiO_2 nanoparticles through dip coating. Then they are taken out and dried in a vacuum-drying oven at 80°C for one hour. Figure 4.38a shows the FTIR spectra of the films with or without modification. Compared with nonmodified films, the CH_3– in modified films decreases, while the intensity of –OH– peak increases. In other words, the hydrophilic groups in original nonmodified films are replaced by the hydrophobic groups; thus the CA is further improved.

Water bouncing experiment is recorded with high-speed video. Water droplets of radius R (about 1.40 mm) impact the modified type VI surface from a height of 30 mm. The water droplets deform after touching the surface and eventually bounce off the surface. There are no adhesions or residues on the surface throughout the process (Figure 4.38e). The actual contact time (t) of the droplets is 15.6 ms, and the short t_c provides favorable evidence for the good hydrophobic properties of the

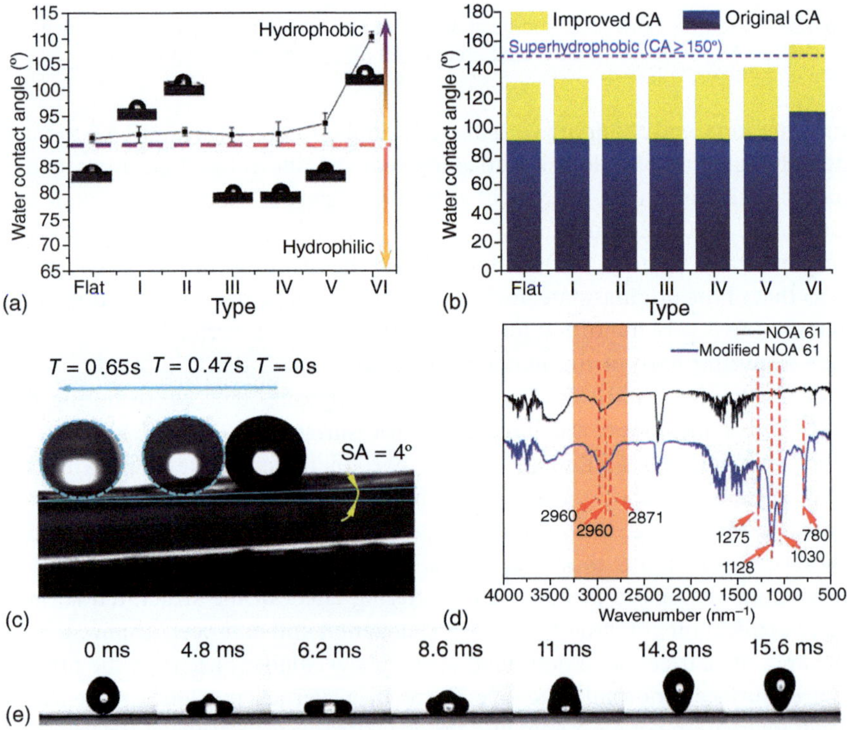

Figure 4.38 (a) Water CA of flat substrate and type I–VI replicas. (b) Improved water CA after modification for all the surfaces. (c) Water SA of modified type VI replica. (d) FTIR spectra of type VI replica before and after modification. (e) Water droplet bouncing on the modified type VI surface.

surface. For further investigation, the theoretical contact time of water droplets hitting the surface is calculated [58]:

$$t_c = 2.6\sqrt{\rho D_0^3/(8\gamma)} \tag{4.10}$$

Water droplets of density $\rho = 1000\ \text{kg}\cdot\text{m}^{-3}$ and surface tension $\gamma = 72\times10^{-3}\ \text{N}\cdot\text{m}^{-1}$ are formed from calibrated needles, and the diameter D_0 is 2.80 mm. It is worth emphasizing that surface theoretical contact time (t_c) will not be affected by droplet impact velocity, but it has a linear relationship with droplet size. According to Eq. (4.10), the theoretical contact time t_c is 16.02 ms, which is very close to the actual time t = 15.6 ms. It indicates the excellent superhydrophobic properties of the modified type VI surface. Above all, both materials and structures are important for superhydrophobicity of modified structural films.

4.4.3.2 Preparation of Bioinspired Light-Trapping AR Materials

To make better use of light-trapping performance of structured AR films, further investigation was carried out. These AR films were plated with gold nanoparticles to improve the absorption of light (Figure 4.39a). Stephan et al. and Matthias et al. have done many detailed and valuable works on the glancing-angle deposition growth and optical properties of gold nanoparticles, confirming plasmon resonances of gold nanoparticles [59, 60]. Since the impact of different deposition rates on the growth regimes has been investigated deeply [61], the sputter parameters must be set rigorously.

Here, the films are coated by an ion sputter (Hitachi E-1045). In fact, the structural change caused by suitable sputtering operation is negligible relative to the bionic structure itself if the sputtering time is 120–150 seconds, and the morphology of the arrays will be changed if the sputtering time is too long. The reflectivity curves of the Au-plated AR films are displayed in Figure 4.39b. The reflection of type II arrays is relatively strong, while the other closely arranged arrays reflect little. The Au-plated type VI AR films shows the best antireflection performance. Reflectivity of Au-plated type VI film is much lower than that of untreated type VI film, and it is caused by the plasma effects of metal particles.

For untreated films, low reflection is the result of high transmission. However, it is no longer applicable to the Au-plated AR films. The gold nanoparticles on treated films absorb the scattering light from different directions perform better AR property. The results in Figure 4.39c indicate the angle-dependent antireflection of the film. When the angle of the incidence changes from normal to 60°, the reflection also increases. Besides, average reflectance is no more than 3% if the AOI is smaller than 30°. In addition, the reflected light paths of the unstructured Au-plated surface and the structured Au-plated surface are compared in Figure 4.39d. The incident light source is a laser pen, and the incident angle during the test remains the same. When the light beam falls on the unstructured Au-plated surface, there is a bright light spot on the screen. However, the light spot disappears when it comes to the structured Au-plated surface, indicating its adequate light-trapping effect.

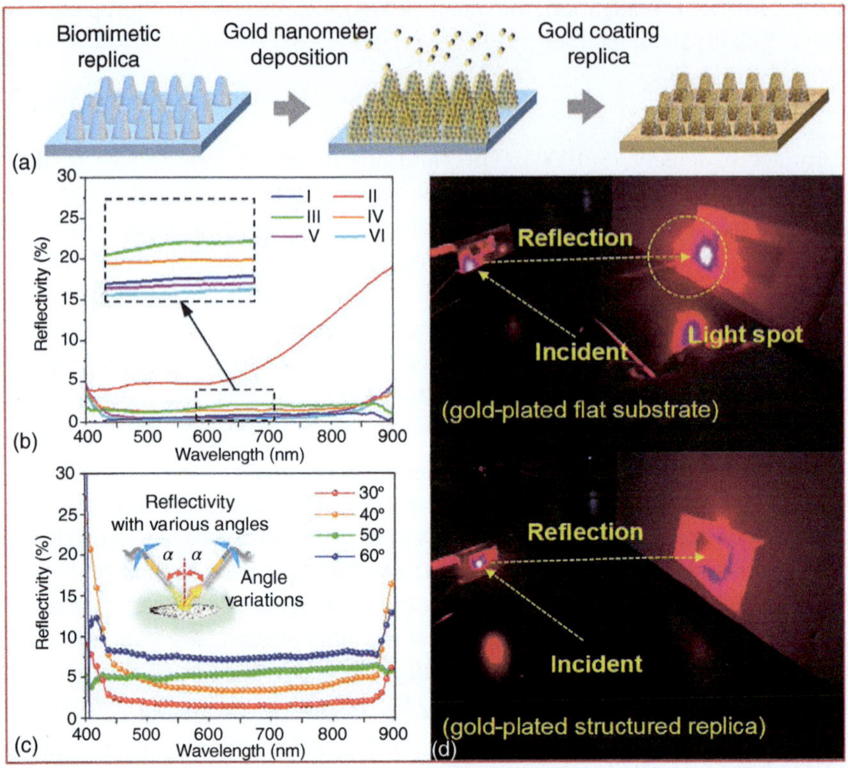

Figure 4.39 Light-trapping performance of the arrays. (a) Schematic illustration of the deposition process. (b) Reflectivity spectra of the replicas after coating with nano-gold particles and the enlarged drawing (inset) for clearer comparison. (c) Reflectivity spectra of the type VI replica measured with varied angles (30°, 40°, 50°, and 60°). (d) Images showing the laser beam path reflected by gold flat substrate and gold nanocone array. Source: Reprinted with the permission from Wang et al. [55]. Copyright 2021, American Chemical Society.

Besides, the Au-plated type VI film in Figure 4.40a takes on light back, indicating its potential for light-absorbing antireflection layer design. The inset in Figure 4.40a displays the morphologies of the Au-plated type VI arrays. It is obvious that the sputtering treatment hardly changes the original structures. Besides, Au-plated films still possess excellent flexibility (Figure 4.40b).

As mentioned, the Au-plated AR film exhibits excellent plasma effect; the light absorption of the Au-plated AR film may also be enhanced. To verify this point, the AR films plated with or without Au nanoparticles were exposed under a stable light source for two minutes, respectively. Figure 4.41a records their infrared thermal imaging pictures. The temperature of the Au-plated AR film is much higher than that of non-plated AR film. It is obvious that Au nanoparticles are crucial to enhance light absorption and photothermal transformation. Subsequently, the Au-plated AR film and the Au-plated flat film are exposed under vertical light to investigate the influence of bioinspired AR structures on energy conversion process. As shown in

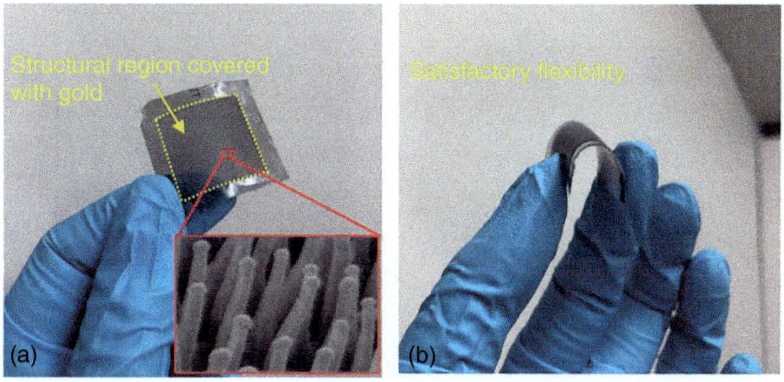

Figure 4.40 (a) Photographs showing the structured replica covered with gold and SEM image (inset) of the gold plating type VI arrays. (b) Photographs showing outstanding flexibility of the replica. Source: Reprinted with the permission from Wang et al. [55]. Copyright 2021, American Chemical Society.

Figure 4.41b, the original temperatures of the Au-plated AR film and the Au-plated flat film are consistent, and the temperature curves change when light is applied. The temperature of the Au-plated AR film raises from 21.5 to 58.3°C in one minute. Besides, the temperature–time curves of surfaces are fitted and analyzed, and the fitted curve and the original data curve show a near-perfect match to the original, indicating that there is an obvious linear correlation between the illumination time and the surface temperature in a short time. The angles between the fitted curves and the horizontal direction are θ_1 and θ_2, respectively. $\theta_2 > \theta_1$ means that the temperature–time curve slope of Au-plated AR film is higher than that of the flat film. In other words, the existence of bioinspired AR structures makes the surface temperature rise faster; as a result, the utilization of light energy is also higher.

Subsequently, the Au-plated AR film and the Au-plated flat film are attached with a thermochromic material (change color at 31°C). The color-changing processes are recorded with a mobile phone, and the major changes are shown in Figure 4.41c. The Au-plated AR film could complete the color-changing process in a shorter time, which implies that it has significantly higher photothermal conversion efficiency than that of the flat surface. On the one hand, the bioinspired AR structures prevent light from escaping from the material surface. On the other hand, the bioinspired AR structures also increase the specific surface area of the material. In this method, the spreading area of gold nanoparticles also expands, and the plasma effect might be more pronounced. The above results indicate that the micron-scale AR structures can make more efficient use of light energy under of plasma effect, which is beneficial to the development and utilization of solar energy.

4.4.3.3 Preparation of Bioinspired Reversible AR Materials

To investigate the effect of coating on the optical properties of bionic reversible antireflection materials, the transmission spectrum, total transmission, and fog of material before and after coating were measured. The surface without coating modification was recorded as OF (Original Film), the surface firstly modified by

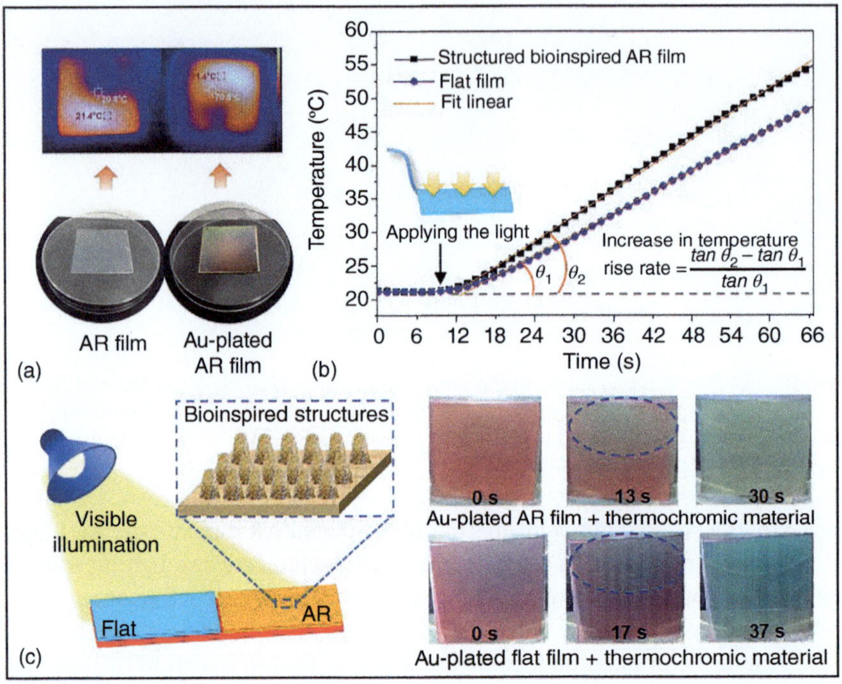

Figure 4.41 Light absorption and photothermal conversion of the Au-plated AR film. (a) Infrared thermal imaging pictures of AR film and Au-plated AR film irradiated for two minutes. (b) Temperature–illumination time curves of Au-plated structured AR film and Au-plated flat film. (c) Test of discoloration response time. Source: Reprinted with the permission from Wang et al. [55]. Copyright 2021, American Chemical Society.

poly(3,4-ethylenedioxythiophene) (PEDOT): poly(styrenesulfonate) (PSS) coating once was recorded as SP-1 (Surface Modified by PEDOT: PSS), the surface modified two times was recorded as SF-2, and the surface modified three times was recorded as SF-3. The average surface transmittance after single coating was about 57%, and it was 25% less than untreated, indicating that PEDOT: PSS exerted certain influence on the material surface transmittance. With the increasing coating times, the thickness of PEDOT: PSS material increases, and light absorption increases, which is consistent with the result that transmission of the material decreases gradually with increasing coating times. In addition, when the number of coating layers of PEDOT: PSS increases, overall transmittance decreases and the fog degree increases, indicating that the coating of PEDOT: PSS should not be too thick to impair the overall transparency of the material and the readability of information. Therefore, a thin coating should be completed as far as possible under the premise of ensuring electrical conductivity.

Experimental results confirmed that suitable material ratio endowed the conductivity of the bionic reversible antireflective materials after coating and their light transmittance can be guaranteed. After coating, bionic reversible minus reflection conductive material with anti-reflection was compressed to form fixed shape and its transmittance dropped significantly. On the one hand, it is due to surface structure due to compression deformation. On the other hand, it is caused by local deformation in conductive coating inside that there might be a small crack or material uneven distribution, thus increase and reduce internal scattering light transmittance. After the material is restored by mild heating, its transmittance increases partly, but it cannot be completely restored to the uncompressed state. There are two reasons. The first may be due to the limitation of the shape memory material's properties, its shape memory recovery rate fails to reach 100%. Secondly, in the process of shape recovery after heating, the ductility of PEDOT: PSS conductive layer is insufficient, which leads to defects such as local cracks in the material. In spite of this, the bionic reversible antireflection conductive material has certain characteristics of high reflection reduction and retains shape memory ability in general.

To explore whether the bionic reversible antireflection conductive material prepared can have application potential in touch control, the response ability of the material to deformation stimulation should be characterized. As shown in Figure 4.42, the two ends of the bionic reversible AR conductive material are bonded with the double-guide copper foil tape with wires, and then respectively connected to the positive and negative electrodes of the digital multimeter KEYSIGHT 34456A, and the supporting software is used to record the material resistance change in real time. The results are recorded in Figure 4.42b. When the power is turned on for seven seconds, the material resistance is stable. The material resistance changes immediately when there is a finger press on the surface of the material. After the finger press is lifted up at an interval of two to three seconds, the material still shows the perception of finger press after the process is repeated several times. When a finger press is applied, the material surface is subjected to a small strain, its internal

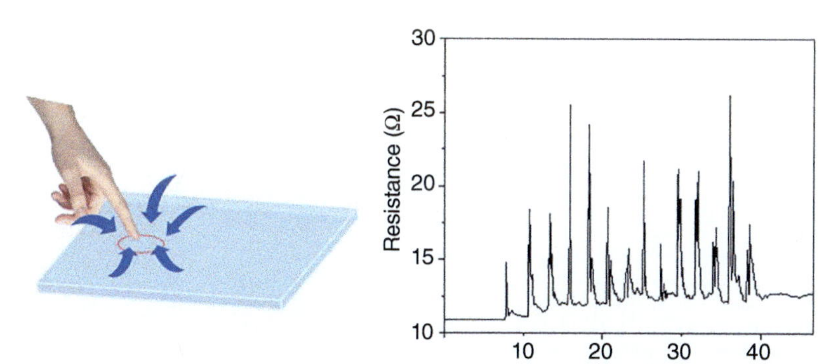

Figure 4.42 Stress–strain response of bioinspired reversible AR conductive material Based on [56].

microstructure and resistance change, then it is collected in the form of electrical signals, and the information is fed back. In these repeated processes, however, the received signal strength differences are apparent, because there is no guarantee that the pressure of the finger every time is exactly the same and every time. As the point of contact is random, the result can only be material surface when stress is to have certain stress–strain response characteristics, but also need to further explore on accuracy.

During the test, it was found that PEDOT: PSS coating had low hardness and was easy to get scratches on the surface, so it needs to be protected on its surface. Transparent polyimide film is generally used as protection material in touch electronic products. It has lightweight, flexibility and scratch resistance properties. It can be used to cover both sides of the bionic reversible anti-reflection conductive material thus to complete the assembly of a capacitor, as shown in Figure 4.43a. Two conductive layers of PEDOT: PSS are, respectively, used as the upper and lower levels of the capacitor, and the shape memory material in the middle is the insulating layer. According to Eq. (4.11),

$$C = \frac{\varepsilon S}{4\pi k d} \tag{4.11}$$

where ε is the dielectric constant of the medium between the plates, S is the opposite area of the capacitor plate, d is the distance of the capacitor plate, and k is the static force constant. When the distance between the two plates changes, the capacitance changes to trigger the electrical signal. After the signal is collected and processed, the results are shown in Figure 4.43b. When the weight is placed on the surface, under

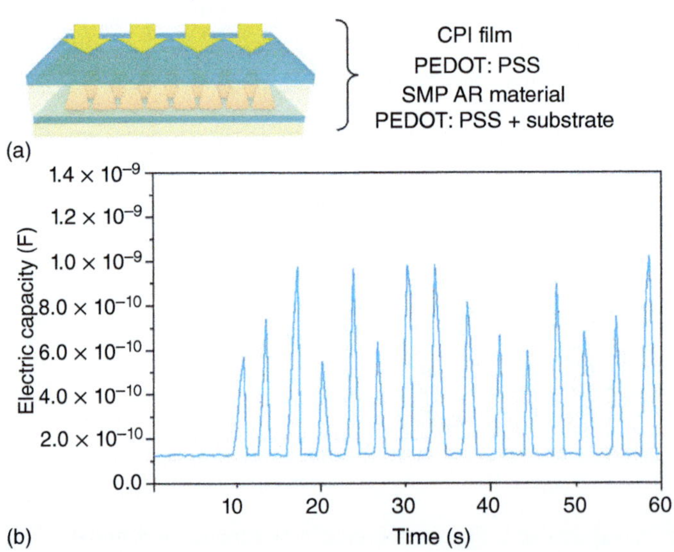

Figure 4.43 Stability of bioinspired reversible reducing reflective conductance materials. (a) Simple assembly diagram of bioinspired reversible AR conductive components; (b) cyclic stability test Based on [56].

the action of gravity, the material is subjected to stress and deformation to cause the distance between the two plates to change, and the capacitance changes to form a peak value. After the weight is removed, the capacitance recovers. The results are still relatively stable after repeated tests, indicating that the capacitor components obtained by this preparation method have certainly cycle stability.

Biology in nature has evolved functional surfaces with excellent performance to adapt to complex and changeable environments, which represent natural wisdom and have guiding significance in actual production and life. The study of biological composite functional surfaces has brought new insights into the field of bioinspired AR materials. However, the service condition of the functional surface is often affected by multiple factors, and a single study of a certain function cannot meet the growing practical demand. This part further explores the design and preparation of bioinspired AR surface under common complex working conditions, using the simple, effective, and universal coating technology. To face different working conditions and needs, different treatment schemes are put forward. Based on the excellent AR performance, multifunctions such as self-cleaning, antireflection, light-trapping antireflection properties have been realized, so that they can be better applied to extensive AR applications. In addition, the bioinspired AR surfaces prepared with shape memory polymers also display reversible AR performance, which endows them with recovery ability after force deformation, and display part of the stress–strain stimulation response through the conductive coating modification, which greatly broadens the application field of bioinspired AR surfaces.

References

1 Lian, Z., Xu, J., Yu, Z. et al. (2020). Bioinspired reversible switch between underwater superoleophobicity/superaerophobicity and oleophilicity/aerophilicity and improved antireflective property on the nanosecond laser-ablated superhydrophobic titanium surfaces. *ACS Applied Materials & Interfaces* 12 (5): 6573–6580.

2 Bacal, D., Lal, N., Jumabekov, A. et al. (2020). Solution-processed antireflective coating for back-contact perovskite solar cells. *Optics Express* 28 (9): 12650.

3 Han, Z., Feng, X., Guo, Z. et al. (2018). Flourishing bioinspired antifogging materials with superwettability: progresses and challenges. *Advanced Materials* 30 (13): 1–32.

4 Wu, L., Jiao, Z., Song, Y. et al. (2017). Water-trapping and drag-reduction effects of fish *Ctenopharyngodon idellus* scales and their simulations. *Science China Technological Sciences* 60 (7): 1111–1117.

5 Han, Z., Fu, J., Wang, Z. et al. (2017). Long-term durability of superhydrophobic properties of butterfly wing scales after continuous contact with water. *Colloids Surfaces A Colloids and Surfaces A: Physicochemical and Engineering Aspects* 518: 139–144.

6 Tan, G., Lee, J., Lan, Y. et al. (2017). Broadband antireflection film with moth-eye-like structure for flexible display applications. *Optica* 4 (7): 678–683.

7 Tan, G., Huang, Y., Li, M. et al. (2018). High dynamic range liquid crystal displays with a mini-LED backlight. *Optics Express* 26 (13): 16572.

8 Wang, Z., Li, B., Feng, X. et al. (2020). Rapid fabrication of bio-inspired antireflection film replicating from cicada wings. *Journal of Bionic Engineering* 17 (1): 34–44.

9 Han, Z., Wang, Z., Li, B. et al. (2019). Flexible self-cleaning broadband antireflective film inspired by the transparent cicada wings. *ACS Applied Materials & Interfaces* 11 (18): 17019–17027.

10 Tan, R., Marzolini, N., Jiang, P., and Jang, Y. (2020). Bio-inspired polymer thin films with non-close-packed nanopillars for enhanced bactericidal and antireflective properties. *ACS Applied Polymer Materials* 2 (12): 5808–5816.

11 Velic, A., Hasan, J., Li, Z., and Yarlagadda, P.K.D.V. (2021). Mechanics of bacterial interaction and death on nanopatterned surfaces. *Biophysical Journal* 120 (2): 217–231.

12 Sun, M., Watson, G.S., Zheng, Y. et al. (2009). Wetting properties on nanostructured surfaces of cicada wings. *Journal of Experimental Biology* 212 (19): 3148–3155.

13 Cordaro, A., Groep, J.V.D., Raza, S. et al. (2019). Antireflection high-index metasurfaces combining Mie and Fabry-Pérot resonances. *ACS Photonics* 6 (2): 453–459.

14 Zhang, F., Wang, H., Wang, C. et al. (2020). Direct femtosecond laser writing of inverted array for broadband antireflection in the far-infrared. *Optics and Lasers in Engineering* 129: 106062.

15 Yoshida, K., Takahashi, L., Takashima, A. et al. (2020). Antireflection in green lacewing wings with random height surface protrusions. *Langmuir* 36 (15): 4207–4213.

16 Pranda, A., Gutierrez Razo, S.A., Fourkas, J.T., and Oehrlein, G.S. (2019). Evolution of photoresist layer structure and surface morphology under fluorocarbon-based plasma exposure. *Plasma Processes and Polymers* 16 (9): 190002.

17 Martínez-Calderon, M., Azkona, J.J., Casquero, N. et al. (2018). Tailoring diamond's optical properties via direct femtosecond laser nanostructuring. *Scientific Reports* 8: 1–9.

18 Bochet-Modaresialam, M., Claude, J.B., Grosso, D., and Abbarchi, M. (2020). Methylated silica surfaces having tapered nipple-dimple nanopillar morphologies as robust broad-angle and broadband antireflection coatings. *ACS Applied Nano Materials* 3 (6): 5231–5239.

19 Jacobsson, R. (1966). Light reflection from films of continuously varying refractive index. *Progress in Optics* 5: 247–286.

20 Southwell, W. (1983). Gradient-index antireflection coatings. *Optics Letters* 8 (11): 584–586.

21 Cheben, P., Halir, R., Schmid, J.H. et al. (2018). Subwavelength integrated photonics. *Nature* 560 (7720): 565–572.

22 Rayleigh, L. (1892). LVI. On the influence of obstacles arranged in rectangular order upon the properties of a medium. *London, Edinburgh, the London, Edinburgh, and Dublin Philosophical Magazine and Journal of Science* 34: 481–502.

23 Bernhard, C. and Miller, W. (1962). A corneal nipple pattern in insect compound eyes. *Acta Physiologica Scandinavica* 56: 385–386.

24 Clapham, P. and Hutley, C. (1973). Reduction of lens reflexion by the "moth eye" principle. *Nature* 244 (5414): 281–282.

25 Halir, R., Bock, P., Cheben, P. et al. (2015). Waveguide sub-wavelength structures: a review of principles and applications. *Laser & Photonics Reviews* 9 (1): 25–49.

26 Ding, L., Yu, Y., Morits, D. et al. (2020). Low loss waveguiding and slow light modes in coupled subwavelength silicon Mie resonators. *Nanoscale* 12 (42): 21713–21718.

27 Halir, R., Cheben, P., Janz, S. et al. (2009). Waveguide grating coupler with subwavelength microstructures. *Optics Letters* 34 (9): 1408–1410.

28 Yang, X., Tian, Z., Chen, X. et al. (2020). Terahertz single-pixel near-field imaging based on active tunable subwavelength metallic grating. *Applied Physics Letters* 116 (24): 241106.

29 Maese-Novo, A., Halir, R., Romero-García, S. et al. (2013). Wavelength independent multimode interference coupler. *Optics Express* 21 (6): 7033.

30 Sun, L., Zhang, Y., He, Y. et al. (2020). Subwavelength structured silicon waveguides and photonic devices. *Nanophotonics* 9 (6): 1321–1340.

31 Vázquez-Guardado, A., Boroumand, J., Franklin, D., and Chanda, D. (2018). Broadband angle-independent antireflection coatings on nanostructured light trapping solar cells. *Physical Review Materials* 2 (3): 1–7.

32 Chattopadhyay, S., Huang, Y., Jen, Y. et al. (2010). Anti-reflecting and photonic nanostructures. *Materials Science & Engineering R: Reports* 69 (1–3): 1–35.

33 Li, J., Wang, W., Mei, X. et al. (2019). Artificial compound eyes prepared by a combination of air-assisted deformation, modified laser swelling, and controlled crystal growth. *ACS Nano* 13 (1): 114–124.

34 Han, G., Nguyen, T., Park, S. et al. (2020). Moth-eye mimicking solid slippery glass surface with icephobicity, transparency, and self-healing. *ACS Nano* 14 (8): 10198–10209.

35 Park, J., Youn, J.R., and Song, Y. (2017). Sustainable antireflection using recoverable nanopattern arrays. *Journal of Materials Chemistry C* 5 (40): 10600–10608.

36 Tseng, H., Chen, Y., Chen, R., and Yang, H. (2020). Reversibly erasable broadband omnidirectional antireflection coatings inspired by inclined conical structures on blue-tailed forest hawk dragonfly wings. *ACS Applied Materials & Interfaces* 12 (9): 10883–10892.

37 Li, P., Han, Y., Wang, W. et al. (2018). Self-erasable nanocone antireflection films based on the shape memory effect of polyvinyl alcohol (PVA) polymers. *Polymers (Basel)* 10 (7): 1–9.

38 Han, Z., Jiao, Z., Niu, S., and Ren, L. (2019). Ascendant bioinspired antireflective materials: opportunities and challenges coexist. *Progress in Materials Science* 103: 1–68.

39 Garlisi, C., Trepci, E., Li, X. et al. (2020). Multilayer thin film structures for multifunctional glass: self-cleaning, antireflective and energy-saving properties. *Applied Energy* 264: 114697.

40 Dudem, B., Leem, J., and Yu, J. (2016). A multifunctional hierarchical nano/micro-structured silicon surface with omnidirectional antireflection and superhydrophilicity via an anodic aluminum oxide etch mask. *RSC Advances* 6 (5): 3764–3773.

41 Mizoshita, N., Ishii, M., Kato, N., and Tanaka, H. (2015). Hierarchical nanoporous silica films for wear resistant antireflection coatings. *ACS Applied Materials & Interfaces* 7 (34): 19424–19430.

42 Bae, S., Lee, Y., Seo, Y., and Jeong, K. (2019). Antireflective structures on highly flexible and large area elastomer membrane for tunable liquid-filled endoscopic lens. *Nanoscale* 11 (3): 856–861.

43 Gu, Z., Kothary, P., Sun, C. et al. (2019). Evaporation-induced hierarchical assembly of rigid silicon nanopillars fabricated by a scalable two-level colloidal lithography approach. *ACS Applied Materials & Interfaces* 11 (43): 40461–40469.

44 Han, Z., Mu, Z., Li, B. et al. (2017). Bioinspired omnidirectional self-stable reflectors with multiscale hierarchical structures. *ACS Applied Materials & Interfaces* 9 (34): 29285–29294.

45 Chen, J., Wei, T., Chang, C. et al. (2014). Fabrication of polymer nanopeapods in the nanopores of anodic aluminum oxide templates using a double-solution wetting method. *Macromolecules* 47 (15): 5227–5235.

46 Ramesh, S., Leen, K., Kumutha, K., and Arof, A. (2007). FTIR studies of PVC/PMMA blend based polymer electrolytes. *Spectrochimica Acta Part A, Molecular and Biomolecular Spectroscopy* 66 (4, 5): 1237–1242.

47 Kim, M., Kang, T., Kim, S. et al. (2019). Antireflective, self-cleaning and protective film by continuous sputtering of a plasma polymer on inorganic multilayer for perovskite solar cells application. *Solar Energy Materials and Solar Cells* 191: 55–61.

48 Chen, W., Li, S., Chu, J. et al. (2018). Fabrication of ordered metallic glass nanotube arrays for label-free biosensing with diffractive reflectance. *Biosensors & Bioelectronics* 102: 129–135.

49 Jayasinghe, R., Perera, G., Zhu, H., and Zhao, Y. (2012). Optical properties of nanostructured TiO_2 thin films and their application as antireflection coatings on infrared detectors. *Optics Letters* 37 (20): 4302.

50 Kong, L., Chiam, S., and Chim, W. (2019). Metal-assisted silicon chemical etching using self-assembled sacrificial nickel nanoparticles template for antireflection layers in photovoltaic and light-trapping devices. *ACS Applied Nano Materials* 2 (11): 7025–7031.

51 Lancellotti, L., Bobeico, E., Castaldo, A. et al. (2018). Effects of different graphene dopants on double antireflection coatings/graphene/n-silicon heterojunction solar cells. *Thin Solid Films* 646: 21–27.

52 Bian, H., Wei, Y., Yang, Q. et al. (2016). Direct fabrication of compound-eye microlens array on curved surfaces by a facile femtosecond laser enhanced wet etching process. *Applied Physics Letters* 109: 221109.

53 Papadopoulos, A., Skoulas, E., Mimidis, A. et al. (2019). Biomimetic omnidirectional antireflective glass via direct ultrafast laser nanostructuring. *Advanced Materials* 31 (32): 1–8.

54 Long, S., Cao, X., Huang, R. et al. (2019). Self-template synthesis of nanoporous VO_2-based films: localized surface plasmon resonance and enhanced optical performance for solar glazing application. *ACS Applied Materials & Interfaces* 11 (25): 22692–22702.

55 Wang, Z., Ding, H., Liu, D. et al. (2021). Large-scale bio-inspired flexible antireflective film with scale-insensitivity arrays. *ACS Applied Materials & Interfaces* 13 (19): 23103–23112.

56 Ross, M., Mirkin, C., and Schatz, G. (2016). Optical properties of one-, two-, and three-dimensional arrays of plasmonic nanostructures. *Journal of Physical Chemistry C* 120 (2): 816–830.

57 Zunke, I., Wolf, S., Heft, A. et al. (2014). Structural properties of zinc oxide deposited using atmospheric pressure combustion chemical vapour deposition. *Thin Solid Films* 565: 45–53.

58 Zhang, C., Zheng, Y., Wu, Z. et al. (2019). Non-wet kingfisher flying in the rain: the water-repellent mechanism of elastic feathers. *Journal of Colloid and Interface Science* 541: 56–64.

59 Roth, S.V., Santoro, G., Risch, J.F.H. et al. (2015). Patterned diblock co-polymer thin films as templates for advanced anisotropic metal nanostructures. *ACS Applied Materials & Interfaces* 7 (23): 12470–12477.

60 Schwartzkopf, M., Santoro, G., Brett, C.J. et al. (2015). Real-time monitoring of morphology and optical properties during sputter deposition for tailoring metal-polymer interfaces. *ACS Applied Materials & Interfaces* 7 (24): 13547–13556.

61 Schwartzkopf, M., Hinz, A., Polonskyi, O. et al. (2017). Role of sputter deposition rate in tailoring nanogranular gold structures on polymer surfaces. *ACS Applied Materials & Interfaces* 9 (6): 5629–5637.

5

Bioinspired Antifogging (AF) Surfaces

Atomization can cause light not to directly penetrate a material surface, which not only has a serious impact on the efficiency of the material but also produces many problems or even dangers to our daily life. For example, fog on the surface of a laparoscopic lens will affect the smooth progress of operation. Sea bridges and ships are exposed to moisture for a long time, which leads to aging and corrosion. Fog on the windshield of aircraft and cars is an important cause of traffic accidents. Fog on the surface of solar panels will reduce the conversion efficiency of photovoltaic cells. Fog on the surface of an agricultural greenhouse will directly affect the crop metabolism. In addition, under some extreme low temperature conditions, which may bring significant safety hazards, such as aircraft, wind turbines, high-voltage transmission, telecommunications equipment, heat exchangers, if frost or excessive uneven water condensation occurs on these surfaces, the operation efficiency of equipment will be reduced, and even the whole system will be paralyzed. Although the problems and losses caused by fog are difficult to be measured by numbers, it is obvious that the efforts to deal with fogging phenomenon will directly or indirectly produce huge economic and social benefits.

Generally, to overcome the problems caused by fog, some methods have been used to remove fog on materials surfaces. The first one is that changing environmental parameters such as the surface temperature of materials, surrounding relative humidity, velocity of airflow. For example, Sui et al. [1] prepared graphene-based electrothermal films/elements on both quartz and flexible polyimide (PI) substrates by using the spin-coating process beginning from graphene oxide (GO) followed by thermal annealing. They prepared AF devices in view of these heating films and observed remarkable efficiencies. In addition, reduced small- or large-size graphene oxide (rSGO or rLGO) sheets were used as a protective layer, and an AgNW hybrid film heater by a simple bottom-up approach was constructed [2]. Particularly, rLGO/AgNW hybrid films showed superior optical transmittance even at 700 °C and excellent heating and defrosting properties at low voltages. An applied single-layer graphene defogger first was fabricated, and graphene films were grown directly on quartz substrates via Ga vapor-assisted chemical vapor deposition (CVD) method [3]. The single-layer graphene defogger rapidly removed fog in less than 15 seconds under a voltage of 20 V and exhibited superior defogging stability after

Nature-Inspired Structured Functional Surfaces: Design, Fabrication, Characterization, and Applications, First Edition. Zhiwu Han.

50 cycles. He et al. [4] reported a scalable production of transparent conductive films (TCs) by combining GO and multi-walled carbon nanotubes (MWNTs). The GO-MWNT inks were deposited on glass by a rod coating method. MWNTs of 0.1 wt% and GO of 0.2 wt% displayed optimum wetting/rheological characteristics. The GO-MWNT films treated with hydroiodic acid turned into rGO-MWNTs, and they demonstrated the potential application of rGO-MWNT films as transparent defrosters. However, these methods have obvious disadvantages such as complex devices, high cost of components, vulnerability, and maintenance issues. So, it is hard to popularize these methods in real-time practice.

The emergence of AF materials exactly makes up for aforementioned disadvantages, and they are gradually becoming a hotspot of the AF research [5, 6]. The AF materials mainly work through the following aspects. The first one is making the fog evaporate quickly. The second one is reducing the quantity of fog produced on material surfaces per unit time. The third one is to prolong the condensation time of fog. The last one is avoiding the condensation of fog droplets on material surfaces. In a word, the fundamental working principle of AF materials mainly relies on the interactions between fog drops and solid surfaces by reasonably designing surface geometries and chemical composition to achieve appropriate wettability.

5.1 Wettability-induced AF Theories

Wettability is a character that shows the affinity between liquid and solid surfaces, and the contact angle (CA) is a parameter that can quantify this concept. Generally speaking, surface tension γ is determined by surface chemical composition, and surface roughness allows it to be modified in a way. On the basis of these prerequisites, some theories are introduced to illustrate the wetting state of solid surface, and these two parameters are considered in a direct way. In fact, to regulate wettability and thus control the AF performance, the surface roughness and surface chemical properties must be adjusted appropriately.

5.1.1 Smooth Surfaces: Young Model and the Static Contact Angle

In the early nineteenth century, Thomas Young first described the static CA using a sessile drop technology on the planar solid surface. When water droplet gets in touch with the solid surface, surface tensions of solid–vapor, solid–liquid, liquid–vapor interfaces are balanced to the minimum energy state, thereby forming a three-phase contact angle (TPCL), which is described by Young's equation [7]:

$$\gamma_{sv} = \gamma_{sl} + \gamma_{lv} \cos \theta \tag{5.1}$$

In this formula, γ_{sv}, γ_{sl}, and γ_{lv} represent the surface tension of solid–vapor, solid–liquid, liquid–vapor interfaces, respectively, as shown in Figure 5.1a. Based on the CA value, the surface features can be classified into superhydrophilic ($\theta \to 0°$) or superhydrophobic ($\theta > 150°$). It should be noted that Young's equation applies to an ideal surface, that is, the solid surface that consists of a homogeneous, smooth,

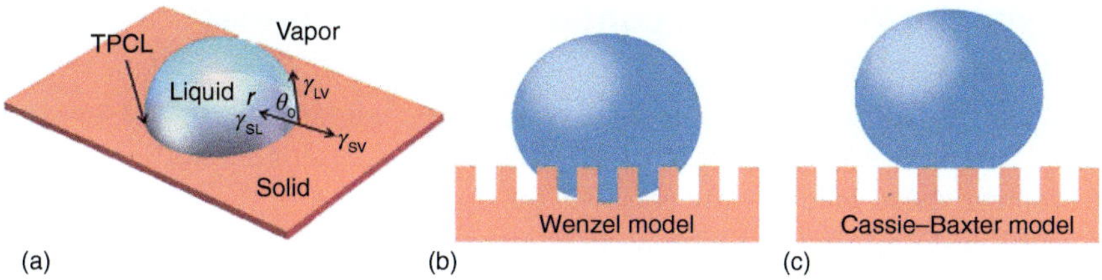

Figure 5.1 (a) Young model; (b) Wenzel model; (c) Cassie–Baxter model.

non-deformable, and isotropic material. Nevertheless, such an ideal surface does not exist in reality. Therefore, in the wettability study, surface roughness should be considered. In general, the relationship between surface roughness and apparent CA is illustrated by two models: Wenzel and Cassie–Baxter [8, 9].

5.1.2 Wenzel Model

When droplets are filled with a void between the solids, a Wenzel state is presented. As shown in Figure 5.1b, the Wenzel model shows that the liquid remains in intimate contact with the surface of the trench structure. The change in CA is described as

$$\cos \theta_W = r \cos \theta \tag{5.2}$$

In this formula, θ_W represents the apparent CA, which is equivalent to the stable equilibrium state, r represents the roughness ratio, which is the estimated value of how the surface roughness affects the uniform surface. It is denoted as the proportion of the actual area to the apparent area of the solid surface. The Wenzel equation shows that hydrophilicity or hydrophobicity of the surface increases with the change of surface microstructures. However, when involving fully wet or nonwet antifogging (AF) surfaces, the distinction between the Wenzel and Cassie–Baxter models is notable.

5.1.3 Cassie–Baxter Model

When droplets do not penetrate into the voids of a rough surface and rest upon a solid surface, a Cassie–Baxter state is presented. Here, a composite state is formed, namely, solid flat protrusions and air cavities, as shown in Figure 5.1c. It is described as

$$\cos \theta_c = f_s \cos \theta_s + f_v \cos \theta_v \tag{5.3}$$

In this formula, θ_c represents the apparent CA between liquid and rough surface, θ_s and θ_v represent the CA in the liquid–solid and liquid–vapor interfaces, and f_s represents the area fraction of the water drop in contact with the top of the solid protrusions, f_v represents the area fraction of the water drop in contact with the air gaps, which is characterized by $\theta_s = \theta$, $\theta_v = 180°$. In view of $f_s + f_v = 1$, Eq. (5.3) becomes

$$\cos \theta_c = f_s (\cos \theta + 1) - 1 \tag{5.4}$$

5.2 Dynamic Wettability: Contact Angle Hysteresis and Sliding Angles

However, whether Wenzel or Cassie–Baxter is used to represent the wetting behavior of droplets, these models only consider the apparent CA of the surface under stable conditions. In fact, the AF property of materials is not only closely related to the static wettability but also more easily affected by dynamic wetting characteristics. The apparent CA is usually between the advancing CA (θ_A) and receding CA (θ_R), as shown in Figure 5.2a. In fact, the difference between θ_A and θ_R is called contact angle hysteresis (CAH) [10]. Moreover, the smooth properties of AF surfaces can also be evaluated by measuring the sliding angle (SA), as shown in Figure 5.2b, when one side of the substrate rises, the droplets begin to move above the horizontal plane, it can be defined according to the Furmidge equation [11]:

$$mg \sin \alpha / w = \gamma_{lv} (\cos \theta_R - \cos \theta_A) \tag{5.5}$$

In this formula, m represents the mass of fog drop, w represents the width of fog drop, g represents gravity acceleration, and α represents SA. According to the difference between $\cos \theta_R$ and $\cos \theta_A$, the tilt angle of the substrate mainly depends on the mass of the droplets and the adhesion force between the droplets and the solid surface.

As shown in Eq. (5.5), the tilt angle of the substrate can cause droplets to glide mainly depending on the mass of the fog drop and on the adhesive force between the fog drop and the solid surface in accordance with the difference of $\cos \theta_R$ and

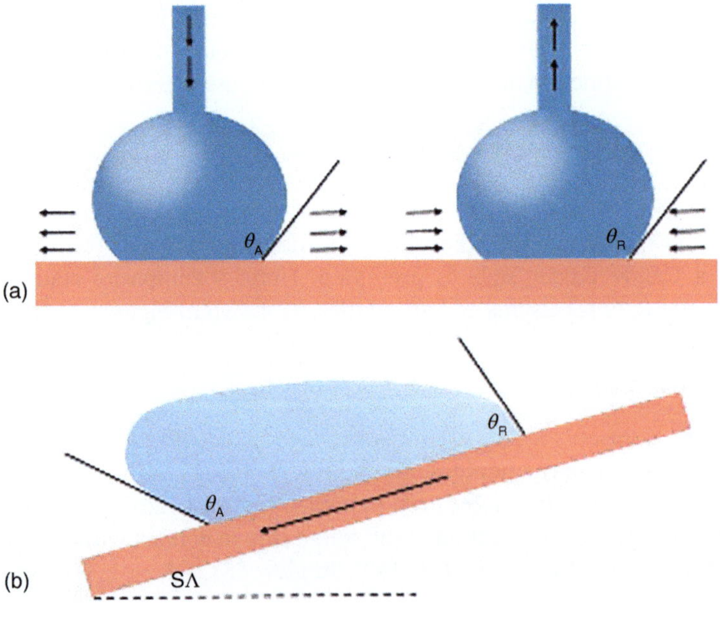

Figure 5.2 (a) Dynamic sessile drop method to measure θ_A and θ_R. (b) The tilting base method.

$\cos \theta_A$. Consequently, AF surfaces may exhibit sticky or slippery behavior according to the various differences.

On the one hand, when the Wenzel model is used to describe the wettability of a solid surface, the actual contact area of the droplets is larger than the projected area on the horizontal plane, because of the close contact between the droplets and the solid surface. Therefore, there is a great difference between θ_A and θ_R. Consequently, the fog drops usually adhere to the surface and require a large angle to slide.

When the Cassie–Baxter model is used to describe the wettability of the solid surface, the actual contact area of the droplets is smaller than the projected area on the horizontal plane. Therefore, the CAH between fog drops and solid surfaces is smaller, because of the low adhesion force between the fog drops and solid surface. Consequently, a small tilt angle is enough to cause the droplet to fall off.

5.3 Definition and Classification of AF Surfaces

Generally speaking, a hydrophilic surface has a strong affinity to water, which is usually well described by the Wenzel model, while a hydrophobic surface tends to reject water and is closer to the Cassie–Baxter state. Unfortunately, the description is too simple to describe the wettability of the AF surface; therefore, a more accurate definition of (super)hydrophilicity and (super)hydrophobicity is needed to provide clear properties. Considering the above reasons, according to the water CA, the AF surface is divided into two categories in this book, superhydrophilic AF surfaces and superhydrophobic AF surfaces [12].

5.3.1 Superhydrophilic AF Surfaces

For superhydrophilic AF surfaces, due to the rapid flow and diffusion of water droplets on the surface, a pseudo film is formed, which leads to low scattering of incident light, thus effectively limits the formation of atomization. Consequently, the transmittance of substrate surface is maintained [13]. In the past few years, there have been many reports on AF surfaces, such as TiO_2 [14–19], SiO_2 [20–29], and TiO_2/SiO_2 [30–40]. Many synthesis methods have been developed to prepare films or coatings with AF performance. Therefore, a large number of superhydrophilic AF materials are obtained.

A CeO_2 film with superhydrophilicity and excellent AF properties was successfully fabricated by one-step electrodeposition without additional modification [41]. The AF mechanism of CeO_2 film surface was elaborated as shown in Figure 5.3a–c. On the one hand, the surface is composed of island-like and microcracks, which leads to rapid diffusion of water droplets on the surface owing to capillary effects, as shown in Figure 5.3d. On the other hand, the surface-active sites due to oxygen vacancies significantly promote the surface free energy and raise the interfacial affinity toward water molecules. In addition, the prepared CeO_2 film exhibits superior corrosion resistance. However, the surface is usually heterogeneous, because of main dependence on the deposition time of CeO_2 on the substrate surface.

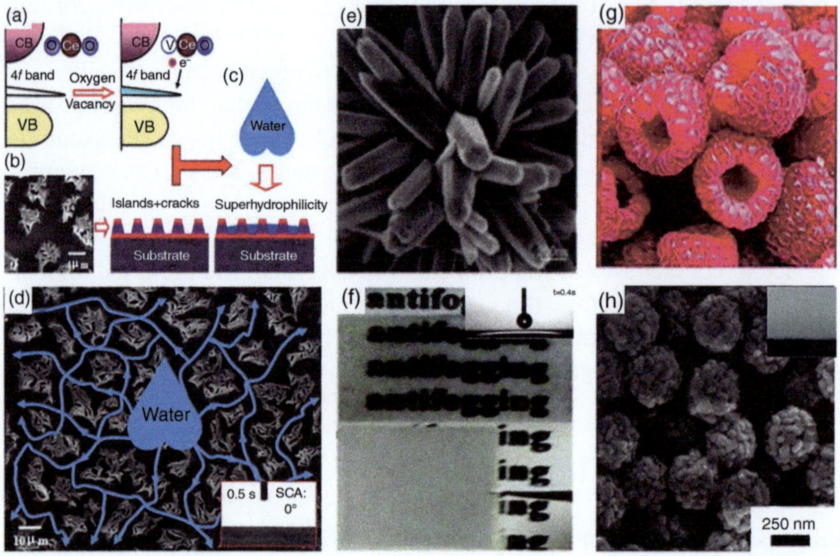

Figure 5.3 (a) Schematic illustration of the valence band of CeO_2 forming an oxygen vacancy. (b) Surface topography of hierarchical islands plus microcracks. (c) Schematic illustration of a water droplet spreading by coupling of the surface topography and oxygen vacancies. (d) Schematic illustration of a water droplet spreading on the surface of CeO_2 with regular islands plus microcracks. Source: Reprinted with permission from Xue et al. [41]. Copyright 2015, Elsevier. (e) FE-SEM images of ZnO nanoflowers prepared at 0.1 M precursor concentrations with high magnification. (f) AF performance of the three-bilayer ZnO nanoflower multilayer. Source: Reprinted with permission from Yuan et al. [42]. Copyright 2013, Royal Society of Chemistry. (g) Photo image of the raspberry in nature. (h) SEM images of the hollow nanospheres obtained at different molar ratios of Co^{2+} ions to Zn^{2+} ions: 3 : 1. Insets: corresponding contact angle images. Source: Reprinted with permission from Peng et al. [43]. Copyright 2011, American Chemical Society.

Another interesting example is that a layered nanostructure similar to flower is constructed by polyelectrolyte-mediated self-assembly, using micro zinc oxide nanoflowers as the base block on different substrate [42]. By contrast, the concentration of the selected precursor is 0.1 M, which can make the graded structure have a larger size and better uniformity, as shown in Figure 5.3e, and the modified glass with graded structure exhibits excellent superhydrophilicity and AF characteristics, as shown in Figure 5.3f. Due to the shortage of samples, the optimum concentration of ZnO needs to be further studied and discussed. Porous superhydrophilic film composed of titanium dioxide hollow spheres was prepared by ion induction technique, as shown in Figure 5.3g [43]. The surface wettability was affected by ion dopants (Figure 5.3h). The AF performance of the film can remain stable for a long time without UV irradiation. However, the preparation procedure is very complicated.

The $La(OH)_3$ nanorod layer was displayed on glass substrate by self-stacking technique [44]. Such a thin film coating significantly reduces the reflection loss in the visible light region, leading to an excellent single-layer AR function. In addition, the

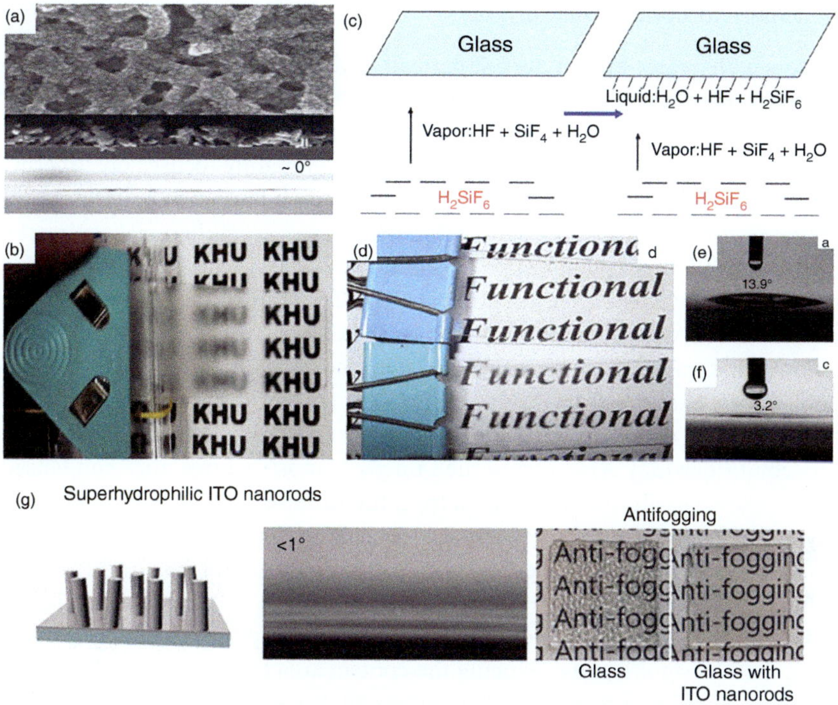

Figure 5.4 (a) FESEM images of surface and cross section and shapes of water droplets on $La(OH)_3/SiO_2$ bilayer. Scale bar: 100 nm. (b) Digital image of a glass slide coated with porous $La(OH)_3/SiO_2$ film (right part) and uncoated glass slide (left part) after being cooled below −5 °C for three hours in a refrigerator followed by exposure to humid laboratory air. Source: Reprinted with permission from You et al. [44]. Copyright 2011, Elsevier. (c) Schematic illustration of the processes occurring during the glass etching. (d) Comparison of the AF behavior of unetched (lower portion) and post-treated etched slide glass (upper portion). Digital images of water droplets after 0.5 second of spreading on the surface of (e) unetched slide glass and (f) slide glass after etching, annealing, and O_2-plasma treatment. Source: Reprinted with permission from Yao and He [45]. Copyright 2013, American Chemical Society. (g) AF properties of the post-annealed ITO nanorod film. Source: Reprinted with permission from Park et al. [46]. Copyright 2013, Royal Society of Chemistry.

silica nanoparticles were deposited on this surface, illustrating superhydrophilic and AF capabilities of the $La(OH)_3/SiO_2$ nanorod film (Figure 5.4a,b). But a relatively large aperture and surface roughness can cause some degree of light scattering. An AF porous surface on a glass substrate was prepared by a vapor-phase etching based on H_2SiF_6, and the transmittance reached 98.95% for an etching time of 30 hours at 15 °C. The schematic of the etching procedure is shown in Figure 5.4c. After annealing and oxygen plasma treatment, the CA of the etched glass was reduced to 3.2°; moreover, the treated surface showed outstanding AF properties (Figure 5.4d–f) [45]. A large-scale, multifunctional, transparent indium tin oxide (ITO) nanorod film was prepared on a glass substrate by a simple sputter deposition method. The performance of the ITO nanorod film was determined by growth time, and superior AF behavior was exhibited at 500 °C for 60 minutes (Figure 5.4g) [46].

5.3.2 Superhydrophobic AF Surfaces

AF surfaces with hydrophilic or even superhydrophilic wettability are widely recognized. Nevertheless, these surfaces may not be able to limit the formation of frost or excessive water condensation under severe atomization circumstances, resulting in inferior optical capability of the surfaces. This section mainly focuses on superhydrophobic AF. Superhydrophobic AF surfaces have low adhesion to water droplets on substrate, which is an advantageous property to repel water, thereby keeping the surface dry [47, 48]. In particular, superhydrophobic surfaces with AF and anti-icing properties have drawn extensive attention [49, 50]. The features of low-temperature superhydrophobic AF have been illustrated by the Cassie–Baxter model [51]. It has been shown that microstructures combined with low-surface energy materials significantly improve the water resistance of surfaces. This means that smaller f_s combined with larger θ may form high-suspended spherical water droplets at low temperature conditions [52]. Also, due to such micro-/nanostructures, tiny, condensed fog drops coalesce easily and then fall off from the surfaces.

Many experiments have been carried out in the fabrication of superhydrophobic AF surfaces. Using polyvinylidene fluoride (PVDF) polymer and zinc oxide to build micro-/nanostructures (ZP-MNs) [53] as shown in Figure 5.5a, Wen et al. developed an outstanding AF surface, in which nanowire arrays can capture air in the gap of the structure, thereby significantly reducing the contact area between water drops and the substrate. Through a breeze or tilt, condensate droplets could almost be eliminated from the ZP-MN surface at −5 °C within 1600 s (Figure 5.5b–d). In addition, by contrast analysis, the A-surface was considered to exhibit excellent performance of AF or anti-icing (Figure 5.5e) [54]. In particular, at −10 °C, the freezing delay time of A-surface was about 9839 seconds (Figure 5.5f). Moreover, the condensed droplets could readily jump from A-surface within 1020 seconds, and the drying region was close to 90.5% (Figure 5.5g).

In recent years, the kinetic behavior of micron droplets on solid surfaces has attracted huge attention. A porous aluminum surface of hierarchical structures was immersed in hot water, and the resulting surface was modified by FAS-17 [55]. The condensate water could be removed from the porous aluminum surface, and self-removal efficiency is closely related to immersion time. Subsequently, the designed micropore array was introduced into the nanostructured superhydrophobic surface to construct the AF/icing surface, and the self-driving transition ability of coalescence droplets can be changed by the size of the micropores, which increases the liquid–vapor interface area, as shown in Figure 5.6a,b [56]. In addition, another porous aluminum surface was prepared, inspiration from the hydrophilic/hydrophobic properties of beetle surface [58, 59], and then modified by polyvinyl alcohol for controllable condensation and microdroplet self-removal, as shown in Figure 5.6c–f. The synergistic effect enabled rapid nucleation of microdroplets, controllable polymerization, and efficient self-removal [57]. Next, the spontaneous wettability transformation of droplets was further studied. Two different tower and conical superhydrophobic surface structures were obtained using a simple hydrothermal reaction (Figure 5.7a–e) [60]. The

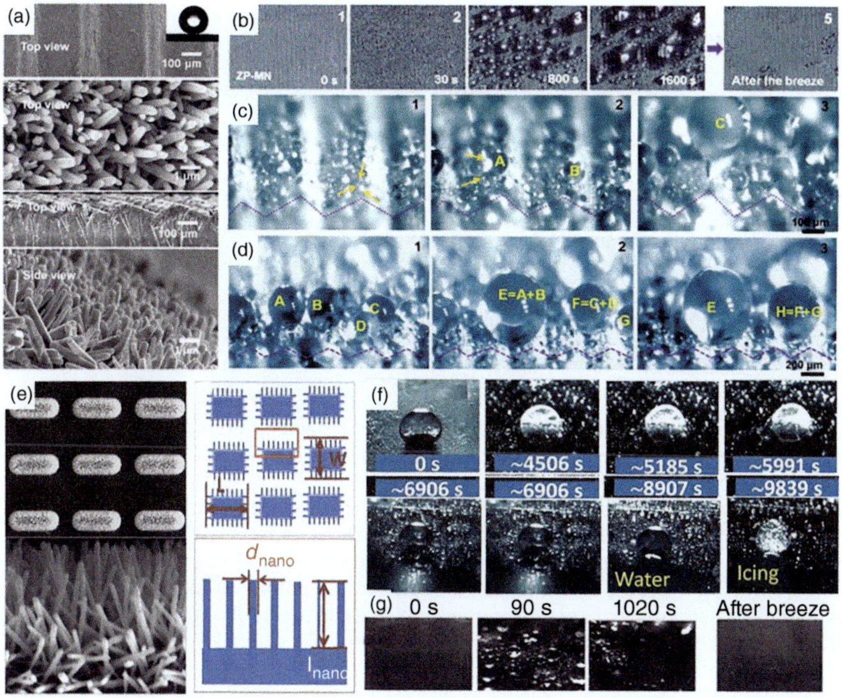

Figure 5.5 (a) SEM images of the ZP-MN surface: Top view and side view. (b) Droplet shedding-off property of the ZP-MN surface. The condensed droplets were formed on ZP-MN over 0–1600 seconds (frames 1–4). The droplets formed on the ZP-MN surface after about 1600 seconds were shed by the applied breeze (frame 5). (c) Magnified high-resolution images captured with a charge coupled device (CCD) camera. (d) Magnified low-resolution images of the interface. Source: Reprinted with permission from Wen et al. [53]. Copyright 2014, American Chemical Society. (e) SEM images of micropillar array surfaces with nanohairs. The A-surface was composed of micropillar arrays covered with nanohairs. (f) Microscopic photographs of the droplets on A-surface at −10 °C with icing delay times about 9839 seconds for droplet frozen than others. (g) Photographs of fog droplet condensation on A-surface, at a low temperature of −5 °C. Within 1020 seconds, fog droplets condensed on the surface, coalesced into bigger ones and some disappeared. Source: Reprinted with permission from Shi et al. [54]. Copyright 2015, John Wiley and Sons.

Laplace pressure of the tower arrays was about 30 times larger than the conical arrays, thereby tiny condensates easily separated from the surface (Figure 5.7f). In contrast, it was difficult for water droplets to detach from the conical array surface (Figure 5.7g).

However, there are many problems with the development of AF surfaces , such as reasonably designing most excellent surface structures with different roughness and developing AF materials with other functions such as wear resistance and high transparency. Fortunately, after millions of years of evolution, nature has created and selected many near-perfect structures to achieve many functions. Inspired by creatures in nature, scientists can copy the natural structures to obtain near-perfect artificial AF surfaces, which are called "bioinspired AF surfaces." So, the bioinspired

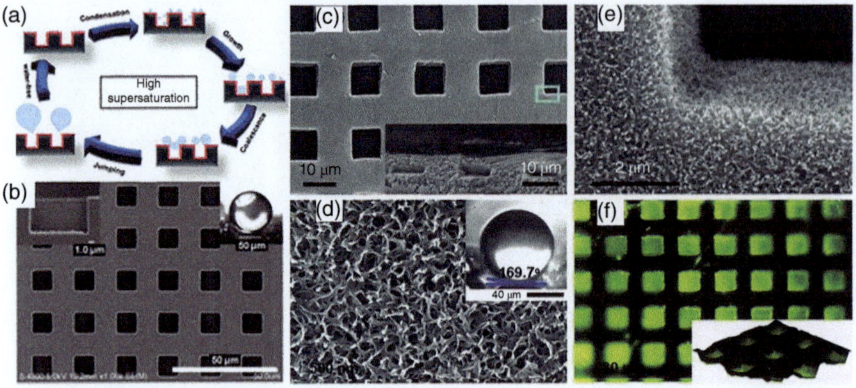

Figure 5.6 (a) Schematic diagram of condensation on the nanostructured superhydrophobic surface with a micropore array at high supersaturation. (b) SEM images of the superhydrophobic surface with a micropore array (left inset), the optical image of condensed microdroplet on the surface (right inset). Source: Reprinted with permission from Zhang et al. [56]. Copyright 2013, Royal Society of Chemistry. (c) Top view of micropores; inset is a side view of micropores. (d) High magnification of nanoporous structures covering all surfaces; inset is the profile of condensed droplets on the porous surface. (e) High-magnification image of area marked with blue open rectangle in (c). (f) Fluorescent micrograph of the superhydrophobic surface with rhodamine B labeled PVA in the bottom of the micropores. Inset is the local 3D presentation of the fluorescence micrograph. Source: Reprinted with permission from He et al. [57]. Copyright 2013, John Wiley and Sons.

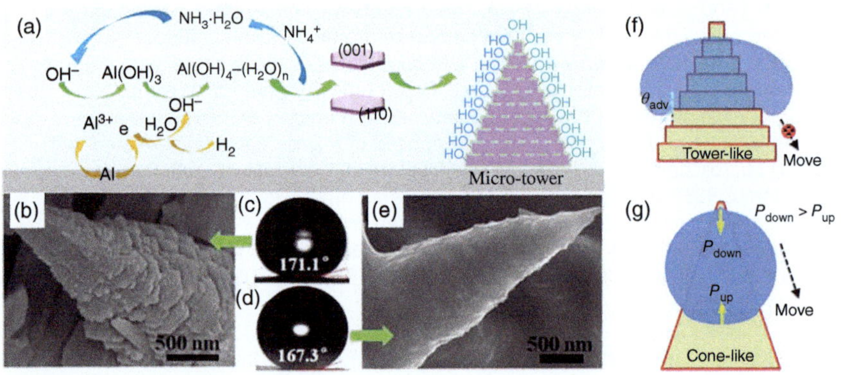

Figure 5.7 (a) Scheme of the reaction of Al with ammonia solution as well as the growth of the microtower. (b, e) SEM images of a single tower-like structure modified by PDMS in 1 and 2 wt% PDMS/*n*-hexane, respectively. It turns into a cone-like structure after being modified by 2% PDMS/*n*-hexane. (c, d) Contact angles of water droplets on the surface in (b, e), respectively. (f) Scheme of a tiny droplet on the sidewall of a tower. (g) Scheme of a tiny droplet on the sidewall of a cone. Source: Reprinted with permission from He et al. [60]. Copyright 2016, American Chemical Society.

AF surfaces with excellent wear resistance, flexibility, transparency, and durability have greatly flourished in recent years.

5.4 AF Surfaces Inspired by Butterfly Wings

Han et al. [61] took butterfly (*Morpho menelaus terrestris*) wings as a bionic prototype and studied the relationship between surface exquisite structure and active AF performance in detail. In addition, a monolayer SiO_2 film with multiscale hierarchical pagoda structures (MHPSs) based on glass substrate was designed and fabricated. More importantly, the underlying active AF strategy gathering initial fog capture and final AF together was revealed.

5.4.1 Characterizations and Analysis of Original AF Butterfly Wings

5.4.1.1 Multiscale Hierarchical Morphologies of Original Butterfly Wings

The multiscale hierarchical morphology of the original wing scale was observed in detail under different magnification by field emission scanning electron microscope (FESEM) (Figure 5.8a–f). Figure 5.8a exhibits these overlapping scales are oval, like tiles on the roof. The dense wing scales are parallel to the scale base and can be divided into two types: transparent cover scales (CSs) and blue ground scales (GSs). CSs are in the upper layer and GSs in the lower layer (Figure 5.8d). CSs and GSs

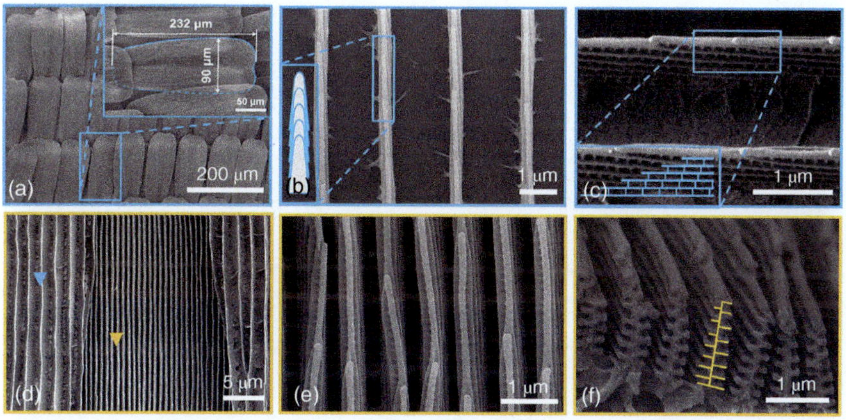

Figure 5.8 FESEM images demonstrate the feature structures of the original cover scales (a–c) and ground scales (d–f). (a) Shapes and distributions of the original scales on the butterfly wing substrate. A single scale (inset) shows its feature sizes in length and width. (b) Periodic parallel ridges arrange along the length of the wing scales. The overlapped lamellae (inset) constitute these narrow ridges. (c) Side view of the ridges shows the ultrafine window-like structures (inset) between two adjacent lamellae. (d) Overlapping region of cover scales (blue triangle) and ground scales (yellow triangle). (e) Denser ridges with similar multiscale hierarchical structures. (f) Cross section of the ridges demonstrates typical pagoda structures (yellow profile). Source: Reprinted with permission from Han et al. [61]. Copyright 2016, American Chemical Society.

overlap each other and ridge–groove composite arrays with different cycle intervals. To be specific, the distance between the two adjacent ridges of CSs is about 2 μm (Figure 5.8b), and it is much wider than that of GSs (Figure 5.8e). Moreover, the ridge width of CSs is about 200 nm, while that of GSs is wider and about 400 nm. Each ridge consists of several layers of parallel slices (Figure 5.8c). Between the two adjacent plates, there are a large number of uniformly distributed micro-ribs, forming rows of window-like pits. Figure 5.8f shows the cross section of a ridge. The cross section of a single ridge has a typical pagoda shape.

5.4.1.2 Chemical Compositions of Original Butterfly Wings

MHPSs are also known as photonic crystals (PCs) [62], which were characterized by X-ray diffraction (XRD) as shown in Figure 5.9a. From the spectrum, two strong Bragg peaks appear at 4.26° and 11.95°, further verifying the periodicity of the MHPSs on original wing scales. In terms of chemical composition, like other insects, the original wing scales of *Morpho* butterfly are mainly composed of water-insoluble organic compounds. The main elements include carbon (C), hydrogen (H), oxygen (O), and nitrogen (N). The characteristic peaks of wing scales

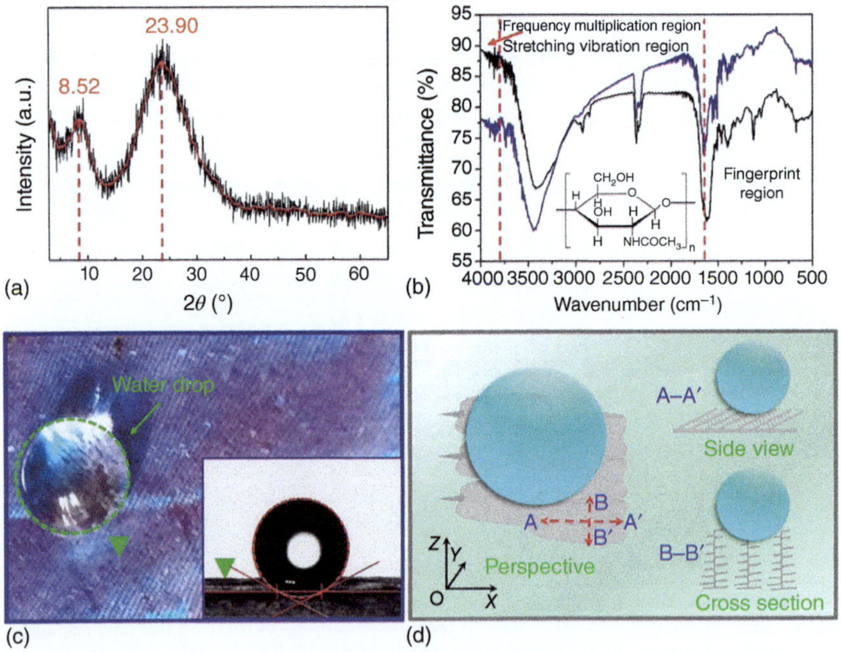

Figure 5.9 (a) XRD spectrum of MHPS presents its quasi-PC structures and corresponding grazing angles. (b) FTIR spectra of original (black line) and treated (blue line) wing scales for comparison indicate the brilliant blue colors result from fine micro/nanostructures in wing scales rather than pigments. (c) Digital photo and CA measurement image (inset) of the wing surface (green triangle) show excellent superhydrophobic property. (d) Simplified illustrations of the static behaviors of a single water drop on a butterfly wing from two different views. Source: Reprinted with permission from Han et al. [61]. Copyright 2016, American Chemical Society.

treated by FTIR are similar to those of the original scales (Figure 5.9b). The sample was treated using a mixture of ether and anhydrous ethanol for 10 minutes. The enhanced absorption peaks appeared at 667 and 2358 cm^{-1}, respectively, due to the bending vibration and asymmetric tensile vibration of CO_2. The absorption peaks at 3428 and 3260 cm^{-1} were attributed to the tensile vibration of hydroxyl (—OH) and N—H bonds, the weak absorption peaks at 2966, 2923, and 2874 cm^{-1} were due to the tensile vibration of C—H bond, and the stretching vibration of C—O bond caused the weak peaks of about 1076 cm^{-1}. XRD combined with FTIR spectra showed that the treated wing scales hardly changed its chemical composition, and the MHPSs were well retained, which offered the basis for the subsequent functional design and bionic manufacturing.

5.4.1.3 Static Wetting Characteristics of Butterfly Wings with MHPSs

The roughness effect can enhance the inherent wettability of the substrate surface, which is an acceptable principle [63]. Because of the advantages of the MHPSs of the wing scales combined with chemical composition, the wing surface exhibits superior superhydrophobicity (CA is about 155.5°), which is beneficial to remove fog drops easily on the surface of wings [64–66] (Figure 5.9c,d). Nevertheless, the fact that most artificial surfaces work in a static state seems to be not considered. Consequently, the development of excellent AF surfaces under static conditions has become an important topic. Superhydrophilic strategy is a well-known effective method to realize static AF surfaces. Based on the rough effect of MHPSs on wing surface, a bionic AF surface by means of hydrophilic materials has been designed and implemented.

5.4.2 Fabrication of Butterfly-inspired AF Surfaces

The manufacturing process of wing scale biological template is shown in Figure 5.10a–d. Firstly, the pretreated wing (1 cm × 2 cm) was clamped between two

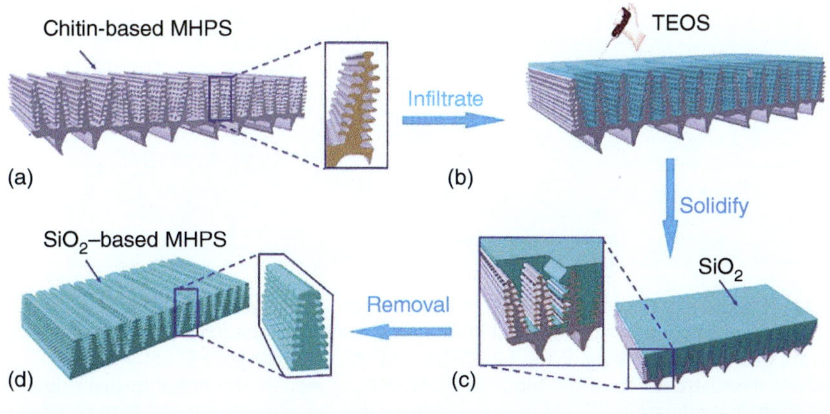

Figure 5.10 Biomimetic fabrication process of the MHPS-based structural surface and the internal chemical synthesis mechanism of 3D SiO_2 networks in MHPSs. Source: Reprinted with permission from Han et al. [61]. Copyright 2016, American Chemical Society.

clean glass slides, which formed an assembly. The precursor solution volume was 10 μL, which consisted of tetraethyl orthosilicate and hydrochloric acid (volume ratio 3 : 1), and then the prepared solution was infused to the assembly edge with a microsyringe. Under the action of capillary force, the precursor solution spontaneously permeated into the microstructure of wing scales. The assembly was baked for 30 minutes at 120 °C, which not only can solidify the sample but also eliminate the residual solution. Subsequently, the assembly was immersed in an acid mixture of nitric acid and perchloric acid (volume ratio 1 : 1) at 130 °C for 30 minutes, to remove the original wing scale. So far, the biomimetic monolayer film (BMF) was retained on the glass slide.

5.4.3 Characterizations and Analysis of Biomimetic AF Monolayer Film (BMF)

5.4.3.1 Multiscale Hierarchical Morphologies of the BMF

The results of AFM and FESEM show the typical ridge valley shape on the surface of BMF (Figure 5.11a). From the top view, each ridge is similar to thorns (Figure 5.11b). The characteristic size of the bionic ridges is measured (Figure 5.11c); the width of

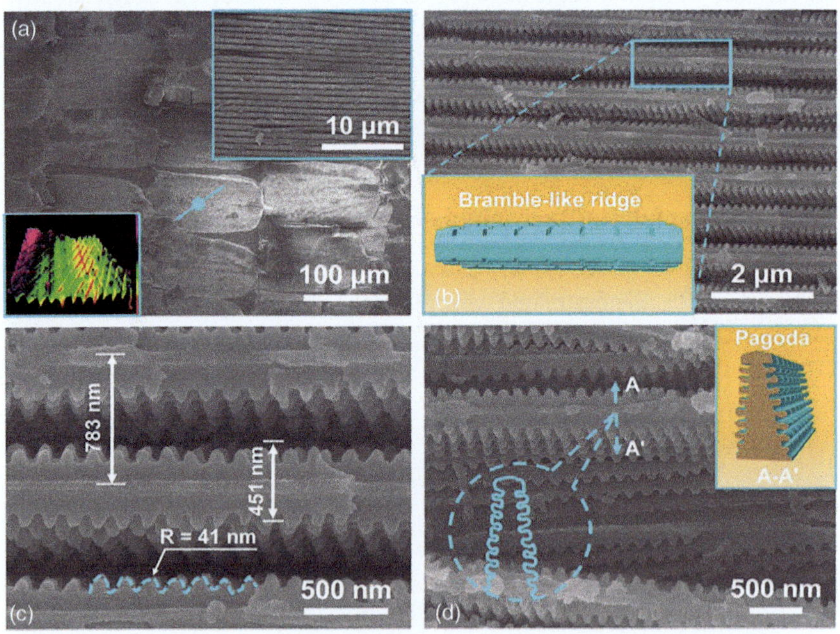

Figure 5.11 FESEM images of the fabricated MHPS-based BMF under different magnifications. (a) Low-magnification FESEM images and AFM image (color inset) demonstrate overall morphology of the MHPS-based BMF. (b) Enlarged FESEM image demonstrates microstructures of the biomimetic MHPSs, especially the bramble-like ridge (inset). (c) High-magnification FESEM image demonstrates the nanostructures' feature sizes of the biomimetic MHPSs, including ridge and papilla. (d) FESEM image of the biomimetic pagoda structure (inset) from the side view under high magnification. Source: Reprinted with permission from Han et al. [61]. Copyright 2016, American Chemical Society.

a single ridge is 451 nm, and the distance between two adjacent ridges is 783 nm. The radius of the nipple structure on both sides of the lamina is 41 nm. It is distinct that the cross section of a single ridge is similar to a typical pagoda structure (Figure 5.11d). Each pagoda ridge is a micro-multilayered system with 5–8 alternating silica and air layers in the vertical direction. Multiscale hierarchical interdependent assemblies, including pagoda ridges, thorns lamellae, and papilla structures, interact to construct the whole MHPS system.

5.4.3.2 Static Wetting Characteristics of the BMF

Here, the static CA of the glass substrate and BMF was obtained separately (Figure 5.12a). It is obvious that the BMF's wettability is much lower than that of the flat glass substrate (Figure 5.12b). The BMF surface shows favorable superhydrophilicity (CA = 6.2°). In addition, the average static CA values for the BMF and flat substrate are 7.0° and 38.3°. These results further illustrate that the roughness of BMF surface increases significantly, providing excellent wettability to the glass substrate.

5.4.3.3 Chemical Compositions of the BMF

The FTIR spectra of the original wing scale and BMF are compared in Figure 5.12c. It is obvious that the chemical composition is distinct. Also, the energy-dispersive

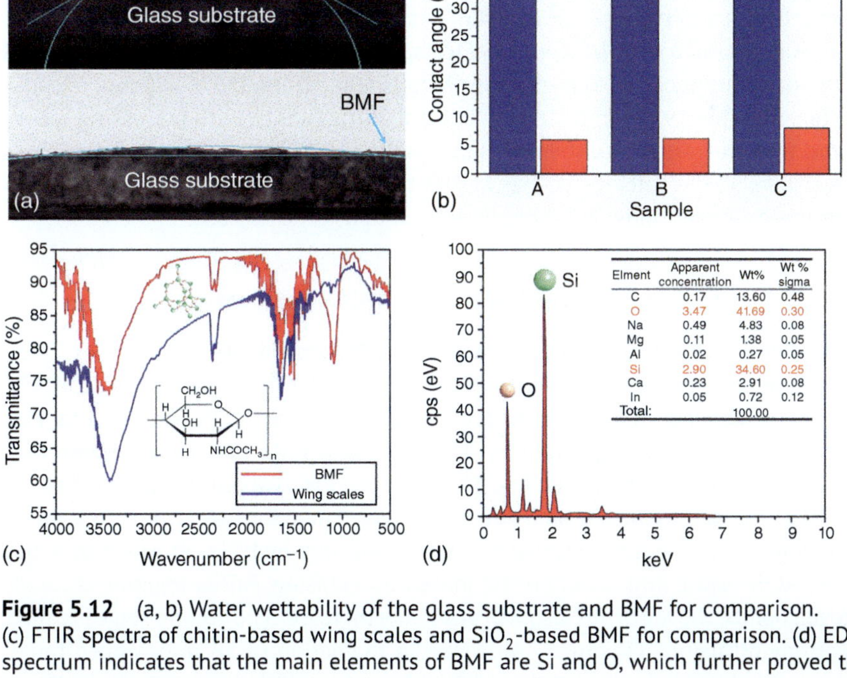

Figure 5.12 (a, b) Water wettability of the glass substrate and BMF for comparison. (c) FTIR spectra of chitin-based wing scales and SiO_2-based BMF for comparison. (d) EDS spectrum indicates that the main elements of BMF are Si and O, which further proved that the nanoscale SiO_2 was obtained successfully. Source: Reprinted with permission from Han et al. [61]. Copyright 2016, American Chemical Society.

X-ray spectroscope (EDS) spectrum indicates that the prepared BMF consists of silicon (Si) and oxygen (O), which are 34.60 and 41.69 wt% (Figure 5.12d). In short, the hydrophilic nano-SiO_2 layers provide a premise for the formation of superhydrophilic BMF. On the other hand, the biomimetic MHPS system further enlarges the hydrophilic effect, thereby achieving the superhydrophilic behavior, which plays an important role in active AF performance.

5.4.4 AF Performance of Butterfly-inspired MHPSs Surfaces

AF performance of BMF was characterized by a self-built spray simulation system (Figure 5.13a). The change in transmittance for the flat substrate and BMF were obtained in the wavelength range of 350–1000 nm. As shown in Figure 5.13b,c, under fog-free conditions low transmittance was exhibited by both flat substrate (82%) and BMF (58%). Meanwhile, it also can be seen that the maximum transmittance for BMF was 95% and flat substrate was 90%. The results illustrate that BMF significantly improved the transmittance of the flat substrate. At a constant flow rate of 0.5 mL/min, stable, continuous fog was produced and controlled by an ultrasonic humidifier. Here, mist flow was consecutively sprayed on BMF surface at an interval of 10 seconds for six times (Figure 5.13d). For the first 10 seconds, the transmittance of BMF surface reached 84% over a range of wavelength from 350 to 480 nm, eventually reduced to 27%. In the wavelength range of 480–1000 nm, it is interesting that the maximum transmittance was always maintained at 95% at different spray times. It was proved that BMF had high AF performance in the whole wavelength range. In addition, to characterize the AF recovery performance, the transmission spectrum of BMF was obtained after 60 seconds of fog spraying (Figure 5.13e). Similarly, BMF had low transmittance (15%) in the wavelength range from 350 to 480 nm, but had high transmittance (95%) from 480 to 1000 nm. Surprisingly, with increasing recovery time, the transmittance also increased rapidly and finally returned to the original fog-free state. The results show that MHPS-based BMF has excellent AF recovery performance (five seconds), which is beneficial to its repeated use.

5.4.5 AF Mechanism of Butterfly-inspired MHPSs Surfaces

5.4.5.1 Dynamic Antifogging Behaviors of the MHPS-based BMF

To illustrate the AF performance of BMF, a simplified three-dimensional MHPS model was established. The whole active AF behavior is shown in Figure 5.14a. The first stage is called "fog drop capture." The MHPSs in BMF may be filled with tiny water droplets from the resulting fog. Then, the fog drops spontaneously adhere to the MHPSs surface to form a hydrone monolayer. The adhesive layer in the interface region attracts an adjacent water mass of multiple hydrides to moisturize the MHPSs. At the same time, as more and more water clusters gather together through hydrogen bonds, the fog drops also will gradually increase. The second stage is called "anisotropy diffusion" (Figure 5.14b). Due to the action of an unbalanced force, droplets spread in space along the x axis, y axis, and z axis at different speeds. Obviously, the diffusion speed along the y axis is quicker than that along the x axis. The

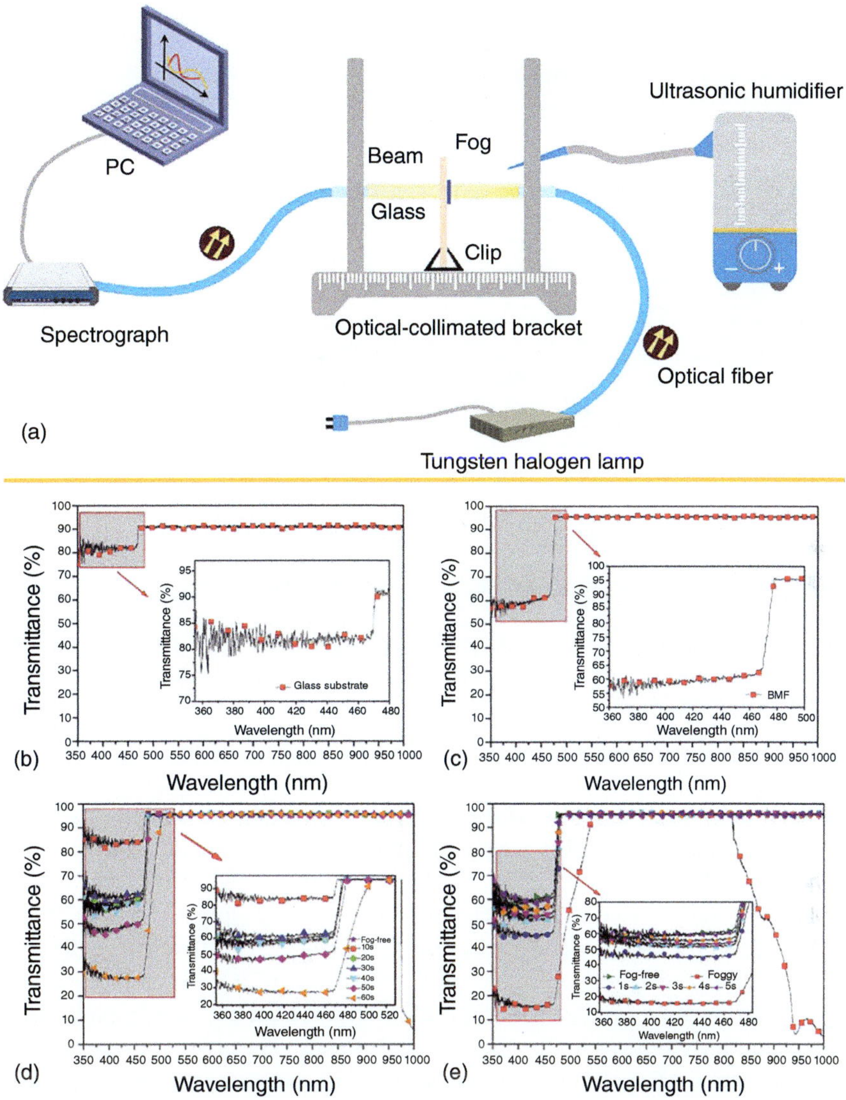

Figure 5.13 Core components of the self-built spray simulation system and transmittance spectra of the glass substrate and MHPS-based BMF. (a) Schematics of the self-built spray simulation system. (b) and (c) Initial transmittance spectra of fog-free glass substrate and BMF before spray simulation. (d) Transmittance of the BMF changes at intervals of 10 seconds during gradient spray simulation for 60 seconds. (e) Transmittance recovers to initial fog-free state after continuous spray simulation for 60 seconds. Source: Reprinted with permission from Han et al. [61]. Copyright 2016, American Chemical Society.

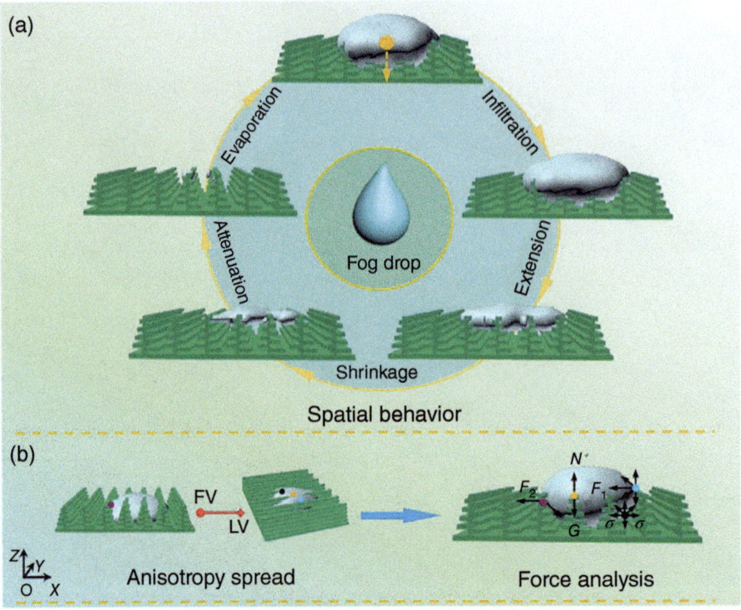

Figure 5.14 Active antifogging behaviors of the MHPS-based BMF. (a) Spatial behaviors of the fog drop on the BMF surface. (b) Static microforce analysis of the fog drop. Here, FV and LV represent front view and lateral view, respectively; N is normal force, G is gravity, F_1 is intermolecular force, F_2 is an attractive force, and σ is surface tension. Source: Reprinted with permission from Han et al. [61]. Copyright 2016, American Chemical Society.

root cause of anisotropy diffusion is that MHPSs form a narrow-layered channel. Droplets spread rapidly along the y axis without obstacles, while along the x axis due to the physical obstacles of MHPSs, the droplets diffuse slowly. Due to the interaction of gravity and surface tension, droplets diffuse slowly along the z axis. Then the entire droplet quickly is divided into several tiny fog drops of different sizes by the MHPS system. Surprisingly, the large specific surface area of MHPSs plays a crucial role in accelerating rapid evaporation of these fog drops. In the end, the droplets vanish entirely and BMF returns to the original fog-free state. The above process can be profiled as an active AF cycle, namely "droplet capture–anisotropy diffusion–rapid evaporation."

5.4.5.2 Active Antifogging Mechanism of the BMF with MHPSs

To further reveal the internal active AF mechanism of BMF, a reasonable explanation is the collaborative effect of chemical composition and physical structure (Figure 5.15). It is a common strategy to construct superhydrophilic surface by using hydrophilic components to form rough structures. It is interesting that the MHPSs of the BMF not only played an important role in enhancing the intrinsic wettability of glass substrate but also significantly improved the surface roughness of the glass substrate. Obviously, the hierarchical amplification effect of MHPSs produced great rewards to realize superhydrophilic surface. Moreover, based on the principle of minimum energy, hydrogenates tend to stay in the nanostructure of BMF to

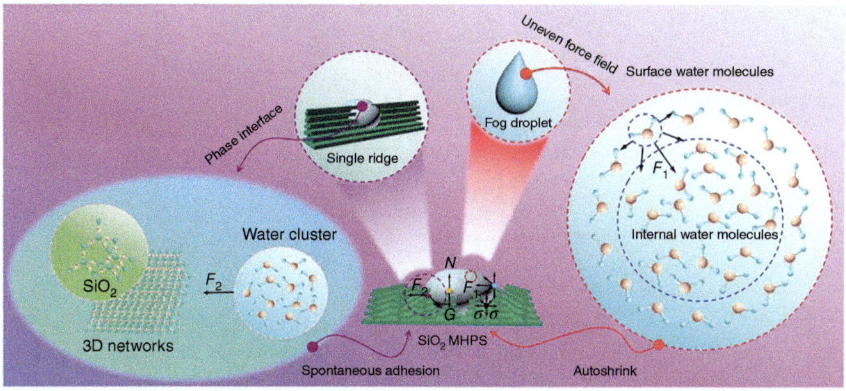

Figure 5.15 Collaborative effect of chemical compositions and MHPSs of the BMF, realizing the active antifogging property. Source: Reprinted with permission from Han et al. [61]. Copyright 2016, American Chemical Society.

reach the lowest energy state, which may lead to the transition of droplets from a metastable Cassie–Baxter wetting state to a stable Wenzel state. Because of the active capture and anisotropy diffusion, the BMF was soon covered with a thin water film; meanwhile due to the increase in the effective contact area between fog drops and MHPSs, the water film can be quickly evaporated. Therefore, the transmittance of BMF would return to fog-free state. These results were in good agreement with the above transmission spectra.

5.5 AF Surfaces Inspired by Mayfly Compound Eyes

Furthermore, Han et al. [67] took mayfly (*Ephemera pictiventris McLachlan*) compound eyes as a bionic prototype, and the relationship between compound eyes structure and excellent AF properties were studied in detail. A bioinspired AF surface was designed and fabricated successfully and also the underlying AF mechanism was revealed.

5.5.1 AF Mechanism of Original Compound Eyes

5.5.1.1 Microstructure Morphologies of Original Mayfly Compound Eyes

The SEM images show that the compound eye is ellipsoidal, which consists of hundreds of ommatidia; as shown in Figure 5.16a,b, these ommatidia are evenly distributed and stacked closely. As shown in Figure 5.16c,d, the top half of ommatidia is spherical, and the bottom half is cylindrical, with a diameter of 20 μm. Additionally, the ommatidia surface is with nonexistent nanostructure. Therefore, it can be concluded that it is the microstructure that finally leads to the AF behavior.

5.5.1.2 Antifogging Behavior of Original Mayfly Compound Eyes

The spray experiment indicated that tiny fog drops were hard to retain on the surface of the compound eyes. After the spray, the surface of the compound eyes still

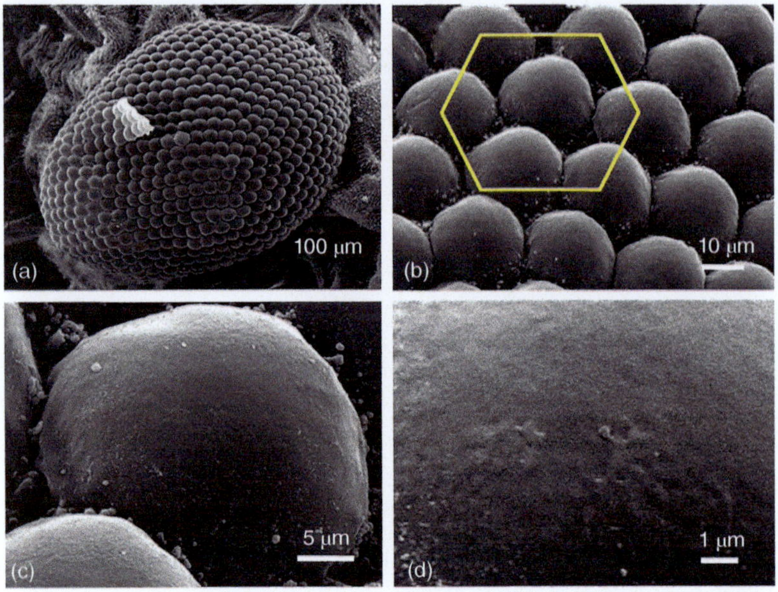

Figure 5.16 SEM images of compound eyes. (a) The compound eyes are ellipsoidal, composed of hundreds of ommatidia. (b) Ommatidia are uniform and tightly arranged. (c) The upper part of ommatidia is spherical, the lower is cylindrical, with a diameter of 20 µm or so. (d) There is no more tiny structure on the ommatidia surface. Source: Reprinted with permission from Han et al. [67]. Copyright 2014, Springer.

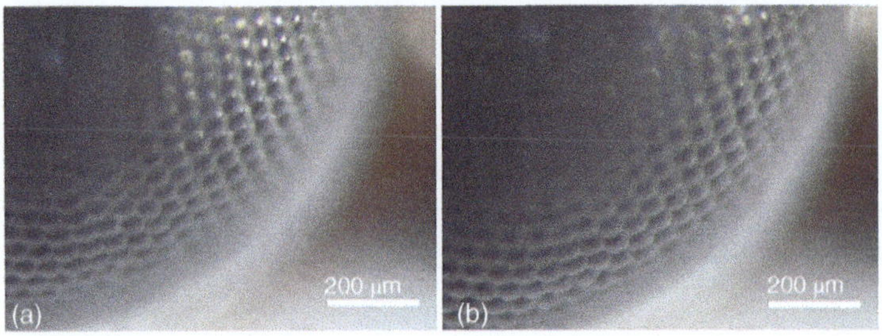

Figure 5.17 An optical microscope image of compound eyes of the sample (a) before blowing fog and (b) after blowing fog. Source: Reprinted with permission from Han et al. [67]. Copyright 2014, Springer.

remained dry and clear. Accordingly, the compound eyes showed significant superhydrophobic AF performance (Figure 5.17).

5.5.1.3 Antifogging Mechanism of Original Mayfly Compound Eyes

The ommatidia are tangent to each other, and the diameter of the tangent circle is about 3 µm. Due to the fog drops that fall on the surface of the compound eyes are randomly distributed, here, the AF process is divided into two situations: at the top and the gap of the ommatidia. Under the influence of gravity, the fog drops may

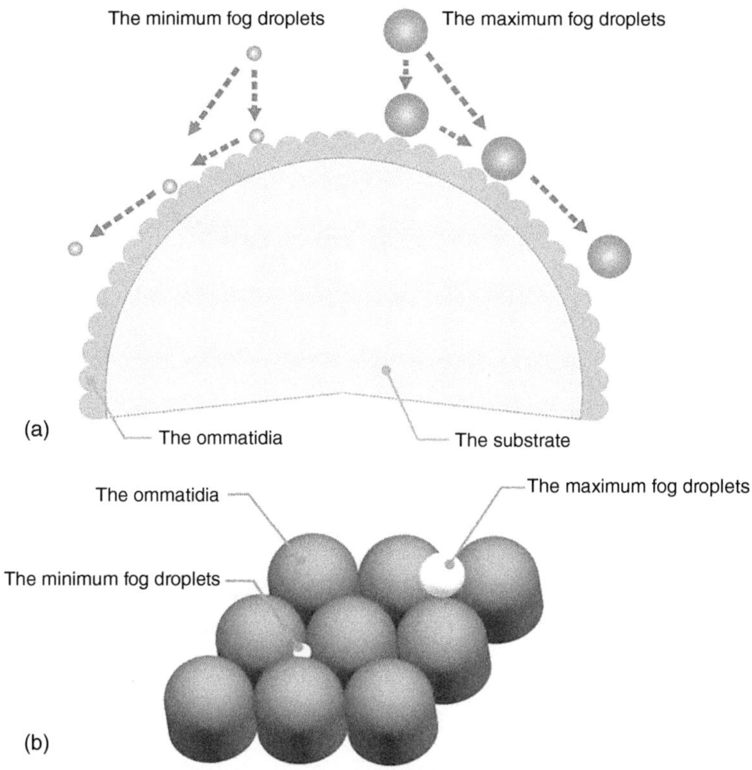

Figure 5.18 A 3D schematic illustrating the point contact between ommatidia and fog droplets. Source: Reprinted with permission from Han et al. [67]. Copyright 2014, Springer.

drop in the gap or drop off the surface, as shown in Figure 5.18a; therefore, it is only necessary to analyze that fog drops fall on the gap. It is well known that the contact mode between two adjacent spheres is only one point. And it just happens that the fog drops and ommatidia both are spherical. Therefore, their contact mode can be seen as point contact. Nevertheless, in a real situation, microscale fog droplets are prone to deformation origin from their motion. However, on account of the effect of air film on fog drops, the deformation is negligible. Therefore, the contact mode between fog drops and ommatidia almost is point contact (Figure 5.18b), namely the actual contact area of the solid–liquid interface is close to zero, which prevents the fog drops from condensing on the surface of the compound eyes.

5.5.2 Fabrication of Superhydrophobic Antifogging Surfaces (SSASs)

5.5.2.1 Preparation Process of Circular Micro-holed Array (CMHA)

The steps as shown in Figure 5.19a,b: First, the slide (4 cm × 3 cm) was cleaned in an ultrasonic bath with acetone, ethanol, and deionized water, and the substrate was baked for 30 minutes at 200 °C, and the SU-8 photoresist was assigned uniformly on

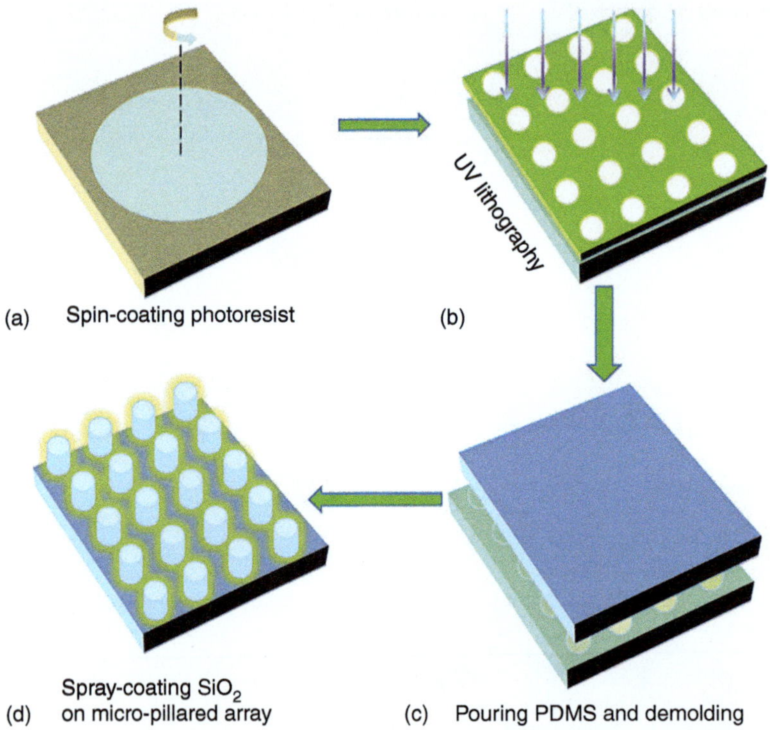

Figure 5.19 The fabrication process of the bioinspired PDMS coupled with sprayed SiO_2 coatings and the MPA. Source: Reprinted with permission from Han et al. [68]. Copyright 2018, Royal Society of Chemistry.

the glass slide. The substrate was heated on a horizontal hot plate for six minutes. Subsequently, the substrate was handled by ordinary ultraviolet lithography and exposed to a mask perpendicular to the circular array pattern (the exposure energy was 158 mJ/cm^2 and the exposure time was 10 seconds). Next, the substrate was baked for six minutes at 95 °C to harden the SU-8 layer. Finally, the substrate was soaked in the developer for 30 seconds, and tweezers were used to stir the solution. Thus, the circular micro-holed array (CMHA) on the glass substrate was obtained.

5.5.2.2 Preparation Process of the Sprayed SiO_2 Coatings on Micro-pillared Array (MPA)

The specific steps as shown in Figure 5.19c,d: First, the prepolymer was evenly mixed with a curing agent (weight ratio 10 : 1) and stirred in a beaker. Then, the mixture was gently poured on the photoresist layer, and the component was put into the vacuum chamber until the bubbles completely disappeared, and then baked in a drying box for one hour to completely solidify the film. Finally, the MPA was smoothly transferred to the cured polydimethylsiloxane (PDMS) film (PF). A commercial SiO_2 coating agent was sprayed on the prepared PF surface. The particles were deposited through an airbrush driven by high-pressure air flow. Then the sample was baked in an oven at 80 °C for 20 minutes.

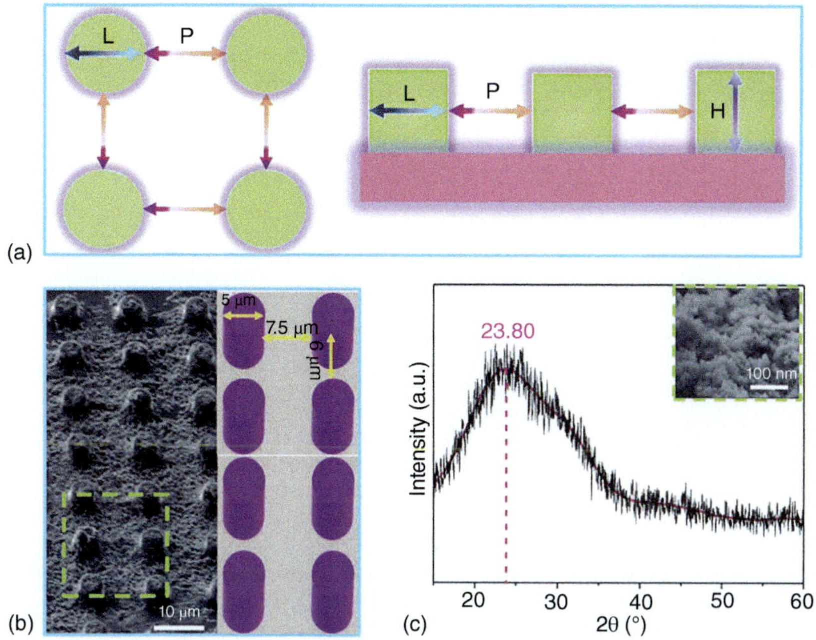

Figure 5.20 (a) The top view (left) and side view (right) of the MPA. This pattern was repeated periodically across the PDMS surface. (b) The FESEM image of the fabricated SiO_2-sprayed MPA under low magnification, demonstrating that the coated PDMS has a negligible effect on the global structure of the MPA. (c) The XRD spectrum of the coating and high-magnification FESEM image shows the morphology of the nanoscale SiO_2 (inset). Source: Reprinted with permission from Han et al. [68]. Copyright 2018, Royal Society of Chemistry.

5.5.3 Wettability and Composition Analysis of the SSASs

5.5.3.1 The Measurements of Surface Wettability for the SSASs

As illustrated in Figure 5.20a, the side length of the pillar (L) is about 5 μm, the pitch between neighboring pillars (P) is about 7.5 μm and the nominal height of the pillar (H) is about 6 μm. The quality and uniformity of the pillars were inspected using FESEM. The FESEM sample was titled at 45° to reveal the actual structures. Figure 5.20b and c show that the PDMS coating has a negligible effect on the global structure of the MPA even though the individual micro-pillar was coated with nanoscale SiO_2. The FESEM results indicate that the as-prepared PF possessed a rough surface and contained many bumps.

The wettability of three samples was tested by CA meter, respectively, flat PF, PF with MPA, PF coupling SiO_2 and MPA, and PF coupling SiO_2 and MPA (superhydrophobic antifogging surfaces [SSASs]). As shown in Figure 5.21, the static CA of the flat PF and PF with MPA reached 114° and 133°, respectively. In particular, it was found that PF coupling SiO_2 and MPA was close to 158°. The results illustrate that the surface structure change caused by MPA remarkably amplified the surface hydrophobicity, and the Cassie–Baxter model was used to explain the SSASs [69],

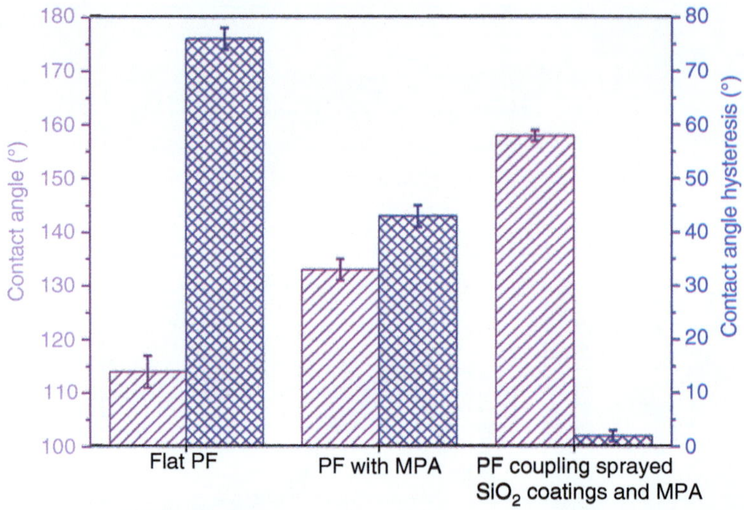

Figure 5.21 CA and CAH measurements of three PF surfaces. The error bars denote standard deviations, which were obtained from distinct measurements on the three different PFs and at least at five different locations on each. Source: Reprinted with permission from Han et al. [68]. Copyright 2018, Royal Society of Chemistry.

which indicated that the microstructure combined with low surface energy materials were favorable for the promotion the surface superhydrophobicity.

The effective CA of the droplets can be regulated by the equation $\cos\theta_c = f_s \cos\theta_s + f_v \cos\theta_v$, where θ_s and θ_v are CAs of the liquid contacting with solid and vapor parts, f_s and f_v are the area fractions of the solid and vapor on the surface. If rough structures on a surface can generate entrapped air pockets, in such circumstances, $f_s + f_v = 1$, $\theta_v = 180°$. The θ_c can be calculated by Eq. (5.4).

In addition, superhydrophobic surface with low adhesion to droplets is a crucial index of AF materials. We found that water adhesion on both flat PF and PF with MPA was high. Therefore, the starting surface was hydrophobic and had high adhesion. Amazingly, PF coupling sprayed SiO_2 coatings and MPA had a low rolling-off angle of less than 2°. From the thermodynamic point of view, eliminating a liquid from its solid substrate requires energy to overcome the adhesion [70]. The basic relation between the work of adhesion and surface wettability is given by the Dupre–Yong equation:

$$W_e = \gamma_{lv}(1 + \cos\theta_c) \tag{5.6}$$

where W_e is the work of adhesion at the equilibrium state and γ_{lv} is the surface tension of the liquid–vapor interfaces. At a large static CA, it requires a small amount of work to remove the droplets. When the SiO_2 coatings were applied on the top of the MPA by spraying techniques, their morphology changed dramatically. Specifically, the pillars retained their microscale geometrical characteristics and also exhibited nanometer roughness due to the presence of the SiO_2 particles. Moreover, the increase in the CA of the SiO_2-sprayed MPA was followed by a noticeable decrease in water adhesion. Since the water drops roll off easily on the patterned surface, it can be assumed that the water drops stay on the top of the MPA without penetrating the gap between the neighboring pillars. Thus, by simple sprayed SiO_2 on the PF with MPA, we created a "nonsticky" superhydrophobic surface. The very low water adhesion is due to the SiO_2 coatings combined with the geometrical features of the rough surface, which further verified the superhydrophobic AF effect of both adding SiO_2 and creating MPA on PDMS surface. In brief, on the one hand, the foundation of hydrophobic SiO_2 nanometer coatings achieved the superhydrophobic PF chemically. On the other hand, the MPA further amplified the hydrophobic effect to realize the superhydrophobic effect (CA = 158° and SA = 2°) physically, which played a crucial role in achieving the AF property.

5.5.3.2 Water Droplet Bounce Behaviors on the SSASs

The water droplet bounce behaviors on as-prepared PF were recorded with a high-speed video camera, when a water droplet of 15.6 μL was dropped from a height of 54.8 mm. The dropping height was determined by the maximum height avoiding the droplet fragmentation when impacting the surface, ensuring maximum droplet momentum. The droplet volume (≈16 μL) was found to be optimum as the droplet could be replicated easily and fell under its weight when dropped from a 23-gauge dispensing tip. As shown in Figure 5.22, the water droplet deformed quickly after contacting the as-prepared PF. The initial impact velocity of the droplet was 0.94 ms^{-1}. At 3.1 ms, the spherical water droplet reached a disc-like state. Then, the water drop began to bounce twice and finally completely separated

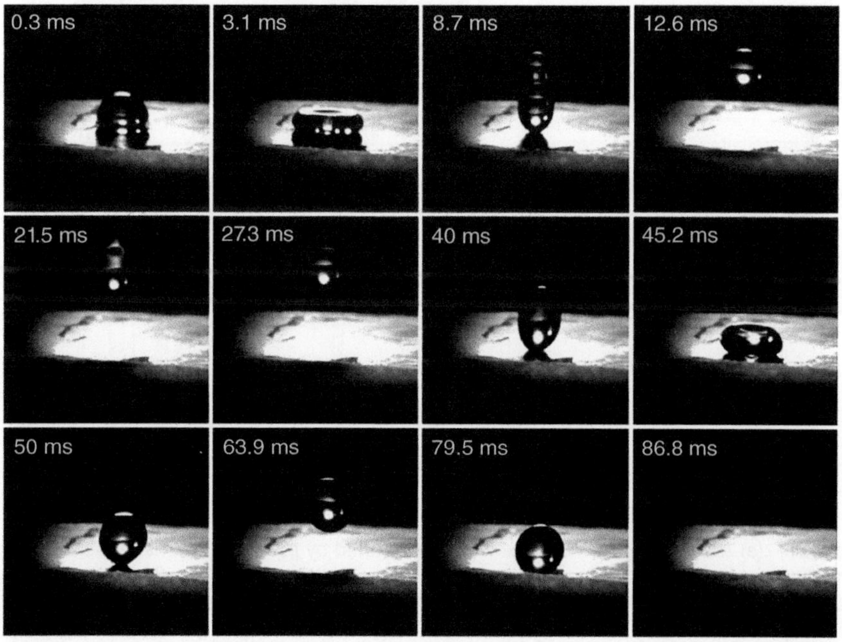

Figure 5.22 Bounce dynamics of a water droplet impacting the surface of SSASs. Source: Reprinted with permission from Han et al. [68]. Copyright 2018, Royal Society of Chemistry.

from the surface of the as-prepared PF within 86.8 ms, which further illustrated excellent superhydrophobic and low adhesion properties.

5.5.3.3 Chemical Compositions of the SSASs

To clarify the reasons that SSASs possessed superhydrophobic–antifogging function, chemical compositions of as-prepared PF surface were investigated. Figure 5.23b shows the EDS spectra of the as-prepared PF. The results indicate that the fabricated PF is composed of three elements, carbon (C), silicon (Si), and oxygen (O). FTIR spectra of commercial SiO_2 coating agent, the PF with MPA and PF coupling sprayed SiO_2 coatings and MPA were obtained, respectively. The FTIR spectrum of the commercial SiO_2 coating agent is shown in Figure 5.23c. The peak at 810 cm^{-1} is assigned to the Si–O–Si symmetric stretching, and Si–O–Si asymmetric vibration at 1082 cm^{-1}. The peaks of 1264 and 2960 cm^{-1} correspond to symmetric bending vibration of Si–CH_3 and symmetric stretching vibration of Si–CH_3, respectively. The peaks at 850 and 1405 cm^{-1} are due to the vibration of Si—C bending and Si—C stretching bonds, respectively. The Si–CH_2 stretching band at 2850 cm^{-1}, the peaks of 1389 and 1460 cm^{-1} are due to the bending vibration of Si–CH_2 [71]. The FTIR

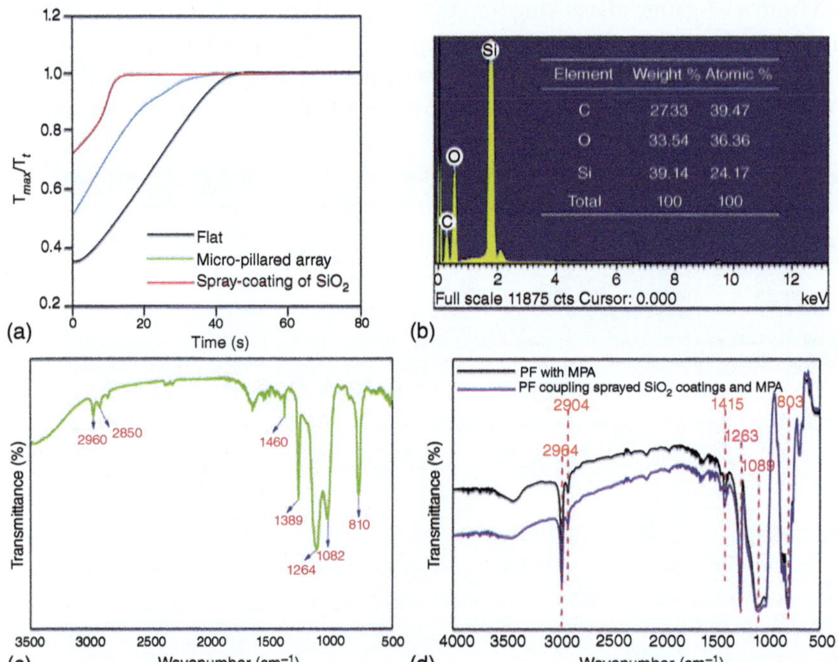

Figure 5.23 (a) Antifogging of the three PFs is quantified by performing time-lapse transmittance measurements (T_t) to determine the required time to restore their original optical properties (T_{max}). The PF coupled with sprayed SiO_2 coatings and the MPA shows a significantly faster recovery from fogging. (b) EDS spectra of the SSASs. (c) FTIR spectra of the commercial SiO_2 coating agent. (d) FTIR spectra of the PF with the MPA (black) and the SSASs (purple). Source: Reprinted with permission from Han et al. [68]. Copyright 2018, Royal Society of Chemistry.

spectrum of the PF with MPA and PF coupling sprayed SiO_2 coatings and MPA is shown in Figure 5.23d. The peak at 1089 cm^{-1} is attributed to Si–O–Si stretching vibration bond. Peaks at around 1263 and 803 cm^{-1} are assigned to the Si—C groups. Other characteristic peaks in the spectrum are assigned to $-CH$, $-CH_2$, and $-CH_3$ groups of the polymer backbone (2964, 2904, and 1415 cm^{-1}, respectively). The FTIR spectrum of the modified PF is consistent with the FTIR spectrum of the untreated PF. Therefore, the above results not only suggest that the commercial SiO_2 coating agent was without any impurities but also that the as-prepared PF was enriched with extreme superhydrophobicity and low adhesion from the hydrophobic functional groups ($-CH_2$ and $-CH_3$) of SiO_2 coating.

5.5.4 AF Performance of the SSASs

5.5.4.1 Time-lapse Transmittance of the SSASs

Furthermore, since fogging results in a certain degree of transmittance loss, we quantified the response of the flat PF, PF with MPA, and SSASs to fogging at regular intervals (T_t), until the original transmittance (T_{max}) was restored, to characterize the AF recovery property, the variation trends of the time-lapse transmittance measurements of three PF were performed after being sprayed by the generated fog. For this purpose, we built a spray simulation system to characterize their AF property [61]. The PF was fixed by a clip and was adjusted to be perpendicular to the light beam. As shown in Figure 5.23a, recovery from fogging was much faster in regard to PF coupling sprayed SiO_2 coatings and MPA, as T_{max}/T_t reached a plateau in far less time than others. It was confirmed that the PF coupling sprayed SiO_2 coatings and MPA possesses superior AF recovery property, especially in wet and humid environments [72].

5.5.4.2 Micro-dynamic Behavior of the Fog Drops Movement on the SSASs

To examine the AF property more intuitively, a 3D ultra-depth stereoscopic microscope was used to observe the micro-dynamic behavior of fog drops on the surface of SSASs. To start with, a great amount of individual spherical fog drops were occurring on the top of the micropillars. As time went on, we found some tiny fog drops began to merge with each other and formed new fog drops. As shown in Figure 5.24, fog drops *A–E* were growing smoothly ($t = 20$ seconds). Subsequently, fog drops *B* and *C* coalesced into a larger fog drop *F* ($t = 30$ seconds); fog drops *D* and E coalesced into a larger fog drop *H* ($t = 45$ seconds); and fog drops *A* and *F* coalesced into a larger droplet *I* ($t = 55$ seconds). Then, fog drops *H* and *I* merged with fog drop *G* and formed fog drop *J* ($t = 90$ seconds). This indicated that the condensed fog drops could be in Cassie's state. We assumed that a dewetting transition phenomenon may occur on the surface [73], and then the released surface energy can propel the fog drop to jump or self-removal from the surface [74]. Furthermore, when this flying fog drop touched another constrained fog drop, a transition would again be stimulated. Due to the occurrence of dewetting transition and self-removal phenomenon, the MPA can get dry and the fog drops can drop from the PF due to low surface adhesion.

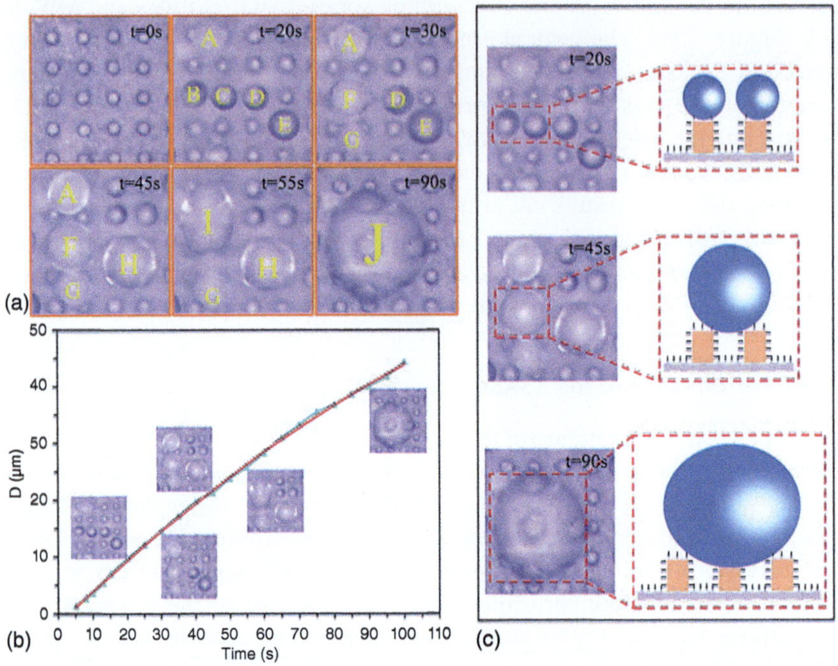

Figure 5.24 (a) Optical images showing the micro-dynamic behavior of the fog drops movement on the surface of SSASs. (b) Time evolution of the diameter of an individual fog drop during the merge process (blue triangles). The insets correspond to $t = 20, 30, 45, 55$, and 90 seconds. (c) Additional details are displayed with the assistance of schematic diagrams. Source: Reprinted with permission from Han et al. [68]. Copyright 2018, Royal Society of Chemistry.

5.5.4.3 Macro-dynamic Process of the Fog Drops Movement on the SSASs

To verify our hypothesis about AF behavior on the surface of SSASs, the macro-dynamic process of fog drops movement was recorded using a 3D ultra-depth stereoscopic microscope. As shown in Figure 5.25a, when the fog spray was applied to the surfaces, some tiny fog drops initially condensed on the surface and subsequently, the fog drops became larger gradually. As time went on, the same phenomenon as in Figure 5.24 happened that some tiny fog drops began to merge with each other and form larger fog drops (see the same color circle from t = 90 seconds to t = 196 seconds). In addition, when the fog drops grew to a certain size, similarly, these fog drops began to roll off suddenly. It is interesting that this roll-off performance removed some circumjacent tiny fog drops, sweeping the surface clean and keeping the area dry (see purple circle at t = 236 seconds). This is consistent with our previous conjecture and demonstrates the excellent AF property of the as-prepared PF surface. To evaluate the AF ability of the SSASs, we estimated the percentage of dry areas versus time, as shown in Figure 5.25b, at t = 50 seconds, the percentage was lower, but the percentage increased suddenly after t = 200 seconds, and the percentage can maintain at ~83% from ~250 to 500 seconds on the whole. This demonstrated the superior AF property of the SSASs.

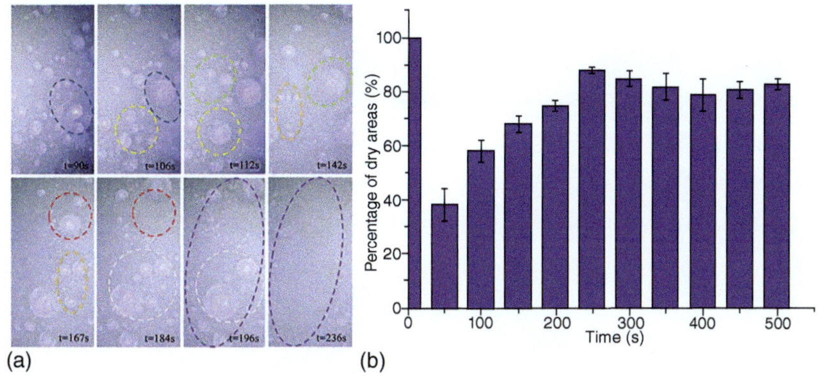

Figure 5.25 (a) Optical images showing the macro-dynamic process of the fog drops movement on the surface of SSASs. (b) The percentage of dry areas versus time. Source: Reprinted with permission from Han et al. [68]. Copyright 2018, Royal Society of Chemistry.

5.5.5 AF Mechanism of the SSASs

To further unfold the internal AF mechanism of the SSASs, one possible reasonable explanation for these findings was that the synergistic effect of droplets intermolecular attraction, chemical compositions, and the key factor MPA. Low surface energy resulted from nanoscale SiO_2 coatings, which further increased the PF surface hydrophobicity and dramatically decreased PF surface adhesion to fog drops. Since water drops rolled off easily on the PF surface, fog drops rested on the top of the MPA without penetrating in the interpillar. Then tiny condensed droplets were induced to merge with each other easily until shed from the surface. Moreover, the evaporation rate of fog drops was also improved [75]. Physically, the fog drops would be repulsive to the MPA because the surface forces were sufficient to suspend liquid against the downward pull of gravity (or other body forces); as a whole, water molecules were affected by the repulsion force of material itself (F_1), intermolecular attraction (F_2), repulsion force of micro-pillared arrays (F_3), surface tension (σ), and its gravity (G), as shown in Figure 5.26.

On the other hand, it is a universal strategy to construct the superhydrophobic surface by creating surface roughness onto low surface energy materials. Specifically, the pillars retained their microscale geometrical characteristics and exhibited also nanometer roughness due to the presence of the SiO_2 coatings. The combination of the micro-/nano-roughness indicated that the true CA of flat hydrophobic surface would be lower with the increase in surface asperities. The very low water adhesion was due to the inherent property of the SiO_2 coatings in combination with the geometrical features of the MPA. Apparently, it was the hierarchical amplification effect of SiO_2-sprayed MPA. In addition, with the increase of the hydrophobic specific surface, the SSAS surface free energy was obviously reduced. According to the energy minimization theory [76], once the droplet coalesces with the adjacent droplets, the released surface energy will overcome droplet adhesion, which may induce the fog drops from a Wenzel state to a Cassie–Baxter state, so that coalescing fog drops can

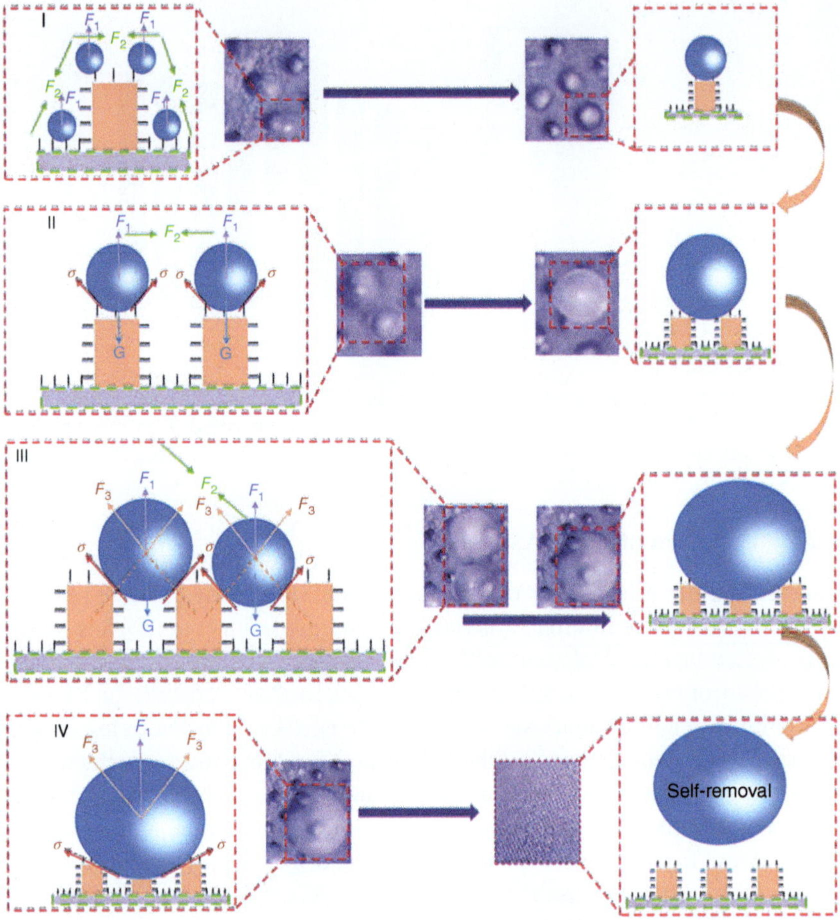

Figure 5.26 The antifogging behaviors of the PF coupled with sprayed SiO_2 coating and the MPA. Here, F_1 is the repulsion force of the material itself, F_2 is the intermolecular attraction, F_3 is the repulsion force of the micro-pillared arrays, σ is surface tension, and G is gravity. Source: Reprinted with permission from Han et al. [68]. Copyright 2018, Royal Society of Chemistry.

undergo self-removal from the surface of SSASs. Consequently, the transmittance of the SSASs will recover to the initial state. These are exactly consistent with the results of the transmittance spectra.

References

1 Sui, D., Huang, Y., Liang, J. et al. (2011). Flexible and transparent electrothermal film heaters based on graphene materials. *Small* 7 (22): 3186–3192.

2 Zhang, X., Yan, X., Chen, J., and Zhang, J. (2014). Large-size graphene microsheets as a protective layer for transparent conductive silver nanowire film heaters. *Carbon* 69: 437–443.

3 Tan, L., Zeng, M., Wu, Q. et al. (2015). Direct growth of ultrafast transparent single-layer graphene defoggers. *Small* 11 (15): 1840–1846.

4 He, L. and Tjon, S. (2016). Aqueous graphene oxide-dispersed carbon nanotubes as inks for the scalable production of all-carbon transparent conductive films. *Journal of Materials Chemistry C* 4 (29): 7043–7051.

5 Han, Z., Feng, X., Guo, Z. et al. (2018). Flourishing bio-inspired antifogging materials with superwettability: progresses and challenges. *Journal of Materials Chemistry C* 30 (13): 1704652.

6 Duran, I.R. and Laroche, G. (2019). Current trends, challenges, and perspectives of anti-fogging technology: surface and material design, fabrication strategies, and beyond. *Progress in Materials Science* 99: 106–186.

7 Young, T. (1805). An essay on the cohesion of fluids. *Philosophical Transactions of the Royal Society of London* 95: 65–87.

8 Wenzel, R.N. (1949). Surface roughness and contact angle. *Journal of Physical and Colloid Chemistry* 53 (9): 1466–1467.

9 Cassie, A. (1948). Contact angles. *Discussions of the Faraday Society* 3: 11–16.

10 Wang, S., Liu, K., Yao, X., and Jiang, L. (2015). Bioinspired surfaces with superwettability: new insight on theory, design, and applications. *Chemical Reviews* 115 (16): 8230–8293.

11 Furmidge, C. (1962). Studies at phase interfaces. I. The sliding of liquid drops on solid surfaces and a theory for spray retention. *Journal of Colloid Science* 17 (4): 309–324.

12 Tzianou, M., Thomopoulos, G., Vourdas, N. et al. (2020). Tailoring wetting properties at extremes states to obtain antifogging functionality. *Advanced Functional Materials* 31 (1): 2006687.

13 Lee, H., Alcaraz, M.L., Rubner, M.F., and Cohen, R.E. (2013). Zwitter-wettability and antifogging coatings with frost-resisting capabilities. *ACS Nano* 7 (3): 2172–2185.

14 Wang, J., Wang, J.J., Sun, Y.L., and Wang, C.W. (2013). The preparation of superhydrophilic surface of TiO_2 coating without ultraviolet irradiation through annealing treatment. *Journal of Sol-Gel Science and Technology* 68 (1): 75–80.

15 Wang, K., Janczarek, M., Wei, Z. et al. (2019). Morphology- and crystalline composition-governed activity of titania-based photocatalysts: overview and perspective. *Catalysts* 9 (12): 1054.

16 Wong, W.S., Nasiri, N., Rodriguez, A.L. et al. (2014). Hierarchical amorphous nanofibers for transparent inherently super-hydrophilic coatings. *Journal of Materials Chemistry A* 2 (37): 15575–15581.

17 Maino, G., Meroni, D., Pifferi, V. et al. (2013). Electrochemically assisted deposition of transparent, mechanically robust TiO_2 films for advanced applications. *Journal of Nanoparticle Research* 15 (11): 2087.

18 Zuo, Y., Zheng, L., Zhao, C., and Liu, H. (2020). Micro-/nanostructured interface for liquid manipulation and its applications. *Small* 16 (9): 1903849.

19 Xiong, Y., Lai, M., Li, J. et al. (2015). Facile synthesis of ultra-smooth and transparent TiO_2 thin films with superhydrophilicity. *Surface and Coatings Technology* 265: 78–82.

20 Chen, P., Hu, Y., and Wei, C. (2012). Preparation of superhydrophilic mesoporous SiO_2 thin films. *Applied Surface Science* 258 (10): 4334–4338.

21 Zhou, G., He, J., and Xu, L. (2013). Antifogging antireflective coatings on Fresnel lenses by integrating solid and mesoporous silica nanoparticles. *Microporous and Mesoporous Materials* 176: 41–47.

22 Park, T., Kim, J.H., and Lee, D. (2014). Excellent anti-fogging dye-sensitized solar cells based on superhydrophilic nanoparticle coatings. *Nanoscale* 6 (13): 7362–7368.

23 Lu, X., Wang, Z., Yang, X. et al. (2011). Antifogging and antireflective silica film and its application on solar modules. *Surface and Coatings Technology* 206 (6): 1490–1494.

24 Thompson, C., Fleming, R., and Zou, M. (2013). Transparent self-cleaning and antifogging silica nanoparticle films. *Solar Energy Materials & Solar Cells* 115: 108–113.

25 Wu, G., Yang, Y., Lei, Y. et al. (2020). Hydrophilic nano-SiO_2/PVA-based coating with durable antifogging properties. *Journal of Coatings Technology and Research* 17 (5): 1145–1155.

26 Zhang, X., Lan, P., Lu, Y. et al. (2014). Multifunctional antireflection coatings based on novel hollow silica-silica nanocomposites. *ACS Applied Materials and Interfaces* 6 (3): 1415–1423.

27 Wang, Y., Li, X., Li, M. et al. (2014). Preparation of silica nanoparticles catalysed by small organic tertiary amine and applications in surface coating with antireflective/antifogging properties. *Journal of Sol-Gel Science and Technology* 70 (1): 1–5.

28 Chen, J., Zhang, L., Zeng, Z. et al. (2016). Facile fabrication of antifogging, antireflective, and self-cleaning transparent silica thin coatings. *Colloids and Surfaces A: Physicochemical and Engineering Aspects* 509: 149–157.

29 Sarkar, S. and Pradhan, S.K. (2014). Tailoring of optical and wetting properties of sputter deposited silica thin films by glancing angle deposition. *Applied Surface Science* 290: 509–513.

30 Xi, R., Wang, Y., Wang, X. et al. (2020). Ultrafine nano-TiO_2 loaded on dendritic porous silica nanoparticles for robust transparent antifogging self-cleaning nanocoatings. *Ceramics International* 46 (15): 23651–23661.

31 Wang, J., Wang, D., Wang, J. et al. (2011). High transmittance and superhydrophilicity of porous TiO_2/SiO_2 bi-layer films without UV irradiation. *Surface and Coatings Technology* 205 (12): 3596–3599.

32 Fardo, F., Ribeiro, R., Strauss, J. et al. (2020). Double layer SiO_2-TiO_2 sol-gel thin films on glass for antireflection, antifogging, and UV recoverable self-cleaning. *Applied Optics* 59 (25): 7720–7725.

33 Li, X. and He, J. (2013). Synthesis of raspberry-like SiO_2-TiO_2 nanoparticles toward antireflective and self-cleaning coatings. *ACS Applied Materials and Interfaces* 5 (11): 5282–5290.

34 Yuan, J., Yan, S., Xiao, Z., and Hu, J. (2020). Multifunctional closed-pore silica/titania composite antireflective coatings based on self-assembly process. *Materials Chemistry and Physics* 250: 123023.

35 Tao, C. and Zhang, L. (2020). Fabrication of multifunctional closed-surface SiO_2-TiO_2 antireflective thin films. *Colloids and Surfaces A: Physicochemical and Engineering Aspects* 585: 124045.

36 Faustini, M., Nicole, L., Boissiere, C. et al. (2010). Hydrophobic, antireflective, self-cleaning, and antifogging sol-gel coatings: an example of multifunctional nanostructured materials for photovoltaic cells. *Chemistry of Materials* 22 (15): 4406–4413.

37 Eshaghi, A., Aghaei, A.A., Zabolian, H. et al. (2013). Transparent super-hydrophilic SiO_2/TiO_2/SiO_2 tri-layer nanostructured antifogging thin film. *Ceramics-Silikáty* 57 (3): 210–214.

38 Xu, L., He, J., and Yao, L. (2014). Fabrication of mechanically robust films with high transmittance and durable superhydrophilicity by precursor-derived one-step growth and post-treatment. *Journal of Materials Chemistry A* 2 (2): 402–409.

39 Saxena, N., Naik, T., and Paria, S. (2017). Organization of SiO_2 and TiO_2 nanoparticles into fractal patterns on glass surface for the generation of superhydrophilicity. *Journal of Physical Chemistry C* 121 (4): 2428–2436.

40 Soklič, A., Tasbihi, M., Kete, M., and Štangar, U.L. (2015). Deposition and possible influence of a self-cleaning thin TiO_2/SiO_2 film on a photovoltaic module efficiency. *Catalysis Today* 252: 54–60.

41 Xue, M., Peng, N., Ou, J. et al. (2015). Hierarchically nanostructured CeO_2 films with superhydrophilicity and corrosion resistance by coupling of surface topography and oxygen vacancies. *Materials Chemistry and Physics* 160: 406–412.

42 Yuan, W., Lu, Z., and Li, C.M. (2013). Self-assembling microsized materials to fabricate multifunctional hierarchical nanostructures on macroscale substrates. *Journal of Materials Chemistry A* 1 (21): 6416–6424.

43 Peng, B., Tan, L., Chen, D. et al. (2011). Programming surface morphology of TiO_2 hollow spheres and their superhydrophilic films. *ACS Applied Materials & Interfaces* 4 (1): 96–101.

44 You, J.-H., Lee, B.-I., Lee, J. et al. (2011). Superhydrophilic and antireflective $La(OH)_3$/SiO_2-nanorod/nanosphere films. *Journal of Colloid and Interface Science* 354 (1): 373–379.

45 Yao, L. and He, J. (2013). Multifunctional surfaces with outstanding mechanical stability on glass substrates by simple H2SiF6-based vapor etching. *Langmuir* 29 (9): 3089–3096.

46 Park, H.K., Yoon, S.W., Chung, W.W. et al. (2013). Fabrication and characterization of large-scale multifunctional transparent ITO nanorod films. *Journal of Materials Chemistry A* 1 (19): 5860–5867.

47 Jin, Y., Qamar, A., Shi, Y., and Wang, P. (2018). Preferential water condensation on superhydrophobic nano-cones array. *Applied Physics Letters* 113 (21): 211601.

48 Yoon, J., Ryu, M., Kim, H. et al. (2020). Wet-style superhydrophobic antifogging coatings for optical sensors. *Advanced Materials* 32 (34): 2002710.

49 Chen, X., Li, B., Hou, K. et al. (2019). Microflower-decorated superhydrophobic copper surface for dry condensation. *Langmuir* 35 (49): 16275–16280.

50 Boreyko, J.B. and Collier, C.P. (2013). Delayed frost growth on jumping-drop superhydrophobic surfaces. *ACS Nano* 7 (2): 1618–1627.

51 Leese, H., Bhurtun, V., Lee, K.P., and Mattia, D. (2013). Wetting behaviour of hydrophilic and hydrophobic nanostructured porous anodic alumina. *Colloids and Surfaces A: Physicochemical and Engineering Aspects* 420: 53–58.

52 Duran, I. and Laroche, G. (2019). Water drop-surface interactions as the basis for the design of anti-fogging surfaces: theory, practice, and applications trends. *Advances in Colloid and Interface Science* 263: 68–94.

53 Wen, M., Wang, L., Zhang, M. et al. (2014). Antifogging and icing-delay properties of composite micro- and nanostructured surfaces. *ACS Applied Materials and Interfaces* 6 (6): 3963–3968.

54 Shi, W., Wang, L., Guo, Z., and Zheng, Y. (2015). Excellent anti-icing abilities of optimal micropillar arrays with nanohairs. *Advanced Materials Interfaces* 2 (18): 1500352.

55 He, M., Zhou, X., Zeng, X. et al. (2012). Hierarchically structured porous aluminum surfaces for high-efficient removal of condensed water. *Soft Matter* 8 (25): 6680–6683.

56 Zhang, Q., He, M., Chen, J. et al. (2013). Anti-icing surfaces based on enhanced self-propelled jumping of condensed water microdroplets. *Chemical Communications* 49 (40): 4516–4518.

57 He, M., Zhang, Q., Zeng, X. et al. (2013). Hierarchical porous surface for efficiently controlling microdroplets' self-removal. *Advanced Materials* 25 (16): 2291–2295.

58 Parker, A.R. and Lawrence, C.R. (2001). Water capture by a desert beetle. *Nature* 414 (6859): 33–34.

59 Thickett, S.C., Neto, C., and Harris, A.T. (2011). Biomimetic surface coatings for atmospheric water capture prepared by dewetting of polymer films. *Advanced Materials* 23 (32): 3718–3722.

60 He, M., Ding, Y., Chen, J., and Song, Y. (2016). Spontaneous uphill movement and self-removal of condensates on hierarchical tower-like arrays. *ACS Nano* 10 (10): 9456–9462.

61 Han, Z., Mu, Z., Li, B. et al. (2016). Active antifogging property of monolayer SiO_2 film with bioinspired multiscale hierarchical pagoda structures. *ACS Nano* 10 (9): 8591–8602.

62 Chung, K., Yu, S., Heo, C.J. et al. (2012). Flexible, angle-independent, structural color reflectors inspired by morpho butterfly wings. *Advanced Materials* 24 (18): 2375–2379.

63 Dou, X., Zhang, D., Feng, C., and Jiang, L. (2015). Bioinspired hierarchical surface structures with tunable wettability for regulating bacteria adhesion. *ACS Nano* 9 (11): 10664–10672.

64 Sambles, J. (2012). Biophotonics: blue butterflies feel the heat. *Nature Photonics* 6 (3): 141.

65 Li, P., Zhang, B., Zhao, H. et al. (2018). Unidirectional droplet transport on the biofabricated butterfly wing. *Langmuir* 34 (41): 12482–12487.

66 Niu, S., Li, B., Mu, Z. et al. (2015). Excellent structure-based multifunction of Morpho butterfly wings: a review. *Journal of Bionic Engineering* 12 (2): 170–189.

67 Han, Z., Guan, H., Cao, Y. et al. (2014). Antifogging properties and mechanism of micron structure in Ephemera pictiventris McLachlan compound eyes. *Chinese Science Bulletin* 59 (17): 2039–2044.

68 Han, Z., Feng, X., Jiao, Z. et al. (2018). Bio-inspired antifogging PDMS coupled micro-pillared superhydrophobic arrays and SiO_2 coatings. *RSC Advances* 8 (47): 26497–26505.

69 Lv, C., Zhang, X., Niu, F. et al. (2017). From initial nucleation to Cassie-Baxter state of condensed droplets on nanotextured superhydrophobic surfaces. *Scientific Reports* 7: 42752.

70 Pan, Z., Zhang, W., Kowalski, A., and Zhao, B. (2015). Oleophobicity of biomimetic micropatterned surface and its effect on the adhesion of frozen oil. *Langmuir* 31 (36): 9901–9910.

71 Zou, X., Tao, C., Yang, K. et al. (2018). Rational design and fabrication of highly transparent, flexible, and thermally stable superhydrophobic coatings from raspberry-like hollow silica nanoparticles. *Applied Surface Science* 440: 700–711.

72 Yong, J., Bian, H., Yang, Q. et al. (2020). Mini-review on bioinspired superwetting microlens array and compound eye. *Frontiers in Chemistry* 8: 57578.

73 Lv, C., Hao, P., Zhang, X., and He, F. (2015). Dewetting transitions of dropwise condensation on nanotexture-enhanced superhydrophobic surfaces. *ACS Nano* 9 (12): 12311–12319.

74 Cao, M., Guo, D., Yu, C. et al. (2015). Water-repellent properties of superhydrophobic and lubricant-infused "slippery" surfaces: a brief study on the functions and applications. *ACS Applied Materials and Interfaces* 8 (6): 3615–3623.

75 Aldhaleai, A., Khan, F., Thundat, T., and Tsai, P. (2020). Evaporation dynamics of water droplets on superhydrophobic nanograss surfaces. *International Journal of Heat and Mass Transfer* 160: 120149.

76 Wen, R., Lan, Z., Peng, B. et al. (2017). Wetting transition of condensed droplets on nanostructured superhydrophobic surfaces: coordination of surface properties and condensing conditions. *ACS Applied Materials and Interfaces* 9 (15): 13770–13777.

6

Structural Color Surfaces Inspired by Butterfly Wings

6.1 Definition of Structural Color Surfaces

There are two main sources of color formation in nature: pigmentation and structural color. Pigmentation is a very common way for plants and animals to produce color. It is caused by absorption and reflection of certain chemicals by certain chemicals. Therefore, the color of a pigment is also known as chemical color. Structural color is a typical physical color, which has long been a research hotspot for scientists. Structural color is generated by the interactions of interference, diffraction, reflection, and scattering between incident light (IL) and regularities and irregularities in photonic nanostructures. This mechanism of interaction generates brilliant colors in nature [1–6]. In fact, the research of structural colors has a long history. However, with the discovery of new structures and the revealing of color formation mechanism, the ancient research topic of structural colors has maintained strong vitality and a vigorous development trend. In particular, the exploration and research on biological structural colors in nature has never stopped, and new discoveries are constantly emerging. Structural colors have many attractive advantages such as high brightness, high saturation, no fading, and iridescent phenomenon with angular discoloration as well as polarization effect. They cannot be available in common pigment colors. Most of structural colors generally originate from different kinds of photonic nanostructures of natural creatures including tail feathers of peacocks [7, 8], butterfly wings [9, 10], and beetles [11, 12], during long evolutionary processes. These natural structurally colored materials are drawing huge attention in various research fields [13–15]. Of all these natural photonic structured surfaces forming structural colors, perhaps butterfly wings are the most famous because of their fine multiscale hierarchical structures (MHSs). The MHS-based structural colors play an important role in many butterfly behaviors including camouflage, courtship display, and aposematic coloration. In the past decades, bioinspired photonic materials with fascinating structures that show variable structural colors have been developed in the world [16–18]. In particular, the MHS-based biomimetic material design mimicking the butterfly wing scales has drawn wide attention and has become a research hotspot [19–22]. In some butterflies, chitin materials and air layer are two important components of the wing multilayer system. The alternating superposition arrangement among them is the main factor in forming surface structural

Nature-Inspired Structured Functional Surfaces: Design, Fabrication, Characterization, and Applications, First Edition. Zhiwu Han.

colors [16]. Moreover, the fine structures of butterfly wing scales are different based on species and gender. Even wings of the same kind of butterfly will have different surface structures. Similarly, the surface of these scales also present various structures in different seasons. The research of photonic structures on wings of butterflies began in 1999. A very constructive report has shown the photonic-crystal-type structures in the scales of the *Morpho rhetenor* butterfly [23]. Subsequently, a series of blue-colored wing scales of the *Morpho* butterflies and other kinds of butterflies have been carefully explored [24–26]. Furthermore, numerous realistic optical models have been reported to explain the mechanism of the structural colors of butterfly wings in detail [27–29]. Inspired by these photonic structures of different kinds of butterfly wings, many bioinspired materials or functional surfaces for structural colors were achieved through different fabrication techniques.

In this section, structurally colored surfaces of *Papilio palinurus* butterfly and the *Morpho* butterflies are first described in detail. Next, the mechanism of the structural colors of these butterfly wings is briefly discussed. The biomimetic materials inspired by different kinds of butterflies and their fabrication techniques are presented in detail. Then, the current applications and the potential value of structurally colored materials in security labels, optical displays, anti-counterfeiting, optical displays, sensors, and bioinspired soft robots are also briefly discussed. Finally, the remaining grand challenges and future development trends of bioinspired structurally colored materials are described.

6.2 Structural Color Surfaces on Butterfly Wings

6.2.1 *Papilio palinurus* Butterfly Wings

6.2.1.1 Multiscale Hierarchical Structures of Original Butterfly Wings

A rare large swallowtail butterfly *Papilio palinurus* lives in the tropical rainforests around the world and it belongs to the family *Papilionidae* [30]. This butterfly can be identified by its unique biological characteristics [31]. As shown in Figure 6.1a, the wingspan of a male butterfly is about 6.6 cm. Some crescent-like green scale areas are orderly located on both fore and hind wings. Shiny rectangle patches in vivid green color between two neighboring nervures can be easily found. Moreover, the biological morphologies and mix-colors (RGB, a general color model) of the original wing scales are characterized by a three-dimensional (3D) ultra-depth stereoscopic microscope. Figure 6.1b demonstrates that the oval wing scales are well aligned on the wing surface along the veins, which resemble tiles on a roof. All these scales on the wings are similar in shape and size. For a single scale, the length and width are about 122 and 61 μm, respectively. Interestingly, there are countless shiny tiny points on each scale, which are distributed across the scale surface along the length. These microscale scale arrays combine action to endow the entire butterfly wings with more vitality on a macroscale. This allows them to be directly observed by naked eyes. To investigate the optical performance properties of the original wing scales, the reflectance spectrum of the green scale regions is characterized by a miniature

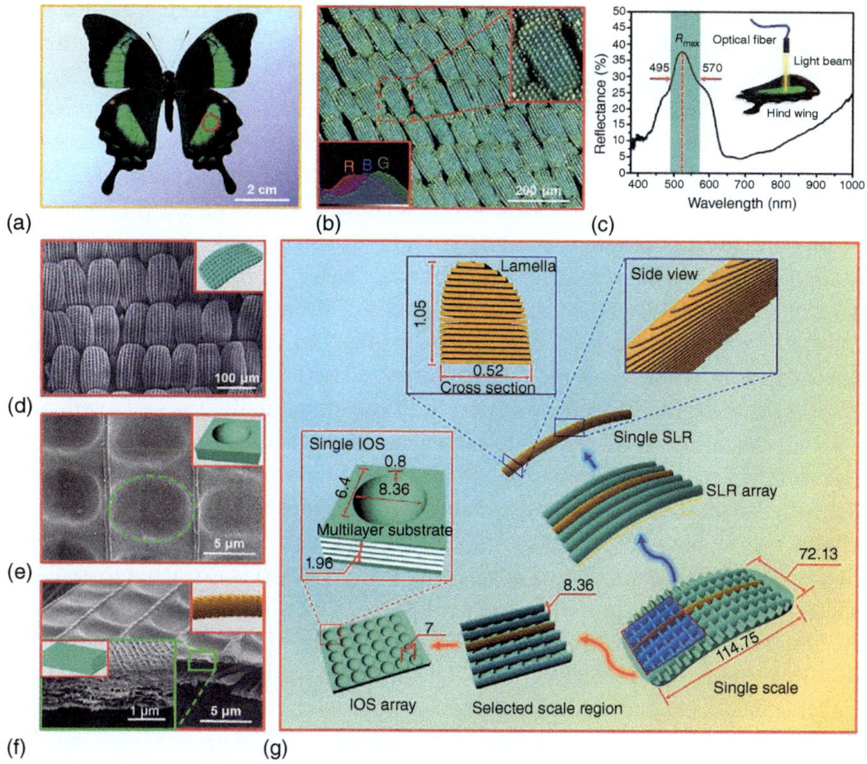

Figure 6.1 (a) An intact male *Papilio palinurus* butterfly showing brilliant axisymmetric green scale areas. (b) Three-dimensional ultra-depth stereoscopic microscope image of the green scale area and its RGB distribution. (c) The measurement of the reflection spectroscopy of the green scale region at normal incidence. (d) Scale array consisting of multitudinous leaf-like scales (inset) that arrange in order. (e) Oval inverse opal-like structures (inset) and narrow stacked lamellar ridges that constitute the primary structures in a single scale. (f) Cross section of the single scale exhibiting the parallel stacked lamellar ridges (inset) and inverse opal-like structures with multilayer (inset). (g) Schematic illustration of structural models presenting different characteristics of structures of the single scale: oval inverse opal-like structure, multilayer substrate, and narrow stacked lamellar ridge. The unit of measurement: μm. Source: Reprinted with the permission from Han et al. [31]. Copyright 2017, American Chemical Society.

fiberoptic spectrometer. As shown in Figure 6.1c, a strong feature peak of maximum reflectance of approximately 38.5% appears at 526 nm. This is consistent with the green color waveband (495–570 nm).

To explore the underlying structural basis of structural colors of the wing, surface structures of a single scale are observed through the ultrahigh-resolution field emission scanning electron microscope (FESEM). Figure 6.1d shows that large numbers of dense inverse opal-like pits cover the whole scale surface. And the density of these pits reaches about 140 per scale. To an extent, the areas of these pits seem to be closer to the Gaussian distribution. Significantly, these tile-like scales are spontaneously distributed in alternate rows and overlap each other. For each single scale, approximately nine ridges can be found along the length. Figure 6.1e indicates that the rim of the pits is in an oval shape, which is distributed between two neighboring ridges. Additionally, exquisite secondary structures of pits and ridges in a single scale are proposed in detail. The tightly arranged grids are located on the bottom and jambs of the pits. The cross section of the cut-off scale shows that the ridges consist of overlapped narrow lamellae. They are exactly tangent to the oval scales. The concave that consists of multiple layers is approximately 1 μm. Besides, a thin air layer exists between the adjacent layers as shown in Figure 6.1f.

Figure 6.1g displays 3D visual structural models based on the FESEM analysis to clearly explain the MHSs of the original wing scale. The entire single scale is slightly curved for forming an arc, which is approximately 114.75 μm long and 72.13 μm wide. Based on the single-scale model and combined with the actual structural characterization results, the secondary structure in the single scale was extracted, and then 3D visualization structure model was also established. On the one hand, the side view of the single ridge selecting from the curved array of the ridge displays overlapped lamellae that resemble shutters. From the cross section of the single ridge, the multilayer system consists of alternating lamella layers and air layers. On the other hand, a relatively complanate area on the single scale model is extracted to analyze the spatial distribution information and feature size data of the secondary structure.

After removing the stacked lamellar ridge (SLR) array from the single scale, the array of inverse opal-like structure (IOS) is completely exposed. Results show that the spacing between two adjacent ridges is approximately 8.36 μm. The center distance between two adjacent IOSs measured along the length is approximately 7.00 μm. To fully evaluate the IOS array, the single IOS is enlarged and its 3D structural characteristic parameters are presented. The measured results show that the long axis and short axis of the ellipse are 8.36 and 6.40 μm, respectively, and the depth of the single IOS reaches approximately 800 nm. Besides, the bottom of the pit is a multilayer structure composed of alternating layers of chitin and air, with a thickness of approximately 2.03 μm. Obviously, the simple primary structures and delicate secondary structures on the wavy scale base surface together constitute the structure system of the single scale. Therefore, the fundamental system architecture of original wing scales is both hierarchical and multiscale.

6.2.1.2 MHS-Based Structural Colors of Original Butterfly Wings

To verify the color type of the green scale regions of the original wings, a simple discoloration experiment using ethyl alcohol (EtOH, A. R.) was performed. First, the bright green scale regions of the wings gradually became dim. After that, as the entire wing was gradually immersed into the EtOH, the green scale regions turned into dark brown. Finally, they regained initial bright green color after drying in the air for a few seconds. Pigments can usually be easily decomposed into common organic solvents. In these circumstances, the result shows that the green color is determined by structural color, not pigment. That is to say, the bright green color is produced by the interactions between the visible light and MHSs of butterfly wing scales, which further testify that the butterfly wing scales can manipulate light with their exquisite MHSs.

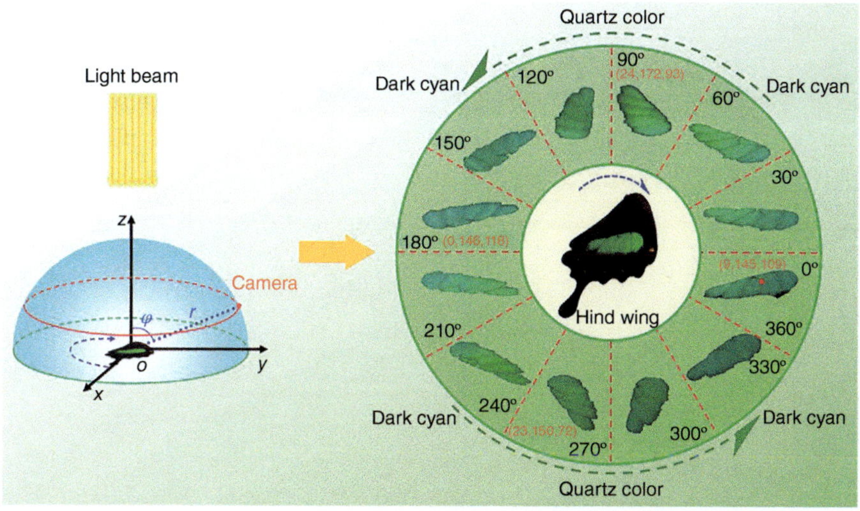

Figure 6.2 A miniature camera was placed at a fixed position (r, φ, 90°) in an observation hemisphere system. Here, r represents the radius of the sphere, φ represents the included angle between the segment and the z-axis, and 90° is azimuth. The left image shows that the parallel light beam illuminates the selected scale region at normal incidence. The right image displays that the hind wing in the inner circle rotates in the clockwise at an interval of 30°. And the corresponding external separated sectors show the variations in color of the hind wing at the same interval within 360°. Source: Reprinted with the permission from Han et al. [31]. Copyright 2017, American Chemical Society.

Besides, the omnidirectional reflective property of original wing scales was carried out through a self-built spherical color observation system. It aims to investigate the relationship between optical performances of butterfly wing scales and different orientations. In the experiment, the reflection characteristic of the original wing scale is explored on a two-dimensional (2D) plane. As shown in Figure 6.2, when the semi-automatic turret rotates from 0° to 180°, the selected area first varies from dark cyan to quartz color, then from 180° to 360°, the color gradually returns to dark cyan. The changes in color are repeated over the period of 180°. When the difference in rotation angle is about one cycle (180°), the RGB values of the two specific positions are similar. A slight change in color of the symmetric observation point is found for the RGB value at 0° versus 180°, and 90° versus 270°. From the above analysis, although the color intensities vary slightly with the hind wing rotating on the x–o–y plane, the color type almost retains itself as an initial green state. This suggests that the bright green color has reflection stability on a 2D plane.

6.2.1.3 Omnidirectional Reflection Property of Butterfly Wings

When a butterfly is flying, its wings cannot always remain horizontal on the 2D plane. We assume that the original wing scale can still maintain reflection stability when expanding from 2D to 3D, meaning all-around stability in the space domain. Here, a clever optical experiment is performed to confirm the omnidirectional reflective self-stable (ORS) characteristics of the original butterfly wing scales. Under the

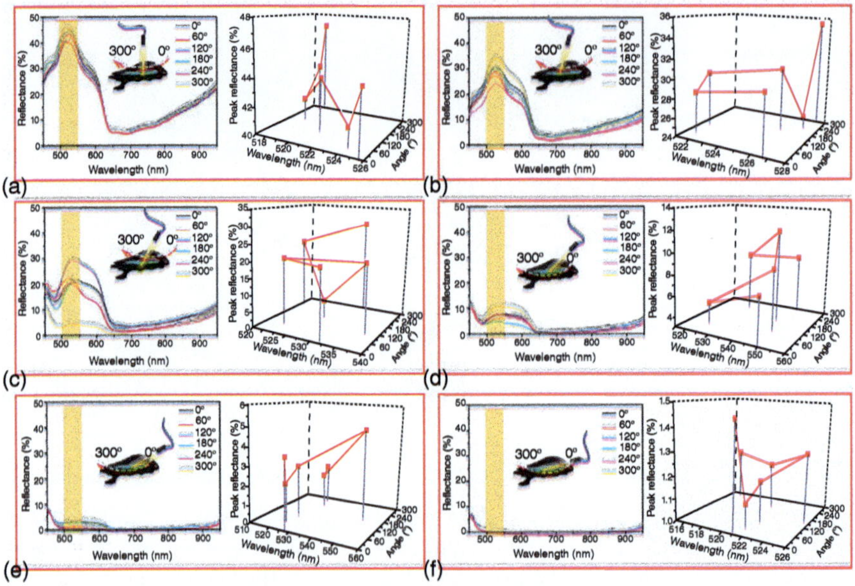

Figure 6.3 Measured reflectance spectra of the ORS characteristics possessing the variable angles at different incident angles: (a) 0°, (b) 15°, (c) 30°, (d) 45°, (e) 60°, and (f) 75°. In terms of each incident angle, six different reflectance curves are presented by varying rotational angles of turntable from 0 to 360° at intervals of 60°, showing the change tendency of the reflectance peaks at different rotation angles for each incident angle. Source: Reprinted with the permission from Han et al. [31]. Copyright 2017, American Chemical Society.

circumstances, the incident angles also vary frequently as the reflectance of the fixed incident angle changes as the turntable rotates. As shown in Figure 6.3a–f (left), a series of reflectance spectrum are obtained by varying the angle of light incidence from 0° to 75° at an interval of 15°, and varying the rotation angle of the turret from 0 to 300° at an interval of 60°. The results suggest that these reflection spectra show almost a similar single-peak tendency, indicating that they are independent of the variation of incident angles and rotation angles of the turntable, and the characteristic peaks always tend to appear at the waveband of green color. By comparison analysis, the characteristic peak intensities of the reflectance spectrum of the original wing scale vary from the lowest 1.8% to the highest 47.5%. In addition, the peak reflectance of the original wing scale possessing variable incident angles and rotation angles is also attained, as shown in Figure 6.3a–f (right). The reflection intensity changes significantly as the angle of rotation varies from 0° to 300°. However, the reflectance peaks shift slightly at the range of 520–550 nm, which is a typical green waveband (495–570 nm). And the characteristic peak positions are basically stable at around 526 nm. This is in agreement with the reflectance spectrum of the original wing scales at an initial angle. Based on these findings, it follows that the original wing scales maintain the ORS performance firmly, regardless of how the angles of incidence and rotation vary in 3D space. Based on previous relevant studies [32, 33], it is speculated that the optical interaction between light and the structural surfaces of original butterfly wings may be the main reason for the formation of novel ORS characteristics.

6.2.1.4 Mechanism Analysis of the ORS Characteristics

On the basis of the previous discussion, we believe that natural MHSs may endow the butterfly wings with ORS properties. To plenarily explore the underlying formation mechanism of the ORS characteristics, the physical interactions between light and the single scale with fine and complex structures of SLR arrays and IOS arrays are schematically elaborated. As shown in Figure 6.4a, for multi-wavelength natural light (NL) incidence, the selected scale region of the original wing exhibits highly selective reflection, resulting in direct observation of natural green structural colors. Actually, according to the previous microscopic characterization of the micro-/nanostructure of green wing scales, the natural MHSs of the scales can be divided into two important basic optical components, that is, the multilayer IOSs and periodic SLRs. Figure 6.4b shows a detailed spatial position relationship between IOS and SLR on the cross section of the selected scale region.

On the one hand, before the multi-wavelength incident light interacts with the bottom and the edge of the IOS, they will first pass through the raised ridges. It

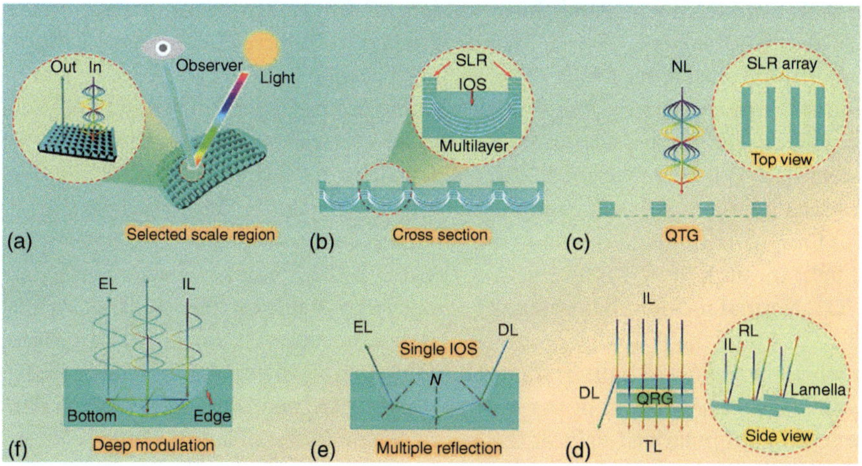

Figure 6.4 (a) The selected scale region of an original wing scale can convert incident multi-wavelength light to a special wavelength light received by the observer, which serves as a light filter that causes a highly color selective. (b) Cross section of the selected scale region showing its basic structural components: the periodic SLRs and multilayer IOSs (inset). (c) The natural light (NL) interacts with the SLR array that acts as a QTG from the top view. (d) Overlapped lamellae serving as the QRG can manipulate the incident light (IL) and divided them into the reflected light (RL) on the surface of lamella, and the transmission light (TL) in the multilayer, as well as the diffraction light (DL) beside the SLR array. (e) When DL from the nearby SLR array gets into the single IOS, the multiple reflection produces on the inner surface of the IOS. (f) In terms of the normal IL, deep modulation effect can also form on the inner surface of the IOS, which results in a special exit light (EL). And it can significantly turn the multi-wavelength IL into the special EL. Source: Reprinted with the permission from Han et al. [31]. Copyright 2017, American Chemical Society.

is interesting to note that these periodic SLRs generate a ridge array from the top view, which actually constitutes a quasi-transmission grating (QTG). As shown in Figure 6.4c, the SLR array serves as a multi-slit diffraction screen. From a side view, more interestingly, the overlapped lamellae of the single ridge constitute a quasi-reflection grating (QRG) (Figure 6.4d). Because both QTG and QRG can change the amplitude of the incident light, they are also called "amplitude grating," which is a kind of diffraction grating. Actually, the multi-slit diffraction gratings are usually based on the Fraunhofer diffraction theory. The relationship between the grating structure, the diffraction angle, and the incident light can be quantitatively expressed by the following formula:

$$d \sin \theta = n\lambda$$

$$d = a + b \tag{6.1}$$

where d is the grating constant, indicating the spacing between two adjacent slits, a represents the width of the slit, b represents the width of light-tight shield, n represents the diffraction series that could be any positive integer, and λ represents the wavelength of the incident light.

In general, d represents the physical quantity at the nanoscale. However, the spacing between two adjacent ridges is approximately 8.36 μm, which is much wider than d. It can effectively avoid the possibility of diffraction on the surface of the scales to a certain extent. Most of the incident light is reflected by the surface of the lamella, and some pass through the alternating layers of the chitin–air, and others beside the ridge is diffracted by the QTG to enter the nearby pits. In general, the incident light can be more easily diffracted by the QTG when the wavelength of the incident light is infinitely close to the width of light-tight shield ($\lambda \to b$). Then, the diffracting light is multi-reflected on the surface of the pit before it escapes the pit in the form of exiting light, as shown in Figure 6.4e. On the other hand, most natural light through the QTG interacts directly with the bottom and edge surface of the IOS. Normal incident light entering the bowl bottom is directly reflected, while the one reaching the IOS edge produces double reflection (Figure 6.4f). It is proved that when the polarized light irradiates the IOS surface, the double reflection in the pit will lead to the rotation of the polarized light. The main reason is the geometrical rotation of polarized light at the interface due to the plane reflections. Thus, the ORS characteristics of original wing scales can be attributed to the interaction of these integrated optical effects. We also believe that the MHSs establish the basis of the novel ORS characteristics physically.

6.2.2 *Morpho* Butterfly Wings

6.2.2.1 Structural Features of Original *Morpho* Butterfly Wings

Morpho butterflies displaying brilliant blue color are widely known for their typical structural color in the most representative animals [34]. Interestingly, this natural color is closely related to the hierarchical ultrastructure of the wing scales [35]. Therefore, it is more necessary to describe the topographical features of the *Morpho* butterflies before carefully explaining the various structural colors. The typical

Figure 6.5 Typical male *Morpho* butterflies. Source: Reprinted with the permission from Kinoshita et al. [36]. Copyright 2008, IOP Publishing.

male *Morpho* butterflies, which belong to the family *Nymphalidae* on the basis of recent classification, are presented in Figure 6.5 [36]. For most of the male *Morpho* butterflies, their dorsal wings always exhibit brilliant blue, while the ventral sides display usually dark brown. More interestingly, the female *Morpho* butterflies usually emit little or no light. Therefore, for the blue coloring in the male, three biological explanations have been summarized. In other words, these structural colors result from the long-term adaptation environment for the butterfly wings [37]: (i) The shining blue can pass a mating signal to a female butterfly of the same species inside. For example, the color of the butterfly wings would become brighter and more sparkling during the breeding season, thereby attracting the opposite sex for mating. In particular, it has been confirmed that a piece of blue metal paper could attract male *Morpho* butterflies [36]. (ii) The second one is to frighten predators such as birds by giving strong blinking flashes to make them feel dizzy. For example, some eye-like structures distributed on the surface of butterfly wings would scare birds and other animals against various attacks, which gives a territorial purpose against other males. (iii) The third is mimicry to protect themselves. For instance, patterns and colors of *Kallima inachis* butterfly wings closely resemble the withered leaves, which helps to escape from predators. The butterfly scales also have excellent heat absorption capacity, which can regulate their body temperature.

The most noticeable characteristic of the *Morpho* butterfly is that the wing color is viewing-angle dependent. For example, when one observes a specimen of the *Morpho* butterfly that is placed horizontally from the direction perpendicular to the scale, they can notice a strong blue color. However, under the premise that the viewing direction remains unchanged, the initial color of the scale does not change when the viewing-angle is tilted, but the wing color suddenly changes to violet or dark blue as the viewing-angle becomes large enough. On the contrary, when one views from the direction parallel to the wing veins, the blue color will disappear and turn into black (Figure 6.6e–f) [36]. Therefore, the viewing direction plays an important role in the color change for butterfly wings. Another typical

Figure 6.6 The change in color of the *Morpho didius* butterfly wings observed when they are placed at different angles and in different mediums. (a) Color of the *Morpho didius* butterfly wing observed from the frontal view in air. (b) Color of the *Morpho didius* butterfly wing observed from the frontal view in liquid ethanol. (c) Color of the *Morpho didius* wing observed from an oblique view in air. (d) Color of the *Morpho didius* wing observed from an oblique view in liquid ethanol. The change in color of the *Morpho cypris* wing observed from the direction (e) perpendicular and (f) parallel to the wing veins. Source: Reprinted with the permission from Kinoshita et al. [36]. Copyright 2008, IOP Publishing.

characteristic is that the color changes when the wings are immersed in different liquids. For example, when the wings are added to an alcohol, the color changes to green possessing slightly dull shining and it displays blue instead of violet from the oblique view, as shown in Figure 6.6a–d [36]. It suggests that the color change depends on the refractive index of the liquid used. Herein, it is worth emphasizing that the reason for not using water for testing is that the wing surface has nonwetting properties.

The fine architectures producing structural colors should also be expounded after clearly describing the morphological features of the *Morpho* butterflies. Structurally, the *Morpho* butterfly wings contain two types of chitin scales. These tile-like scales distribute orderly on the dorsal and ventral wing surfaces. The basal scales distribute on the surfaces of wing lamina, on which cover scales overlay. The main ultrastructure on both scales contains periodically arranged ridges and shelves. Normally, the shining blue color of the wing scales mainly arises from the interaction of light reflection and interference between the shelves and incident light [36, 38]. Actually, some researchers have further elaborated on the detailed regular and irregular structural features of the butterfly scales [13]. Based on the analysis of most previous studies, it can be confirmed that the reflected color is mainly attributed to the cover scales. But there are some exceptions in this regard. For some *Morpho* species, for example, the dorsal wing color is the result of the combined action of the basal and cover scales [23, 39]. For the *Morpho* butterfly wings, the structures on the scale surface show complex variability in morphologies and optical multifunction. For example, these patterns present a state of the reflectors with a multilayer system. In addition, it also works as the diffraction gratings. As shown in Figure 6.7, a kind of typical multilayer structure in the scales of *Morpho* butterflies wings is presented.

Figure 6.7 (a) Digital image of the *Morpho* butterfly. Source: Reprinted with the permission from Watanabe et al. [38]. Copyright 2005, The Japan Society of Applied Physics. (b) Optical image of the scales locating on the magnified region of *Morpho rhetenor* butterfly wing. Scale bar: 100 μm. Source: Reprinted with the permission from Vukusic et al. [23]. Copyright 1999, Royal Society. (c) Scanning electron microscope (SEM) image showing the cross section structure of a single scale. (d) Schematic of the optical mechanisms in the ridge array of multilayer structure system. Source: Reprinted with permission from Yu et al. [40]. Copyright 2013, Elsevier.

This fine structure is often considered the "Christmas-tree" multilayer structures at the nanoscale (Figure 6.7c). In general, iridescence is closely related to angles. Therefore, one can observe a sharp iridescent color over a relatively small range of angles. The *Morpho* butterflies, however, have strong and wide-angle iridescence owing to the existence of well-designed microstructures. The strong blue iridescence could display over the wide-angular range. This is another fascinating optical property of the *Morpho* butterfly wing scales.

It is well known that both the colors of the cover and basal scales of butterfly wings possess the features of orientation dependence. Figure 6.8 displays an optical photograph of the *Morpho peleides* butterfly wings, which presents the surface colors and appearances of different scales observed from different directions. Comparing the results of different spectra of the original wing and the cover scales, it shows that the cover scales play a major role in producing the structural color [29]. But the mechanism of various functions of the cover and basal scales are still controversial. Meanwhile, most related research provides a reference for the direct use of butterflies as biological templates to create smart photonic structures [29].

Figure 6.8 (a) An optical photograph of a male *Morpho peleides* butterfly. (b) An image of the wing scales that are observed by the microscope at the vertical direction. (c–f) Optical photographs of the cover and ground wing scales that are observed under low and high magnifications at the vertical direction, respectively. (g and h) Optical photographs of the cover and ground wing scales that are illuminated from different directions. Interestingly, it suggests that surface colors of the wing scales change with the illuminating direction. Source: Reprinted with the permission from Ding et al. [29]. Copyright 2009, AIP Publishing.

6.2.2.2 Mechanism of the Structural Color of Original *Morpho* Butterfly Wings

6.2.2.2.1 Structural Parameters of the Original Morpho Butterfly Wings

By carefully analyzing the structural color observed on some creatures including peacock feathers and butterfly wings, effective interaction between various parameters can be inferred. The brilliant colors are produced by the joint action between the periodically arranged 3D surface structures, chemistry, and the photonic bandgap, with structural parameters as much as the wavelength of visible light [41]. 1D photonic structure that is observed in *Morpho* butterflies arises from the periodical arrangement of cuticles. The wing of male *Morpho didius* contains a slender cover scale and an overlapping basal scale, both of which are consisting of several ridges. It shows a pitch of approximately 0.7–1.4 μm and possesses a lamellar structure with 6–8 alternate layers of cuticle and air. The basal scale has a pigment, imparting brown color to the wings. For the *Morpho rhetenor* butterfly, some of the scales on the wings gradually degenerate into small flakes, and the others possess ridges with a spacing of about 0.7–0.8 μm. The wavelength of reflected light (RL) is attributed to the multilayer interference at the cuticle–air interfaces and relies on the refractive index of the cuticle and the periodical spacing [42, 43].

As shown in Figure 6.9, the *Morpho menelaus* butterfly has a wingspan of approximately 12 cm. The ridges are covered on both cover and basal scales, which are arranged longitudinally on each scale. The distance between two adjacent ridges is approximately 2 μm, and the ridge has a thickness of about 0.8 μm. Moreover, the cover scales have five periods of the chitin and air layers with the parameters of 190 nm. In addition, the *M. menelaus* butterfly has the ridges with randomly arranged height, resembling the typical species *M. didius* [5]. In short, the suggested "Christmas-tree" nanostructures of the chitin and air layers on the ridges form a discrete multilayer structure system. Because the characteristic parameters of single ridges and random height are in the micron scale, the diffraction interaction between fine structures and incident light is caused. Thus, the blue color of the wings is clearly displayed in a wide range of angle by diffraction [23]. Besides, the structural colors of the *Morpho* butterfly also depend on the number of chitin layers of the ridges [23]. The more layers of the ridges, the more obvious the blue intensity. Significantly, the appearance of the wings is affected by structural parameters. For example, if the reflectance spectral intensity or angular anisotropy of the reflected light becomes larger, the wings will have more glossy appearance [39].

6.2.2.2.2 Angle Dependence of the Original Morpho Butterfly Wings

The angular dependence is one typical feature for the *Morpho* butterfly wings, which results from the effects of the multilayer interference and grating diffraction. Figure 6.9 shows the cross-scale ultrastructure in the scales of the *Morpho* butterfly wing. And the simplified schematic of the angular dependence feature is presented in Figure 6.7. The angle-dependent features can be described with an ideal model of the multilayer structure. It comprises two alternately superimposed layers of chitin and air, which has a constant thickness and a different refractive index. It is

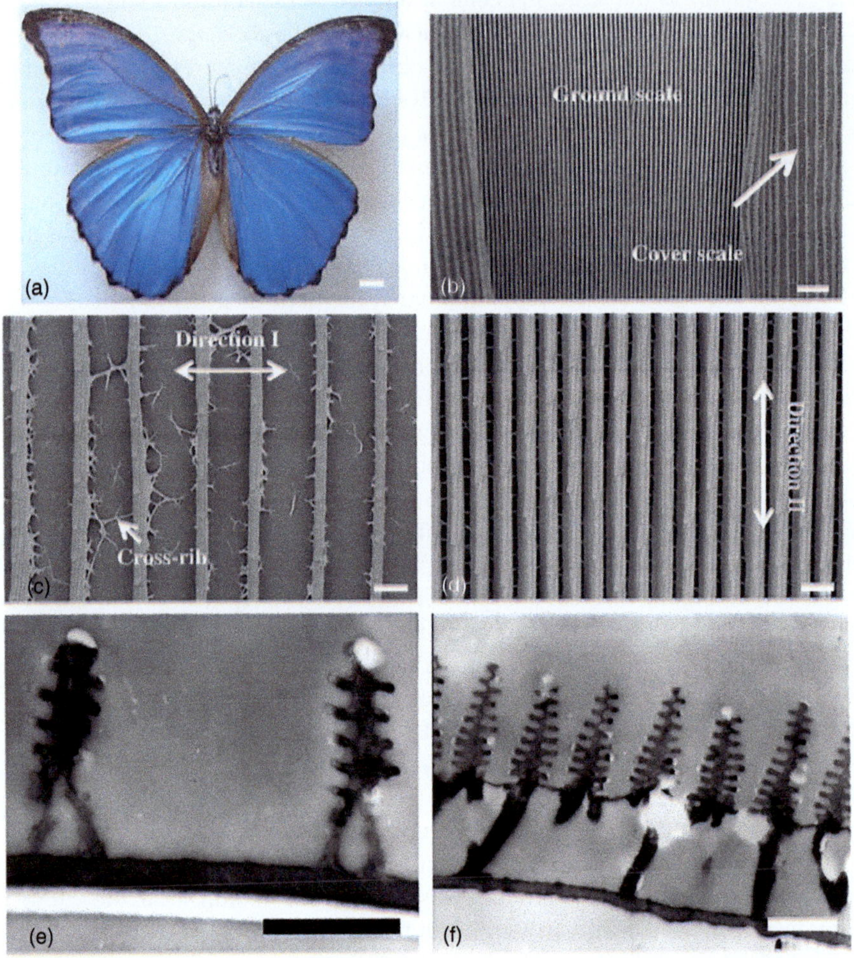

Figure 6.9 (a) Photograph of the *Morpho menelaus* butterfly. (b) SEM image of the wing scales: the basal scale and cover scale. The close-up SEM image of (c) the cover scale and (d) the basal scale from the top view. Transmission electron microscope (TEM) image of transverse cross section of (e) the cover scale and (f) the basal scale. Source: Reprinted with the permission from Liu et al. [44]. Copyright 2011, Elsevier.

well known that optical interference occurs when the thickness of the film material approximates a quarter wavelength of the incident light. This makes the narrow band reflectance of the wings become sharp. In particular, as the angle of incidence increases, the optical length will gradually increase, leading to a blueshift of the reflection peak [40].

In general, both the grating and multilayer structures produce high angular dependence. But the structural color of the *Morpho* butterflies is less dependent on angle. In recent years, the mechanism of the structure-based color features has been carefully summarized [45]. The models based on the ultrastructure of *Morpho* butterfly wings, which explain the mechanism of the structure-based color, have experienced several stages. Firstly, a simple multilayer structure is built up [45]. After that, it turns into the distribution of dielectric multilayers [23]. In the end, a more sophisticated model with a tree-like structure is established [46, 47].

6.2.2.2.3 Isotropic Reflection of the Original Morpho Butterfly Wings

In addition to the above two features, isotropic reflection is a main feature of *Morpho* butterfly wing to achieve structural color. As shown in Figure 6.10, the directions of reflected light and transmitted light on the same plane can better illustrate the isotropic reflection mechanism. When the factors leading to angular broadening are not considered, the intensity ratios of the reflected light from three different directions reach 0.14, 0.54, and 0.03, respectively. Herein, the reflectance spectral intensity is separated into three different directions with an angular separation of 16°. This distribution may be employed to partially interpret the directional ridge-top at the angular dependence [39].

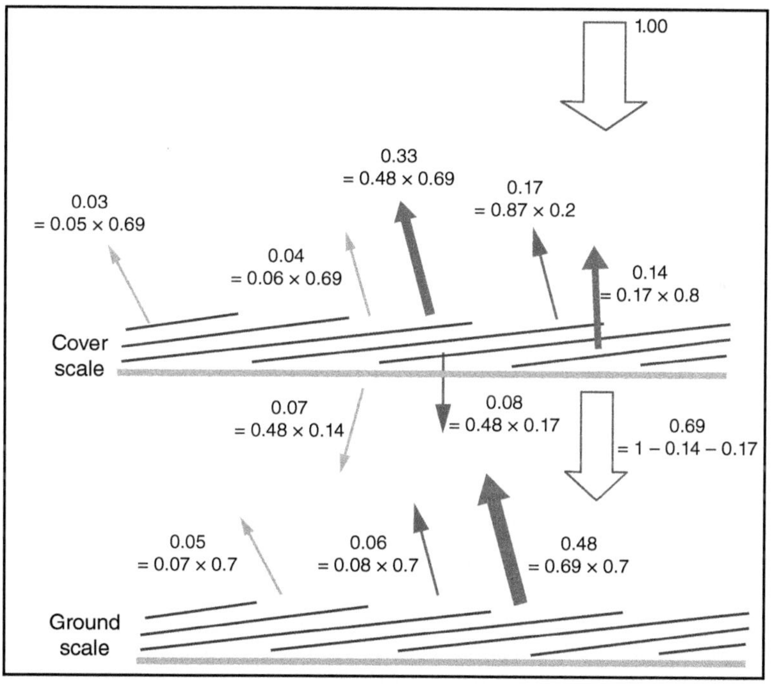

Figure 6.10 A schematic diagram of the isotropic reflection feature of the *Morpho* butterfly wings. It displays the direction and intensity of the reflected and transmitted light. The reflectance spectral intensity divided into three different directions are 0.14, 0.54, and 0.03 respectively. And the rest of the light is considered to be transmitted light. Source: Reprinted with the permission from Yoshioka and Kinoshita [39]. Copyright 2004, Royal Society.

6.3 Fabrication of the Butterfly-Inspired Structure Color Materials

Increasing numbers of detailed investigations demonstrating the microstructures of butterfly wings, especially structural colors, have inspired numerous materials scientists to fabricate structurally colored bioinspired materials. Models and mechanisms of butterfly wings have enhanced our understanding of fine structures, guiding the design and preparation of advanced functional materials. In this section, recent progress in the preparation of novel functional materials inspired by wing scales by using advanced approaches has been summarized with particular emphasis on the sol–gel process.

6.3.1 Sol–Gel Process

The sol–gel method is widely used in bio-templating and bioinspired fabrication [48]. Sol–gel process is mainly divided into the following processes. Firstly, the raw material including an inorganic substance or a metal alkoxide in a liquid state is uniformly mixed and then made into a precursor solution. During the synthesis of the precursor solution, a series of hydrolysis reactions and polymerization reactions will occur. After these two reactions, the precursor solution will gradually become a stable state, in which the precursor solution has changed to a transparent colloidal system. After that, polymerization occurs between the colloidal particles present in the colloidal system after aging, forming a gel of a 3D network structure. Finally, after the gel is dried, sintered, solidified, etc., molecular or nanostructured materials are prepared. Among them, the sol is a colloidal system with liquid characteristics, and the dispersed particles are solid or macromolecules with diameters of 1–100 nm. The sol–gel method helps to regulate the structure of the materials at the nanoscale, starting from the initial stage of processing. It is very suitable for fabricating the sorts of microstructures displayed in butterfly wings, due to its flexibility, facile shape control, and effectiveness at low cost.

Inspired by the MHSs found in the *Papilio palinurus* butterfly wing scales, the SiO_2-based bioinspired color reflectors (BCRs) have been prepared by using the sol–gel method and a selective etching processing [31]. As shown in Figure 6.11, simplified 3D structural models are established to elaborate the key steps of the preparation process of the bioinspired optical film with structural colors in detail.

Taking into account that the existing machining accuracy of mechanical manufacturing cannot meet the bionic manufacturing of natural complex 3D structures, the original wing scales are chosen as the initial bio-template to design and manufacture the bionic structural color surface, as shown in Figure 6.11a. To improve the mechanical strength of the initial bio-template and remove the possible impurities filled in the hierarchical structures, organic solvents such as EtOH or diethyl ether are employed to immerse the selected wing slice for a few minutes (Figure 6.11b). The precursor solution, which is a mixture solution of tetraethyl orthosilicate (TEOS) and EtOH with excellent liquidity, is used to immerse and fill the MHSs of the selected wing slice after the sample is naturally dried in the

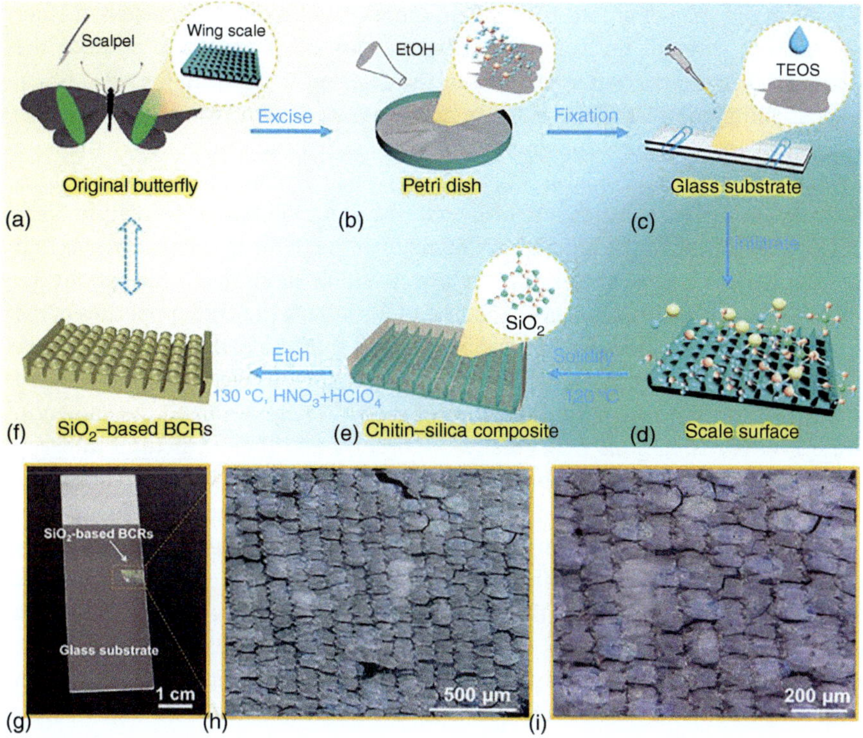

Figure 6.11 (a) A selected wing scale as a bio-template was cut from the original butterfly. (b) The bio-template was completely soaked into an organic solvent. (c) TEOS was poured into the gap between two glass substrates, in which the bio-template was properly fixed. (d) The sol–gel reaction process occurred in the surface of the bio-template with MHSs. (e) SiO_2-based film forming along the MHSs after solidification at 120 °C. (f) SiO_2-based BCRs were achieved after removing the original butterfly wing by etching. (g) A digital photograph of the SiO_2-based BCRs possessing brilliant structural colors. (h and i) Stereoscopic microscope images of the obtained BCRs that resemble overlapped tiles on a roof at low and medium magnifications. Source: Reprinted with the permission from Han et al. [31]. Copyright 2017, American Chemical Society.

air (Figure 6.11c). After that, the sol reaction occurs on the surface of wing scales (Figure 6.11d). Then, it is heated in the drying baker to gradually solidify the sol that is filled in the MHSs to achieve the SiO_2-based 3D networks (Figure 6.11e). Actually, the 3D networks are responsible for the framework formation of the BCRs, which contains a two-step hydrolytic polymerization reaction of TEOS in the precursor solution. Herein, the formation process of the mechanism of the 3D networks is briefly interpreted. Under acidic or at low pH conditions, TEOS undergoes the hydrolysis reaction to form hydrated silica [$(EtO)_2Si(OH)_2$] and EtOH at first. At the same time, TEOS undergoes a condensation reaction with $(EtO)_2Si(OH)_2$ to generate the colloid mixture consisting of a transparent and homogeneous suspension. Next, a condensation reaction between $(EtO)_2Si(OH)_2$ and SiO_2 clusters occurs, and cross-linked SiO_2 clusters and H_2O are produced. The aforementioned chemical reactions play a crucial role in the formation of the main body of the sol–gel processing. At last, these cross-linked SiO_2 clusters further linked together through the polymerization reaction of oligomers to produce the SiO_2-based 3D networks. After removal of the selected wing slice, the SiO_2-based BCRs possessing complex structures are achieved, as shown in Figure 6.11f. The SiO_2-based BCRs on a glass substrate display brilliant structural colors, which are structure-based at the macro level (Figure 6.11g). Furthermore, the dynamic images of the BCRs on a glass substrate are also achieved at an interval of 30° within 360°. It is found that although the reflectance intensity of the BCRs is not as strong as that of the selected wing slice, they exhibit relatively stable light green throughout the whole rotation period. This indicates that the BCRs inherit the ORS properties of the original bio-template. Moreover, it can be inferred that the SiO_2-based BCRs may also be multiscale and hierarchical, according to the obvious view structures. Based on this, the morphologies and dimensions of the BCRs are observed. As shown in Figure 6.11h, the squamous BCRs distribute orderly and their structural integrity is also completely maintained, showing that the BCRs inherit well the squamous structures from the bio-template. Figure 6.11i indicates the structural appearance of the BCR amplification, which consist of abundant, neatly arranged artificial scales. It is interesting that all these artificial scales also show brilliant structural colors at the micro level. It further proves the SiO_2-based BCRs are endowed with the MHSs to a great extent.

As shown in Figure 6.12a,b, a plenty of bump structures that look like smooth colorful soap bubbles are distributed on the BCR surface, which is arranged with a regular array. Actually, most of the serried textures are covered on each single bump surface, originating from the fine hierarchical structures inside the original pits of the bio-template. Besides, there are lots of groove structures with folded textures on both sides of the bump structures (Figure 6.12c). In addition to the main groove structures, it can be clearly seen that the microgrooves are parallel to the main groove at about 45° (Figure 6.12d). To further explain the overall morphology of the BCRs, 3D visual models were built to briefly show the structural characteristics of a single BCR observed from three different views (Figure 6.12e). These feature structures are formed by the original MHSs that are filled with the SiO_2 sol during the biomimetic manufacturing process. In short, the hierarchical structures including bump structures and groove structures on the glass substrate jointly constitute the main body of the SiO_2-based BCRs.

By the finite-difference time-domain (FDTD) method, the optical characteristics of the SiO_2-based BCRs are investigated. FDTD method is a good solution to solve optical problems and can simulate and analyze the interaction between light and structural surfaces [49]. A simplified optical model of the BCRs was employed for local reflectance simulations through the FDTD method. Furthermore, the bright structural colors directly originate from the bump structures instead of the groove structures. Hence, the bump structure is chosen as the simulation object. Then, the simulated light reflectance at three different positions is attained, which is a function of wavelength (Figure 4.12f–h). Results show that BCRs have the simulated maximal reflectance values of around 30% near 520 nm, which are in good agreement with the experimental result of the original wing scale. Also, it

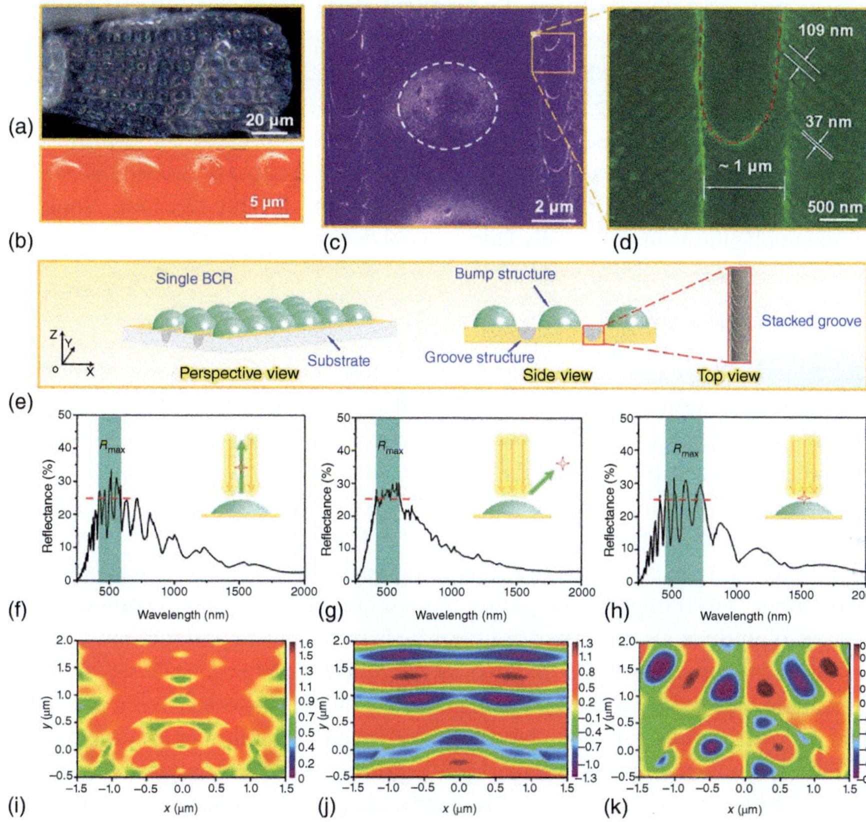

Figure 6.12 (a) A single BCR with biomimetic structures showing brilliant structural colors. (b) Highly ordered array of the bump structure. (c) Single bump structure possessing the oval shape. (d) Groove structure possessing the regular patterns. (e) A schematic diagram of two typical periodic structures including the bump structure and groove structure in the SiO_2-based BCR. The simulated reflectance of special monitoring sites near a single bump structure of the BCR: (f) paralleling to the incident light (0°), (g) making 45° with the incident light, and (h) at the bump surface. (i) The electric field (*E*) profile of the bump structure at $z = 0$ plane when the plane wave light gets into the structure from top to bottom. (j) X-component of the electric field of the BCR (E_x). (k) Y-component of the electric field of the BCR (E_y). Source: Reprinted with the permission from Han et al. [31]. Copyright 2017, American Chemical Society.

can provide further insights into the optical interaction between incident light (350–1000 nm) and MHSs for the electric field profiles at selected wavelengths [50]. The simulated results show that the intensive electric field enhancement occurs around the surface of the bump structures (Figure 4.12i). In other words, these optimal structures produce strong light reflectance in a wide space. Moreover, after incident light interacts with them at normal incidence, the changes of dynamic electric field also show significant light–matter effect produced near the bump structures. In particular, the obvious warps of the electric field are found when the incident light draws near the bump structures (Figure 4.12j). Because of the symmetry of the bump structures, the excellent space symmetry can be observed from the intensity of the electric field, as shown in Figure 4.12k. From this discussion, it can be confirmed that both fine MHSs and novel ORS characteristics of the natural butterfly wing are well inherited by the SiO_2-based BCRs. Thus, the bioinspired SiO_2-based BCRs prepared using the sol–gel process could provide a promising strategy to duplicate complex 3D structured surfaces possessing ultrafine MHSs. Furthermore, novel bioinspired materials can also broaden the range of potential applications including displays, optical devices, and so on.

6.3.2 Soft Lithography

Most advanced approaches including semiconductor microfabrication, soft lithography, or transfer printing are one of the most fundamental and effective process strategies to achieve industrially viable high-end manufacturing and creating low-cost structures at the nano–microscale. To some extent, we are very skilled to create functional patterns on planar surfaces using these sequential operations, which allows us to easily create a series of 2D patterns possessing resolutions of a few nanometers. In addition to the ability to replicate the aforementioned exquisite 2D structures, the strategy of integrating these sequential processing methods even allows us to produce 3D complex structures by layer-by-layer construction, such as functional structured materials inspired by the *Morpho* butterfly scale [38, 51, 52]. For example, inspired by the *Morpho* butterfly, bionic materials with photonic crystal microstructures have been fabricated by using the soft lithography technique [53]. The results confirmed that the morphology and optical properties of the bionic replica inherit those of the original butterfly wing with a reasonable degree of accuracy. However, the drawback of this replica is that the ridge characteristics exhibit a greater width than that of the original scale, and the scales consist of fewer ridge structures. Although this method has the advantages of simplicity and low cost, it requires corresponding master molds involving complicated manufacturing processes [54]. Besides, there are still significant limitations to these integration methods. For example, a reasonable processing sequentiality and a good balance are required between resolution, structural complexity, high-throughput large-area processing and restrictions are formed in interwoven hierarchical and 3D material structures [55].

6.3.3 Layer Deposition Techniques

Layer deposition technology is a common combination method used in the manufacture of 3D refined complex structures in recent years. It mainly includes colloidal self-assembly, sputtering, and atomic layer deposition (ALD) techniques. These combined approaches provide greater efficiency, versatility, and scalability for the preparation of 3D micro-/nanoscale structures, particularly those based on structural colors inspired by butterfly wings. It is well known that the complex structures found on the butterfly wing scales are difficult to replicate, which is particularly challenging. As researchers continue to explore, bionic replicas were successfully prepared by using the ALD technique employing the *Papilio blumei* butterfly wing scales as a template [56]. However, this method has a certain drawback that it may not be suitable for accurate fabrication of the periodically shaped inner multilayer architectures over a large-area surface. To achieve precise manufacturing of this complex 3D structure, the photonic structures that imitate the color mixing effect discovered on the *P. blumei* butterfly wings have been fabricated by the combination of layer deposition techniques including colloidal self-assembly, sputtering, and ALD [32]. Practically speaking, the replication processes of the multilayer architecture of the wing scales involve five simple steps. Firstly, the polystyrene colloids having a diameter of 5 μm were deposited on a gold-plated silicon substrate to form regularly arranged concavities in an appropriate size. Then, a 2.5 μm-thick layer of platinum or gold, which was equal to the radius of the microsphere, was grown into the interstices of the colloidal array by using the electroplating approach to generate a negative replica. Next, the sample immersed into dimethylformamide or acetone was subjected to ultrasonic oscillations to remove the colloids from the substrate, which formed a template of metal concavities arranged in the hexagonal shape. Subsequently, the thin carbon film with a thickness of about 20 nm was sputtered onto the gold surface. Finally, a conformal multilayer of thin quarter-wave TiO_2 and Al_2O_3 layers was grown by using the ALD technique. Significantly, a thin carbon film between the gold (or platinum) and the multilayer stack was employed to adsorb the light passing through the multilayer stack. It can decrease the specular reflections and unwanted destructive interferences, which will severely limit the optical performance of the mimicking structures. Thus, the square-centimeter-sized bioinspired samples were achieved by means of a facile and scalable process. Importantly, the square-centimeter-sized structures and pictures with microscale resolution can be encoded in photonic structures. The method can be used in a wide range of applications, such as security labels or dynamic manufacture, vivid paints and coatings. Also, such an amazing effect may provide valuable preferences for research on other patterned insects and should be further explored in future works [32].

6.3.4 Electron Beam Lithography

In recent years, nanolithography, including extreme UV lithography, electron beam lithography, and nanoimprint lithography, has been the basis for the fabrication of nanostructures. It is also the prerequisite for research in applications such as nanophotonic crystals, metamaterials, structural colors, and nanodevices [57–60]. Recently, by mimicking structural colors caused by multilayer nanophotonic architectures found in *Morpho* butterfly wing scales, Zhang et al. [61] fabricated polymethyl methacrylate (PMMA) films with 3D lamellae layers using the top-down electron beam lithography combined with an alternating development/dissolution process of PMMA/lift-off resist (LOR) multilayers. This method overcomes the shortcoming of electron beam lithography that is difficult to achieve 3D nanostructures, which successfully achieves artificial butterfly wings and their brilliant color effects. It is widely believed that the blue color of the *Morphs* butterfly wings shows multiple reflections from the multilayered membrane structure of the Christmas tree. To simulate these structures of butterfly wings, this team successfully prepared the PMMA/air structures. The color characteristics of the PMMA/air structures were explored by angular resolution spectroscope. The results show that the difference in refractive index between the PMMA/LOR and PMMA/air structures is a key factor in successfully simulating the color display of butterfly wings. Furthermore, in the case of $0°–30°$ incident light, the blue band of the sample possesses a distinct strong reflection peak. The blue reflection peak remains constant in the range of $\pm 10°$, which has better angular independence than the original butterfly wings. This is caused by the bioinspired lamellae layer structures with strong regularity. In addition, the green color can be produced by adjusting the thickness of the PMMA/air structure (15 layers). Compared to the blue sample, the green sample can keep the color unchanged at a wider angle. Therefore, the successful application of nanolithography in the bioinspired 3D layered structures not only allows us to deepen the understanding of the mysterious blue mechanism of the *Morph* butterfly wings but also breaks the bottlenecks in technological development. This technology will be widely used in the field of gas/liquid sensors, coloring media, 3D metamaterials, infrared imaging devices, etc., and provide an important reference value for practical applications.

6.4 Applications of the Structural Color Materials

Structural colors in nature play a vital role in industrial manufacturing and human life. Particularly, structurally colored materials inspired by butterfly wings have been increasingly attracting in recent decades. These bioinspired materials are usually applied for various applications, including security labels [32], bioinspired soft robotics [62], anti-counterfeiting [63, 64], optical display [65, 66], sensors [67, 68], accessories, and furniture decoration.

For example, anti-counterfeiting is a practical technique to identify the authenticity of a marked object, which is critical in many concerns such as economic security, property safety, and human health. However, in practical applications, the development of anti-counterfeit labels is a very complicated process; meanwhile, it is also highly cost-effective process for commodity applications. Fortunately, among many anti-counterfeiting materials, structurally colored materials are well known for their advantages including easy-to-identify and stable optical signals and are widely used. Besides, these structurally colored materials can be designed on different substrates by selecting any patterns, which has the advantages such as relatively low cost and simple operation. The above characteristics, which are combined with their outstanding reaction capabilities, enable structural color materials to be largely used in anti-counterfeiting applications. Very recently, the biomimetic structurally colored polyvinylidene fluoride (PVDF) films with a bilayer inverse heterostructure (BLIHS) mimicking the *Morpho* butterfly wing have been successfully prepared by using template approach [69]. Particularly, the bioinspired PVDF BLIHS is composed of the ordered array layers inverse structure (OALIS) and the quasi-amorphous array layers inverse structure (Q-AALIS). It exhibits different colors under the combined action of these two structures. On the one hand, the film could selectively present iridescent structural color due to the Bragg diffraction of OALIS. On the other hand, the film color could change to non-iridescent color by rotating the angle of the sample, resulting from the effect of Q-AALIS. The biomimetic design of this novel photonic structure makes it easy to switch between hiding and displaying patterns by transforming the viewing angle. This further provides a new technical approach for the practical application of anti-counterfeiting materials.

In addition, structurally colored sensor materials inspired by butterfly wings have drawn attention of numerous researchers in recent years. These bioinspired sensor materials can read the responsive color shifts to further identify or quantify a certain parameter [70, 71]. For example, the robotic tactile flexible sensors inspired by the structural colors found on the *M. menelaus* butterfly wing have been presented by using the transfer technique [67]. Most periodical parallel grooves with structural colors were fabricated on the surfaces of polydimethylsiloxane (PDMS) film to conveniently capture the diffracted light by a charged-coupled device (CCD) camera. This can effectively predict the position and magnitude of the contact force resulting from a change in the diffraction patterns. It suggests that this tactile flexible sensor shows high sensitivity in contact force and resolution. Moreover, the feasibility of the prototype as a tactile flexible sensor was further verified by a practical application on the UR-5 manipulator. Thus, the tactile sensing approach with bioinspired structures can be widely used in the field of robotics through miniaturizing the design in future research work.

Overall, inspired by natural biological materials with bright structural colors, plenty of artificial structurally colored materials with refined structure arrays have been prepared by using various advanced approaches in the past few decades. Moreover, great achievements have been done including the mechanism of interactions between light and fine structures found in creatures, the preparation of bioinspired structure-based color materials, and exploration of their applications in a wide range of fields. These groundbreaking achievements have revolutionized the development of structurally colored materials and have endowed them with amazing performance. Thus, the structural color materials, especially for mimicking butterfly wings, provide great contribution to the development of coloration-mediated man–machine interaction systems and devices, which have been confirmed in practical applications, such as in anti-counterfeiting, security labels, optical displays, sensors, bioinspired soft robots. In this chapter, recent developments in some typical butterfly wings with structural colors caused by the multilayer optical structures were described in detail. Next, several advanced fabrication approaches and color performances of bioinspired structurally colored materials were highlighted. Finally, the current applications of structurally colored materials in many promising applications such as security labels, optical displays, anti-counterfeiting, optical displays, sensors, and bioinspired soft robots were also briefly presented. By exploring recent progress in structurally colored materials especially for the materials inspired by butterfly wings, both opportunities and challenges exist. Some challenges are related to the fabrication of structurally colored materials and their commercial applications. First, although many fabrication processes including combining multiple preparation techniques have been proposed to construct structurally colored materials, they are complicated or expensive, which make them hard to mass produced over a large scale. Second, in addition to a single structural color feature, versatile features including optical, sensing, and wetting properties can be integrated for biomimetic materials.

As scientific research continues to deepen, opportunities also follow. The following research directions may attract more attention in the near future. Firstly, preferable optical structures of butterfly wings in nature with structural color characteristics should be further investigated or discovered in future work to provide a perfect natural template for structurally colored surfaces. Secondly, it is still necessary to prepare structurally colored materials with properties of regulating light propagation, which is the most important and irreplaceable property, on the basis of which other functional properties including wettability and abrasion resistance can also be introduced. Finally, it is necessary to propose facile, cost-effective, and large-scale industrial preparation methods to fabricate structurally colored materials. Particularly, with the development of 3D and 4D printing techniques, we believe these advanced methods will be an effective means to fabricate perfect structurally colored materials in upcoming research.

References

1 Vukusic, P.J., Sambles, J.R., and Lawrence, C.R. (2000). Structural colour–colour mixing in wing scales of a butterfly. *Nature* 404 (6777): 457–457.

2 Mcdonald, L.T., Narayanan, S., Sandy, A. et al. (2020). Brilliant angle-independent structural colours preserved in weevil scales from the Swiss Pleistocene. *Biology Letters* 16 (4): 20200063.

3 Vukusic, P. and Sambles, J.R. (2003). Photonic structures in biology. *Nature* 424 (6950): 852–855.

4 Thayer, R.C., Allen, F.I., and Patel, N.H. (2020). Structural color in Junonia butterflies evolves by tuning scale lamina thickness. *eLife* 9: e52187.

5 Kinoshita, S., Yoshioka, S., and Kawagoe, K. (2002). Mechanisms of structural colour in the morpho butterfly: cooperation of regularity and irregularity in an iridescent scale. *Proceedings Biological Sciences* 269 (1499): 1417–1421.

6 Parker, A.R., Welch, V.L., Driver, D., and Martini, N. (2003). Opal analogue discovered in a weevil. *Nature* 426 (6968): 786–787.

7 Zi, J., Yu, X., Li, Y. et al. (2003). Coloration strategies in peacock feathers. *Proceedings of the National Academy of Sciences of the United States of America* 100 (22): 12576–12578.

8 Li, Y., Lu, Z., Yin, H. et al. (2005). Structural origin of the brown color of barbules in male peacock tail feathers. *Physical Review E* 72 (1): 010902.

9 Liu, X., Wang, D., Yang, Z. et al. (2019). Bright silver brilliancy from irregular microstructures in butterfly *Curetis acuta* Moore. *Advanced Optical Materials* 7 (18): 1900687.

10 Niu, S., Li, B., Ye, J. et al. (2016). Angle-dependent discoloration structures in wing scales of *Morpho menelaus* butterfly. *Science China Technological Sciences* 59 (5): 749–755.

11 Onelli, O.D., van de Kamp, T., Skepper, J.N. et al. (2017). Development of structural colour in leaf beetles. *Scientific Reports* 7 (1): 1373.

12 Kilchoer, C., Pirih, P., Steiner, U., and Wilts, B.D. (2019). Diffusive structural colour in *Hoplia argentea*. *Journal of Experimental Biology* 222 (24): jeb213306.

13 Kinoshita, S. and Yoshioka, S. (2005). Structural colors in nature: the role of regularity and irregularity in the structure. *ChemPhysChem* 6 (8): 1442–1459.

14 Wang, Y., Shang, L.R., Chen, G.P. et al. (2020). Bioinspired structural color patch with anisotropic surface adhesion. *Science Advance* 6 (4): eaax8258.

15 Scarangella, A., Soldan, V., and Mitov, M. (2020). Biomimetic design of iridescent insect cuticles with tailored, self-organized cholesteric patterns. *Nature Communications* 11 (1): 1–10.

16 Zhao, Y., Xie, Z., Gu, H. et al. (2012). Bio-inspired variable structural color materials. *Chemical Society Reviews* 41 (8): 3297–3317.

17 Braun, P.V. (2011). Colour without colourants. *Nature* 472 (7344): 423.

18 Kolle, M., Lethbridge, A., Kreysing, M. et al. (2013). Bio-inspired band-gap tunable elastic optical multilayer fibers. *Advanced Materials* 25 (15): 2239–2245.

19 Lou, S., Guo, X.M., Fan, T.X., and Zhang, D. (2012). Butterflies: inspiration for solar cells and sunlight water-splitting catalysts. *Energy Environmental Science* 5 (11): 9195–9216.

20 Zhao, Q., Guo, X., Fan, T. et al. (2011). Art of blackness in butterfly wings as natural solar collector. *Soft Matter* 7 (24): 11433–11439.

21 Han, Z., Niu, S., Li, W., and Ren, L. (2013). Preparation of bionic nanostructures from butterfly wings and their low reflectivity of ultraviolet. *Applied Physics Letters* 102 (23): 233702.

22 Han, Z., Mu, Z., Li, B. et al. (2016). Active antifogging property of monolayer SiO_2 film with bio-inspired multiscale hierarchical pagoda structures. *ACS Nano* 10 (9): 8591–8602.

23 Vukusic, P., Sambles, J.R., Lawrence, C.R., and Wootton, R.J. (1999). Quantified interference and diffraction in single *Morpho* butterfly scales. *Proceedings of the Royal Society B: Biological Sciences* 266 (1427): 1403–1411.

24 Siddique, R.H., Diewald, S., Leuthold, J., and Hendrik, H. (2013). Theoretical and experimental analysis of the structural pattern responsible for the iridescence of *Morpho* butterflies. *Optics Express* 21 (12): 14351–14361.

25 Prum, R.O., Quinn, T., and Torres, R.H. (2006). Anatomically diverse butterfly scales all produce structural colours by coherent scattering. *Journal of Experimental Biology* 209 (4): 748–765.

26 Wong, T.H., Gupta, M.C., Robins, B., and Levendusky, T.L. (2004). Color generation in butterfly wings and fabrication of such structures. *Optics Letters* 28 (23): 2342–2344.

27 Mejdoubi, A., Andraud, C., Berthier, S. et al. (2013). Finite element modeling of the radiative properties of *Morpho* butterfly wing scales. *Physical Review E* 87 (2): 022705.

28 Steindorfer, M.A., Schmidt, V., Belegratis, M. et al. (2012). Detailed simulation of structural color generation inspired by the *Morpho* butterfly. *Optics Express* 20 (19): 21485–21494.

29 Ding, Y., Xu, S., Wang, Z.L. et al. (2009). Structural colors from *Morpho peleides* butterfly wing scales. *Journal of Applied Physics* 106 (7): 074702.

30 Wang, W., Zhang, W., Fang, X. et al. (2014). Demonstration of higher colour response with ambient refractive index in *Papilio blumei* as compared to *Morpho rhetenor*. *Scientific Reports* 4: 5591.

31 Han, Z., Mu, Z., Li, B. et al. (2017). Bioinspired omnidirectional self-stable reflectors with multiscale hierarchical structures. *ACS Applied Materials & Interfaces* 9 (34): 29285–29294.

32 Kolle, M., Salgard-cunha, P.M., Scherer, M.R.J. et al. (2010). Mimicking the colourful wing scale structure of the *Papilio blumei* butterfly. *Nature Nanotechnology* 5 (7): 511–515.

33 Diao, Y.Y. and Liu, X.Y. (2011). Mysterious coloring: structural origin of color mixing for two breeds of *Papilio* butterflies. *Optics Express* 19 (10): 2325–2331.

34 Smith, G.S. (2009). Structural color of *Morpho* butterflies. *American Journal of Physical Anthropology* 77 (11): 1010–1019.

35 Jiang, X., Shi, T., Zuo, H. et al. (2012). Investigation on color variation of *Morpho* butterfly wings hierarchical structure based on PCA. *Science China Technological Sciences* 55 (1): 16–21.

36 Kinoshita, S., Yoshioka, S., and Miyazaki, J. (2008). Physics of structural colors. *Reports on Progress in Physics* 71 (7): 076401.

37 Niu, S., Li, B., Mu, Z. et al. (2015). Excellent structure-based multifunction of *Morpho* butterfly wings: a review. *Journal of Bionic Engineering* 12 (2): 170–189.

38 Watanabe, K., Hoshino, T., Kanda, K. et al. (2005). Brilliant blue observation from a *Morpho*-butterfly-scale quasi-structure. *Japanese Journal of Applied Physics* 44 (1–7): L48–L50.

39 Yoshioka, S. and Kinoshita, S. (2004). Wavelength-selective and anisotropic light-diffusing scale on the wing of the *Morpho* butterfly. *Proceedings of the Royal Society of London Series B Biological Science* 271 (1539): 581–587.

40 Yu, K., Fan, T., Lou, S., and Zhang, D. (2013). Biomimetic optical materials: integration of nature's design for manipulation of light. *Progress in Materials Science* 58 (6): 825–873.

41 Kinoshita, S., Yoshioka, S., Fujii, Y., and Okamoto, N. (2002). Photophysics of structural color in the *Morpho* butterflies. *Forma* 17 (2): 103–121.

42 Bruzzone, A.A.G., Costa, H.L., Lonardo, P.M., and Lucca, D.A. (2008). Advances in engineered surfaces for functional performance. *CIRP Annals* 57 (2): 750–769.

43 Malshe, A., Rajurkar, K., Samant, A. et al. (2013). Bio-inspired functional surfaces for advanced applications. *CIRP Annals* 62 (2): 607–628.

44 Liu, F., Liu, Y., Huang, L. et al. (2011). Replication of homologous optical and hydrophobic features by templating wings of butter flies *Morpho menelaus*. *Optics Communications* 284 (9): 2376–2381.

45 Zhang, D., Zhang, W., Gu, J. et al. (2015). Inspiration from butterfly and moth wing scales: characterization, modeling, and fabrication. *Progress in Materials Science* 68: 67–96.

46 Lee, R.T. and Smith, G.S. (2009). Detailed electromagnetic simulation for the structural color of butterfly wings. *Applied Optics* 48 (21): 4177–4190.

47 Zhu, D., Kinoshita, S., Cai, D.S., and Cole, J.B. (2009). Investigation of structural colors in morpho butterflies using the nonstandard-finite-difference time-domain method: effects of alternately stacked shelves and ridge density. *Physical Review E* 80 (5): 051924.

48 Zhang, W., Gu, J., Liu, Q. et al. (2014). Butterfly effects: novel functional materials inspired from the wings scales. *Physical Chemistry Chemical Physics* 16 (37): 19767–19780.

49 Butt, H., Yetisen, A.K., Mistry, D. et al. (2016). Morpho butterfly-inspired nanostructures. *Advanced Optical Materials* 4 (4): 497–504.

50 Feng, W., Kim, J.Y., Wang, X. et al. (2017). Assembly of mesoscale helices with near-unity enantiomeric excess and light-matter interactions for chiral semiconductors. *Science Advance* 3 (3): e1601159.

51 Potyrailo, R.A., Bonam, R.K., Hartley, J.G. et al. (2015). Towards outperforming conventional sensor arrays with fabricated individual photonic vapour sensors inspired by *Morpho* butterflies. *Nature Communications* 6: 7959.

52 Poncelet, O., Tallier, G., Mouchet, S.R. et al. (2016). Vapour sensitivity of an ALD hierarchical photonic structure inspired by *Morpho*. *Bioinspiration & Biomimetics* 11 (3): 1–9.

53 Kang, S., Tai, T., and Fang, T. (2010). Replication of butterfly wing microstructures using molding lithography. *Current Applied Physics* 10 (2): 625–630.

54 Han, Z., Jiao, Z., Niu, S., and Ren, L. (2019). Ascendant bioinspired antireflective materials: opportunities and challenges coexist. *Progress in Materials Science* 103: 1–68.

55 Mcdougal, A., Miller, B., Singh, M., and Kolle, M. (2019). Biological growth and synthetic fabrication of structurally colored materials. *Journal of Optics* 21 (7): 073001.

56 Gaillot, D.P., Deparis, O., Welch, V. et al. (2008). Composite organic–inorganic butterfly scales: production of photonic structures with atomic layer deposition. *Physical Review E* 78 (3): 031922.

57 Aryal, M., Ko, D.H., Tumbleston, J.R. et al. (2012). Large area nanofabrication of butterfly wing's three dimensional ultrastructures. *Journal of Vacuum Science & Technology B Microelectronics and Nanometer Structures* 30 (6): 061802.

58 Dumanli, A.G. and Savin, T. (2016). Recent advances in the biomimicry of structural colours. *Chemical Society Reviews* 45 (24): 6698–6724.

59 Bae, W.G., Kim, H.N., Kim, D. et al. (2014). 25th anniversary article: scalable multiscale patterned structures inspired by nature: the role of hierarchy. *Advanced Materials* 26 (5): 675–700.

60 Zhang, S. and Chen, Y. (2015). Nanofabrication and coloration study of artificial *Morpho* butterfly wings with aligned lamellae layers. *Scientific Reports* 5: 16637.

61 Zhang, S., Chen, Y., Lu, B. et al. (2016). Lithographically-generated 3D lamella layers and their structural color. *Nanoscale* 8 (17): 9118–9127.

62 Fu, F., Shang, L., Chen, Z. et al. (2018). Bioinspired living structural color hydrogels. *Science Robotics* 3 (16): eaar8580.

63 Zhang, Z., Chen, Z., Sun, L. et al. (2019). Bio-inspired angle-independent structural color films with anisotropic colloidal crystal array domains. *Nano Research* 12 (7): 1579–1584.

64 England, G., Kolle, M., Kim, P. et al. (2014). Bioinspired micrograting arrays mimicking the reverse color diffraction elements evolved by the butterfly *Pierella luna*. *Proceedings of the National Academy of Sciences of the United States of America* 111 (44): 15630–15634.

65 Zhao, Y., Zhao, Y., Hu, S. et al. (2017). Artificial structural color pixels: a review. *Materials* 10 (8): 944.

66 Kim, H.M., Kim, S.H., Lee, G.J. et al. (2015). Parametric studies on artificial *Morpho* butterfly wing scales for optical device applications. *Journal of Nanomaterials* 16 (1): 451834.

67 Zhu, Y., Hao, J., Li, W. et al. (2019). A new robotic tactile sensor with bio-mimetic structural colour inspired by *Morpho* butterflies. *Bioinspiration & Biomimetics* 14 (5): 056010.

68 Kertesz, K., Piszter, G., Balint, Z., and Biro, L.P. (2018). Optical vapor sensing on single wing scales and on whole wings of the *Albulina metallica* butterfly. *Sensors* 18 (12): 4282.

69 Meng, Y., Qiu, J., Wu, S. et al. (2018). Biomimetic structural color films with a bilayer inverse heterostructure for anticounterfeiting applications. *ACS Applied Materials & Interfaces* 10 (44): 38459–38465.

70 Ge, J. and Yin, Y. (2011). Responsive photonic crystals. *Angewandte Chemie International Edition* 50 (7): 1492–1522.

71 Fenzl, C., Hirsch, T., and Olfbeis, O.S.W. (2014). Photonic crystals for chemical sensing and biosensing. *Angewandte Chemie International Edition* 53 (13): 3318–3335.

7

Bioinspired Oil–Water Separation Materials

7.1 Definition and Classification of Oil–Water Separation Materials

Over the past few years, oil–water separation has become an increasingly important and urgent issue in modern chemical industrial processes and environmental protection due to increasing industrial oily wastewater discharge, as well as frequent oil spill accidents. The rapid growth of interface science and biomimetic materials, especially for special wettability materials [1–4], has offered a cutting-edge method for manufacturing efficient, robust, and recyclable oil–water separation materials that can separate an oil–water mixture. Compared to the conventional separation strategies, special wettability-controlled oil–water separations show dramatic advances in both flux and efficiency of the separation processes. Materials with special wettability, including superhydrophobic–oleophilic, superhydrophilic–oleophobic, superhydrophilic–underwater oleophobic, and switchable wettability, have been successfully applied for efficient separation of the common layered oil–water mixture.

7.1.1 Superhydrophobic–Oleophilic Materials

Superhydrophobic–oleophilic materials are typical oil removing materials. The superhydrophobic and superoleophilic properties make the oil phase spread easily while the water phase will be repelled, thus obtaining oil from an oil–water mixture. As we all know, surface wettability is determined by the combined effects of chemical composition and morphology. Accordingly, superhydrophobic–superoleophilic materials can be manufactured using two methods: the first one is constructing a complex structure on a hydrophobic surface; the second is modifying chemicals with low surface energy on a surface with a large roughness [5]. The ideal oil removing materials are generally considered to possess superhydrophobic and superoleophilic properties, high oil absorption ability and low affinity for water, low density, environmentally friendly nature, self-propelled exhausting external energy, and good recyclability for a wide range of oil–water mixtures. The most universal oil removing material is metallic mesh-based materials, which are discussed in detail in the upcoming sections.

Nature-Inspired Structured Functional Surfaces: Design, Fabrication, Characterization, and Applications, First Edition. Zhiwu Han.

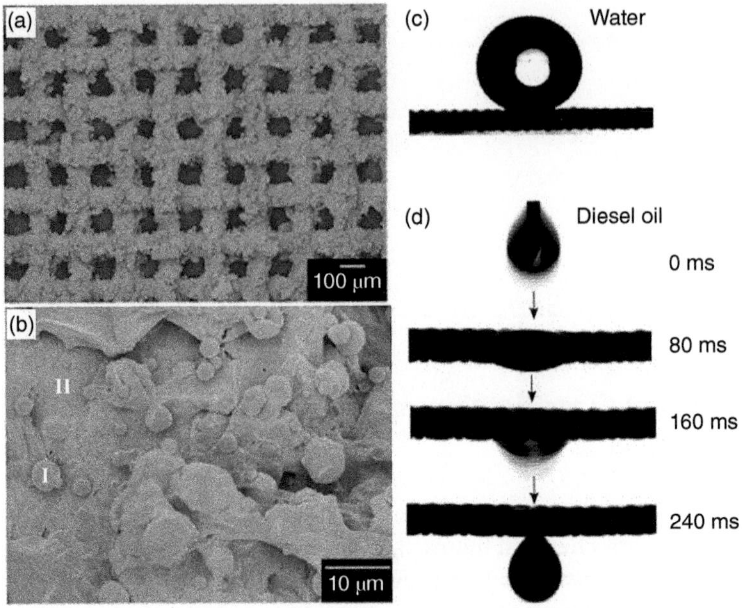

Figure 7.1 (a) Scanning electron microscope (SEM) images of the stainless steel mesh coating (average pore diameter is about 115 μm); (b) High-resolution view of (a). (c) The shape of a water droplet on the as-prepared mesh film with a water contact angle (WCA) of 156.2° ± 2.8°; (d) The diesel oil spreads and penetrates the copper mesh film quickly within ~240 ms, showing an oil contact angle (OCA) of 0°. Source: Reprinted with permission from Feng et al. [8]. Copyright 2004 WILEY-VCH Verlag GmbH.

Superhydrophobicity of a metallic material has been studied over the past few decades [6, 7]. To enhance the superhydrophobic and superoleophilic property of a porous metallic mesh membrane, a filtering mesh with selective permeability to water and oil is fabricated, forming oil/water separating functions. Stainless steel mesh was the first material used for manufacturing superhydrophobic and superoleophilic surfaces via a spray-and-dry method by Jiang et al. in 2004 (Figure 7.1a,b) [8]. The pre-mixed aqueous emulsion containing Teflon, adhesive (polyvinyl acetate), dispersant (polyvinyl alcohol), and surfactant (sodium dodecyl benzene sulfonate) was sprayed evenly on copper mesh with compressed air and was then subjected to a high-temperature drying process (350 °C) to reduce the adhesives, dispersants, and surfactants. The as-prepared mesh is so water-repellent that a water droplet is unstable to stay on the mesh and easily rolls off (Figure 7.1c). However, oil droplet spreads quickly on the mesh and permeates it thoroughly within only 240 ms (Figure 7.1d). It is demonstrated that the as-prepared mesh could separate an oil–water mixture with high efficiency. It could be found that the hydrophobicity of the as-prepared mesh films was mainly determined by the pore diameter of the mesh. Since then, various strategies and methods were used to manufacture superhydrophobic and superoleophilic oil–water separation materials, not only on stainless steel mesh substrates [9, 10] but also on the other metal substrates, which are introduced later in the chapter . Feng et al. [11] reported a stainless steel

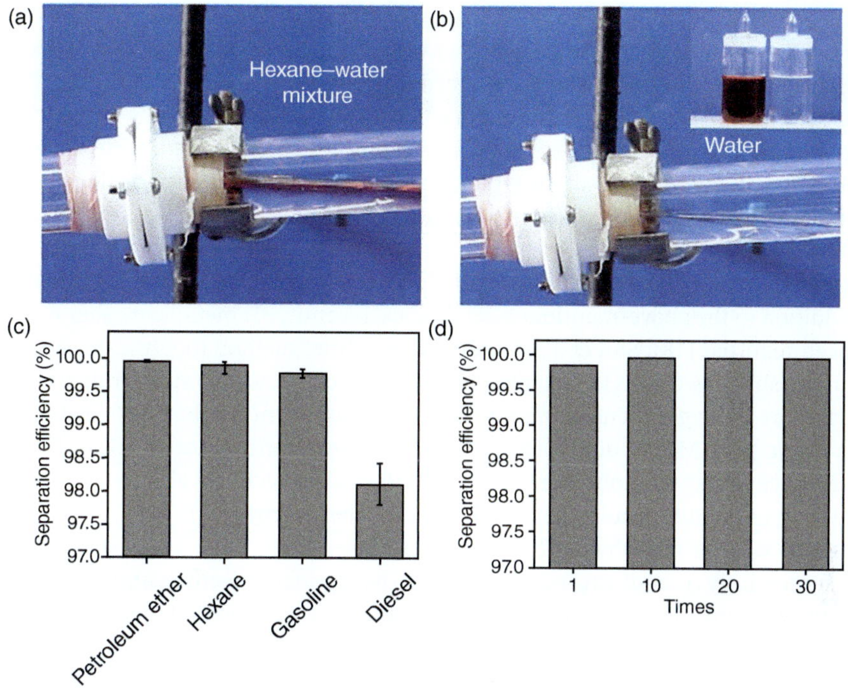

Figure 7.2 (a, b) The oil–water separation process based on the highly hydrophobic PDA-NDM mesh. (c) The separation efficiency of the PDA-NDM mesh for a selection of oil–water mixtures. (d) The separation efficiency keeps high after 30 cycles. Source: Reprinted with the permission from Cao et al. [11]. Copyright 2013 American Chemical Society.

mesh coated with polydopamine through combining mussel-inspired chemistry (self-polymerization of dopamine at pH = 8.5) and a Michael addition reaction of *n*-dodecyl mercaptan (NDM) with polydopamine (PDA). As shown in Figure 7.2, the as-prepared mesh could separate a series of oil–water mixtures. Besides, the separation efficiency was relatively high (99.95% for a hexane–water mixture, Figure 7.2c) after 30 cycles (Figure 7.2d). More importantly, the relatively high intrusion pressure (2.2 kPa) showed great potential in the separation of oil–water mixtures. Thanks to the easy oxidation and polymerization of dopamine under a mild environment, the method is also available to other substrates such as fabric [12] and sponge [13].

Except for stainless steel mesh, the copper mesh is another common substrate for oil–water separation [14, 15]. Surface modification with low surface energy materials is widely used in fabricating a superhydrophobic (or hydrophobic) and superoleophilic (or oleophilic) surface. However, secondary pollution might be caused by modifiers, especially fluorides. Recently, superhydrophobic (hydrophobic) and superoleophilic (oleophilic) stainless steel meshes have been prepared without the modification of low surface energy materials or fluorinated chemicals [16]. Lee et al. [16] synthesized vertically carbon nanotubes on a stainless steel mesh to separate various oil–water mixtures. The multiscale structures (with nanoscale needle-like tubes on the mesh with microscale pores) was coordinated with the low surface energy of carbon, resulting in both enhanced hydrophobicity and oleophilicity. The contact angles for diesel and water were 0° and 163° ± 4°, respectively. Similarly, superhydrophobic and superoleophilic stainless steel mesh coated with carbon nanotubes was manufactured through thermal chemical vapor deposition [17]. Very interestingly, the as-prepared mesh is able to dewater the water–oil emulsion. It is obvious that the modification by low surface energy materials without fluorine is more environmentally friendly since it reduces the usage of poisonous chemicals.

In addition to the above-mentioned 2D meshes, recently, 3D metallic meshes have also been reported [18]. Sun et al. [19] described a brief method to fabricate super-wetting mesh films (SMFs) by engineering commercial stainless steel grids based on a layer-by-layer graphene assembly. Due to the excellent mechanical flexibility of the mesh, the SMFs could be easily folded or bent forming closed 3D-SMF for long-distance transportation. The cubic 3D-SMF could selectively absorb and store the oil from an oil–water mixture in its inner chamber (Figure 7.3).

In short, various superhydrophobic and superoleophilic metallic meshes have been successfully manufactured, which can be applied to efficiently separate oil–water mixtures. However, there are still some problems and challenges. First,

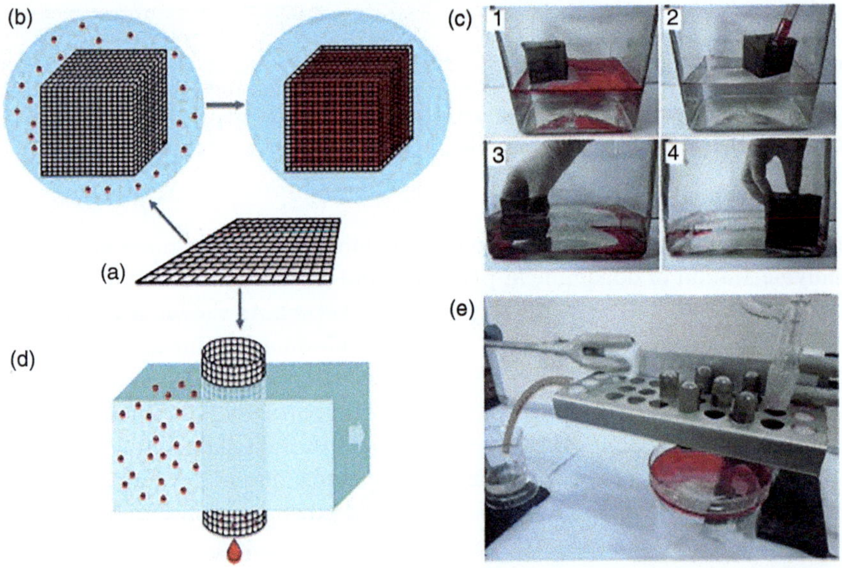

Figure 7.3 (a) Schematic model of an SMF. (b) Schematic model of cubic 3D-SMF prepared using the SMF and the selective absorption process of oil–water mixture (the red spots represent oil). (c) Optical images demonstrating the selective absorption of octane (1, 2) and chloroform (3, 4) from water phase using a cubic 3D-SMF (6 cm × 6 cm × 6 cm). (d) Schematic model of columnar 3D-SMF prepared using the SMF for oil–water separation. (e) Photograph of the equipment assembled for chloroform–water mixture separation, showing the water by-passed the columnar 3D-SMF, while the chloroform permeated and was collected by the container underneath. Source: Reprinted with permission from Sun et al. [19]. Copyright 2013 WILEY-VCH Verlag GmbH.

the rough structures of the superhydrophobic and superoleophilic metallic meshes can be easily destroyed, which results in the loss of superhydrophobicity and oil–water separation function. Second, these superwetting meshes cannot be applied to in situ oil removal process since the oil-contaminated water must be firstly collected and then filtered, which is not suitable for large-scale oil spills. Third, the density of most oils and organic solvents is smaller than that of water. Thus, superhydrophobic/superoleophilic meshes are not suitable for large-scale oil–water separation since oil floats on water, oil cannot come in contact with the mesh surface, which results in low separation efficiency and speed. Fourth, so far, the metallic materials used for oil–water separation are mainly stainless steel meshes and copper meshes, other metallic meshes are seldomly investigated. Therefore, other metallic meshes should be also developed since they may provide unique advantages in complex and specific application scenarios.

7.1.2 Underwater Superoleophobic Materials

Since most of the oil–water separation materials allow the oil phase to penetrate the surface while repelling water, some researchers have tried to develop inverse oil–water separation materials that can make the water pass through freely, yet repelling oil totally. Fish scales can prevent themselves from being contaminated by oil pollution in the sea. Inspired by fish scales, researchers have developed materials with hydrophilic and underwater superoleophobic properties and these materials show promising potential as water-removal materials [20]. This bionic idea overcomes the drawbacks of oil-removal materials, i.e. easily fouled or even blocked by oils because of their intrinsic oleophilicity, thus affecting their recyclability. Moreover, water-removal materials avoid the formation of a water barrier between the mesh surface and the oil phase, which frequently occurs with oil-removal materials because water density is higher than that of oil, thus preventing contact between the oil and substrates [21]. Besides, these materials will waste both oils and oil-removal materials, especially oils with high viscosity. The as-absorbed oils are hard to clear away, which will easily lead to secondary pollution during the post-treatment process of materials [22]. Jiang et al. [22] utilized stainless steel mesh coated with hydrophilic polyacrylamide (PAM) hydrogel (Figure 7.4a,b) to separate water from an oil–water mixture after pre-wetting the as-prepared mesh before the oil–water separation process. The as-prepared mesh possessed both underwater superoleophobic (Figure 7.4c, d, and f) and low oil-adhesion (Figure 7.4e,f) properties in an oil–water–solid three-phase system. The separation efficiency was above 99% for a series of oils, as given in Figure 7.4g. This is the first oil–water separation material that allows water to pass through (Figure 7.4h,i).

7.1.3 Superhydrophilic–Superoleophobic Materials

In addition to the superhydrophilic and underwater superoleophobic materials introduced in the previous sections, another superhydrophilic and in-air superoleophobic material was fabricated, which could also remove the water phase

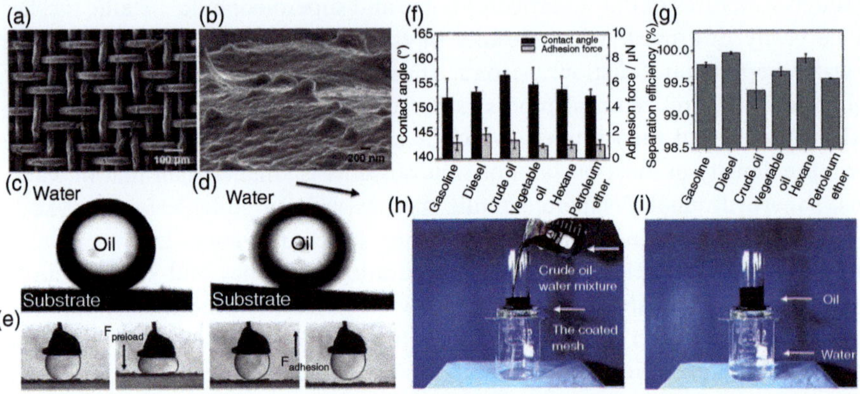

Figure 7.4 SEM images of the PAM hydrogel-coated mesh prepared by a stainless steel mesh with an average pore diameter of ~50 μm. (a) The integral view and (b) The magnified view of one single wire on the hydrogel-coated stainless steel mesh. (c) The still image shows the underwater oil droplet (1,2-dichloroethane, 2 μL) on the hydrogel-coated stainless steel mesh. (d) The underwater sliding angle of oil droplet on the stainless-steel mesh. (e) Optical images of the dynamic underwater oil adhesion process on the mesh. A 5 μL oil droplet was applied as the detecting probe to contact the mesh surface and was released. The mesh shows an ultralow affinity to the water droplet. (f) Underwater oil wettability and oil adhesion of the resultant mesh for various oils in terms of CA and oil-adhesion force. (g) The oil–water separation efficiency of the resultant mesh for a series of oils. (h, i) The oil–water separation process of the resultant mesh. The mesh film was fixed between two glass tubes. (h) A snapshot of the separation process of crude oil and water. (i) A snapshot after the oil-removal process. Source: Reprinted with permission from Xue et al. [22]. Copyright 2013 WILEY-VCH Verlag GmbH.

from oil–water mixtures and followed a quite different mechanism. Usually, a superhydrophilic (hydrophilic) surface exhibits superoleophilicity (oleophilicity) because oils commonly possess lower surface free energy than water [23]. In addition, for a superoleophobic surface, it commonly shows superhydrophobicity [24]. However, some scientists have found that a few stimuli-responsive surfaces can simultaneously possess hydrophilic and oleophobic properties based on a favorable interaction with polar liquids (e.g. water) and an unfavorable interaction with nonpolar liquids (e.g. hexadecane) [25]. To utilize the complexation of a polyelectrolyte and oppositely charged surfactants, polyelectrolyte–fluorosurfactant complexes can be constructed on the mesh surface. The stimuli-responsivity is triggered by the layer of polyelectrolyte–fluorosurfactant complexes that contain both hydrophilic (the polyelectrolyte) and oleophobic (the fluorosurfactant) components. On these materials, oil is repelled by the surface as the fluorosurfactant is exposed outward. However, the interaction between polar water molecules and the surface polyelectrolyte can induce molecular rearrangement so that the hydrophilic moieties are locked at the solid–liquid interface.

Although some amazing hydrophilic–oleophobic films have been synthesized [25], extreme superhydrophilic–superoleophobic films are still a challenge. To combine hydrophilic–oleophobic surfaces with a multiscale rough structure, Zhang et al. [26] first fabricated superhydrophilic and superoleophobic poly

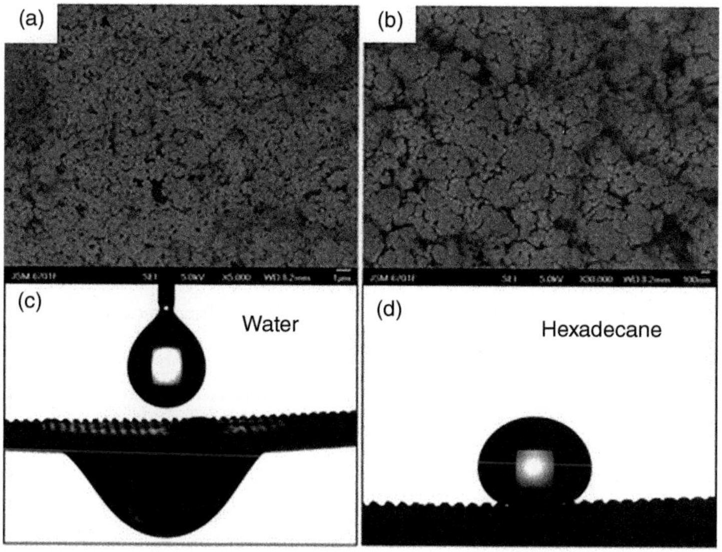

Figure 7.5 (a) SEM images of the PDDA-PFO/SiO_2 coating indicating plenty of protrusions and pores on its surface. (b) An enlarged view of the coating. (c) The water droplet spreads and passes through the resultant mesh. (d) The shape of a hexadecane droplet on the mesh with an OCA of 157° ± 2°. The PDDA-PFO/SiO_2-coated meshes possess both superhydrophilicity and superoleophobicity. Source: Reprinted with permission from Yang et al. [26]. Copyright 2012 Royal Society of Chemistry.

(diallyldimethylammonium chloride)-perfluorooctanoate/SiO_2 (PDDA-PFO/SiO_2) coatings. The high surface concentration of fluorinated groups together with carboxyl and quaternary ammonium groups formed a hydrophilic–oleophobic surface, whose mechanism was similar to previous stimuli-responsive surfaces [25]. Moreover, SiO_2 nanoparticles created multiscale hierarchical structures, which could enhance the intrinsic wettability to obtain superhydrophilicity–superoleophobicity (Figure 7.5a,b). The PDDA-PFO/SiO_2-coated mesh with an average diameter of ~200 μm showed water permeation (Figure 7.5c) and oil repellent (Figure 7.5d) properties, which showed that the coated mesh could be applied to separate oil–water mixtures. The meshes can be washed with water and then dried for reuse.

In addition, hygro-responsive films with both superhydrophilic and superoleophobic properties in air and underwater were investigated by Tuteja et al. [27]. The membranes were manufactured using a dip coating method with fluorodecyl polyhedral oligomeric silsesquioxane (POSS) and cross-linked poly(ethylene glycol) diacrylate (x-PEGDA), forming a POSS + x-PEGDA complex. In the air, the mesh surface was superoleophobic with several fluorodecyl POSS aggregates. When the material was immersed in an aqueous solvent, fluorodecyl POSS aggregates disappeared because of surface reconfiguration induced by water molecules. The film could separate an oil-in-water emulsion (Figure 7.6a,b) and a water-in-oil emulsion (Figure 7.6d,e) by a solely gravity-driven device. The thermogravimetric analysis (TGA) results demonstrated that the separation efficiency was more than 99% (Figure 7.6c).

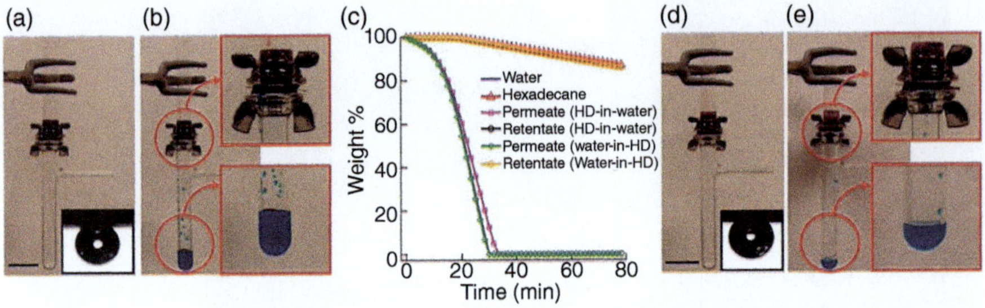

Figure 7.6 (a, b) Photographs showing the separation of a 50 : 50 v : v hexadecane-in-water emulsion. Inset in (a) presents the wettability underwater (with an SDS of 1 mg mL^{-1}) of a hexadecane droplet on a surface spin-coated with a 20 wt% fluorodecyl POSS + *x*-PEGDA blend. (c) TGA data showing the permeates and the retentates. Therein, HD is short for hexadecane. (d, e) The separation process of a 30 : 70 v : v water-in-hexadecane emulsion. Inset in (d) presents the wettability of a hexadecane droplet on the surface spin-coated with a 20 wt% fluorodecyl POSS + *x*-PEGDA blend. Source: Reprinted with permission from Kota et al. [27]. Copyright 2012 Nature Publishing Group.

Excellent superhydrophilic and superoleophobic surfaces have potential in industrial oil-polluted water treatments, cleanup of oil spills, and fuel purification. However, the preparation methods of superhydrophilic-superoleophobic materials are quite complex. So far, superhydrophilic-superoleophobic surfaces and their applications still lack investigation, which remains a great challenge in the field of oil–water separation.

7.1.4 Smart Materials with Switchable Wettability

Materials with stimuli-responsive wettability have attracted increasing interest because of their importance in the fundamental study and industrial applications [28]. Controllable surface wettability can be achieved by applying an external stimulus such as light illumination [29], temperature, electrical potential [30], and pH [31]. Recently, based on an extreme wettability switch strategy, controllable oil–water separation materials have been manufactured, which is significant in the treatment of oil-contaminated water for various purposes [32].

Polymers containing acid or basic functional groups usually possess pH-responsive wetting property because their conformation and charges are dramatically affected by solutions with different pH [33]. Based on this view, Zhang et al. [32] first developed smart materials that could be used as highly controllable oil–water separation devices. These smart materials with switchable superoleophilicity and superoleophobicity in aqueous environments were synthesized by grafting a BCP comprising pH-responsive poly(2-vinylpyridine) and oleophilic/hydrophobic polydimethylsiloxane (i.e. P2VP-*b*-PDMS) on widely used materials (Figure 7.7a), such as fibers (Figure 7.7b,c) and sponges. Therein, the P2VP block could alter its wettability and conformation via protonation and deprotonation upon changing the pH of the aqueous solvent, which provides a controllable and switchable access of oil by the PDMS block, forming surface with switchable oil wettability in aqueous

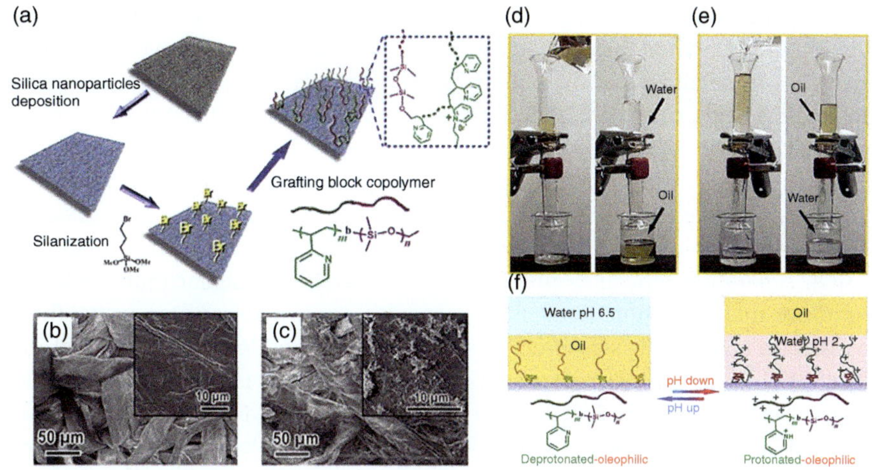

Figure 7.7 (a) A schematic illustration of the preparation process of a surface with switchable water and underwater oil wettability on a nonwoven textile surface. (b) and (c) show the SEM images of the raw textile and the textile after deposition of silica nanoparticles and BCP grafting. The insets in (b) and (c) show the enlarged view of single fibers. (d) and (e) show the controllable oil–water separation using the functionalized textile. (d) Removal of oil from the water phase. (e) Separation of water from the oil phase. Therein, the functionalized textile was pre-wetted with acidic water (pH = 2.0). (f) Schematic diagrams for the switchable oil wettability of the P2VPb-PDMS-grafted textile. Source: Reprinted with the permission from Zhang et al. [32]. Copyright 2012 Nature Publishing Group.

media (Figure 7.7f). Figure 7.7d,e demonstrates a highly controllable oil–water separation process of the resultant textile. When a gasoline–water mixture at pH 6.5 was poured into an upper glass tube, the gasoline quickly penetrated the textile film, but the water did not. However, when the textile film was wetted in advance by acidic water with a pH of 2.0 without subsequent drying and used under the same conditions, the opposite separation process was realized, with water passing through the film.

Light is another widely reported external stimuli source. Photo-responsive materials, whose wettability switch between superhydrophobicity and superhydrophilicity, have been studied and reported recently [29]. Inspired by the extreme wettability switch case, Zhai et al. [34] first fabricated a photo-responsive oil–water separation film, based on switchable superhydrophobicity–superhydrophilicity and underwater superoleophobicity, using a stainless steel mesh coated with an aligned ZnO nanorod array. The ZnO-coated mesh showed good controllability for separating different oil–water mixtures in an oil–water–solid three-phase system.

Thermo-responsive materials have also been developed to control the wettability of solid surface in the early years. Poly(*N*-isopropylacrylamide) (PNIPAAm), as a thermal-responsive polymer with a low critical solution temperature (LCST) of about 32–33 °C, is widely applied to fabricate a material with switchable wettability. Gao et al. [35] reported a polymethylmethacrylate-*b*-PNIPAAm BCP with reversible wettability states (hydrophilicity/oleophobicity and hydrophobicity/oleophilicity) at different temperatures. That is to say, water could permeate the BCP-coated mesh, while oil could not when the temperature was below the LCST. However, oil could penetrate the mesh and water could not when the temperature was above the LCST. Therefore, this film provids promising applications in the controllable separation of oil–water mixtures.

Electric field response was developed by Tuteja et al. [36] to switch the wettability and separate the oil–water mixtures in a smart way. They developed a film-based single-unit separator that could separate all types of oil–water mixtures with high separation efficiency under a high electric field (higher than 1000 V). Using these methods, functional materials with controllable water wettability (responsive to acidic or basic water) in air and oil wettability when immersed in an aqueous environment were expected to be used in many practical applications and help researchers design and develop smart functionalized interfacial materials for both in-air and underwater cases. So far, there are a few reports involving responsive materials for controllable oil–water separation. In addition, the study of stimuli-responsive materials focused on pH-responsive materials, photo-responsive materials, thermal-responsive materials, and high electric-field-responsive materials. Therefore, further studies on smart oil–water separation materials should mainly focus on fabricating new responsive materials, for instance, low electric field-responsive materials, stress-responsive materials, dual- and multi-responsive materials.

7.2 Oil–Water Separation Materials Inspired by Butterfly Wings

After billions of years of evolution, creatures in nature have developed almost perfect hierarchical structures and surface functions. Nature has experimented with various methods to its challenges and has improved the effective ones, which can supply ready answers to scientific and technical problems and inspire us with a series of amazing designs and ingenious structures. Abundant and gorgeous species in nature have been studied and mimicked for many decades to design unique structural materials with excellent functions and extensive practical values [37]. Especially, increasing interest has been directed to biomimetic nonwetting surfaces, such as lotus leaves [38], rose petals [39], water striders [40]. To a certain extent, existing biomimetic surfaces also break the limitation for the oil–water separation of various oil–water mixtures and possess the anti-oil-fouling properties. However, the underlying mechanism is not very clear and unified until now [41]. As a typical prototype of the bionic field recently, buttery wings are endowed with many excellent properties [42]. Butterfly wings exhibit either static or vibrating states in their daily life. Irrespective of situation, butterfly wings should be light-loaded to ensure flexibility of their flying. On the one hand, for butterflies in rainforests, water droplets usually move down on butterfly wings. Therefore, a nonwetting surface is developed on the wings. On the other hand, butterflies also need to prevent oil secreted by trees from adhering on their wing surfaces during their flight. Therefore, oil droplets need to be volatilized as soon as possible. In a word, butterfly wings possess excellent nonwetting properties, which present inspiration and opens up an avenue for realization of oil–water separation processes.

7.2.1 Characterizations of Original Butterfly Wings

To find out the reasons why butterfly wings possess oil–water separation ability, we investigated physical structures and chemical compositions of butterfly wings. The wingspan of a male *Trogonoptera brookiana* butterfly is about 12 cm, and the dorsal wings include overlapping scales with mysterious brilliance (Figure 7.8) [43]. It was clear that these scales were distributed in good order on the substrate at low magnification. The wing scales resembled the rooftop tiles very much in shape with about 160 μm in length and 70 μm in width. The root of each scale was embedded on the wing membrane. More exquisite structures were found at medium magnification. Scale surface consisted of a set of raised longitudinal quasi-parallel lamellae (ridges). The distance between two adjacent ridges was about 1.5 μm. The space between adjacent ridges was filled with a netlike reticulum composed of pores, which was

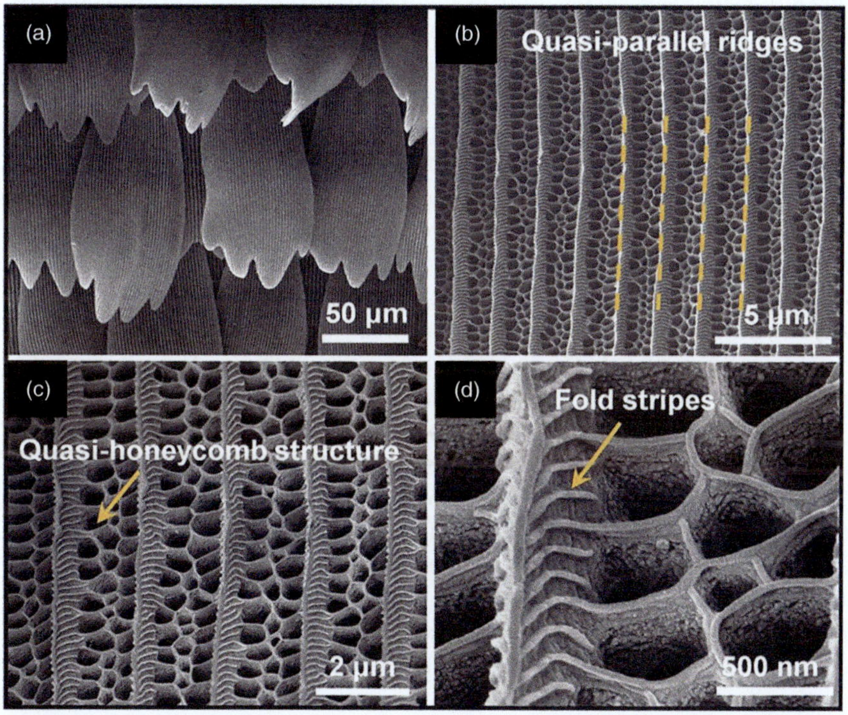

Figure 7.8 Structural characteristics of original butterfly wings. (a) Scales range in good order on wing substrate. (b, c) Scale surface contains raised longitudinal quasi-parallel lamellae (ridges). The space between ridges is filled with quasi-honeycomb structures. (d) Fold stripes distribute on both sides of ridges. Source: Reprinted with permission from Han et al. [43]. Copyright 2013 WILEY-VCH Verlag GmbH.

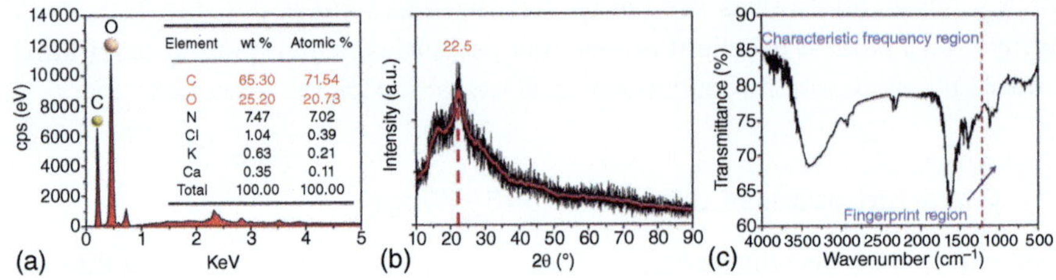

Figure 7.9 Chemical composition analysis of original butterfly wings. (a) EDS spectrum presents the element types of wings. (b) XRD patterns. A strong Bragg peak confirms the periodic feature of micro-/nanostructures in wings. (c) FT-IR spectrum. Butterfly wings are mainly chitin-based and primarily composed of water-insoluble components. Source: Reprinted with permission from Han et al. [43]. Copyright 2013 WILEY-VCH Verlag GmbH.

named a quasi-honeycomb-like structure. There were parallel fold stripes on both sides of the ridges. Ridges and fold stripes formed the structural nonwetting framework (SNWF) of butterfly wings.

The chemical elements and content analysis were characterized by energy dispersive spectrometer (EDS). The main elements were carbon (C), hydrogen (H), oxygen (O), and nitrogen (N), which made up the nonwetting organic framework of butterfly wings as shown in Figure 7.9a. Atomic percentages of C and N elements were 71.54% and 7.02%, respectively. In the X-ray diffraction (XRD) spectrum (Figure 7.9b), a strong Bragg peak was at 11.25°, which further proved the periodic structure of original butterfly wings. For the chemical groups, Fourier transform infrared spectrum (FTIR) was utilized to analyze the chemical bonds of the SNWF in butterfly wing scales. As shown in Figure 7.9c, the enhanced absorption peaks of butterfly sample were at 667 and 2358 cm^{-1}, which corresponded to the bending vibration and asymmetric stretching vibration of carbon dioxide. The broad absorption peaks at 3428 and 3260 cm^{-1} were assigned to the stretching vibration of hydroxyl groups (—OH) and the N—H bond, respectively. The weak absorption peaks at 2966, 2923, and 2874 cm^{-1} were related to the stretching vibration of the C—H bond. The multipeaks at 1655, 1550, and 1310 cm^{-1} around the fingerprint region were attributed to amide I, amide II, and amide III, which were the functional groups of chitins. The weak peak around 1076 cm^{-1} was assigned to the stretching vibration of the C—O bond and the weak peak around 886 cm^{-1} to the stretching vibration of the hexatomic ring, which was the backbone of chitin. In a word, the original scales in butterfly wings are mainly made up of chitin and primarily contain water-insoluble organic components.

Trogonoptera brookiana butterfly was selected as the biological prototype because of its typical humid habitat. They are distributed across a lot of tropical rainforests around the world, especially in Southeast Asia. Figure 7.10a,b shows the overall view of the original butterfly [43]. It can be found clearly that triangular green patches and gorgeous metallic black areas are distributed on both fore and hind wings using a digital camera. With an optical stereo microscope, the micro morphologies of wing scales were observed as shown in Figure 7.10c,d. These scales were placed in alternate rows and overlapped each other. This kind of arrangement ensured that these

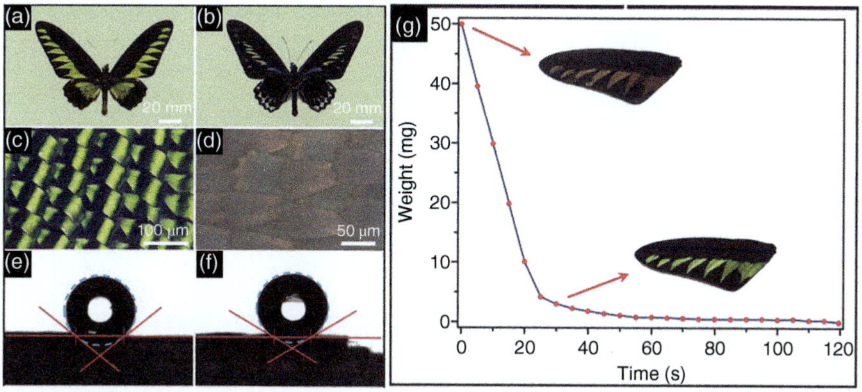

Figure 7.10 Wettability and self-recovery anti-oil-fouling property of butterfly wings. (a, b) Digital images of female butterfly. (c, d) Stereoscopic microscope images of green and black parts. (e, f) Water contact angles of green and black parts. (g) Self-recovery anti-oil-fouling property of butterfly wings. Butterfly wings return to initial oil-free state within ≈ 30 seconds. Source: Reprinted with permission from Han et al. [43]. Copyright 2013 WILEY-VCH Verlag GmbH.

scales form a dense membrane on butterfly wing surface. Wetting behaviors of water on butterfly wings were further examined using an optical contact angle (CA) measuring device. The water droplets were spheroidal on the wing surface. The water CAs of green and black areas were 148.4° (Figure 7.10e) and 152.7° (Figure 7.10f), respectively. Therefore, the hydrophobic property of black region was better than that of green region. In particular, the butterfly wings not only possessed the superhydrophobic feature but also showed self-cleaning property. They were found to be responsible for the waterproof capability. The wetting behavior of oil droplets on the front butterfly wing surface was also examined carefully. About 50 mg oil was added dropwise onto the wing surface through a glass capillary. Oil droplets spread out over the entire surface evenly and the oil CA was approximately 0°, which proved the superoleophilic property of butterfly wings. During the abovementioned processes, butterfly wings were fixed on a high-sensitivity electronic balance. A high-speed camera was used to record the readings. Figure 7.10g shows the curve of the weight of residual oil as a function of time t. After dried for about 30 seconds, oil droplets were completely released and butterfly wings recovered to the original state. It was confirmed that butterfly wings possess superior anti-oil-fouling property, which enables butterflies to keep their wings light-loaded.

To examine liquid wetting behaviors of butterfly wings in more detail, a high-speed camera system was used to record bounce dynamic processes of water droplets (radius $R = 1.58$ mm, velocity $U = 0.87$ m s^{-1}) and spreading processes of oil droplets (radius $R = 1.16$ mm, velocity $U = 0.89$ m s^{-1}) as shown in Figure 7.11 [43]. Water droplet impacts on the wing surface (up of Figure 7.11a,b) and spreads to form a film. Then, the film retracts to recover as a drop and lifts off within just 12.33 and 21.33 ms, respectively. Little contact time indicated that butterfly wings had excellent ability to resist water. Besides, the overall contact time of the black part was much less than

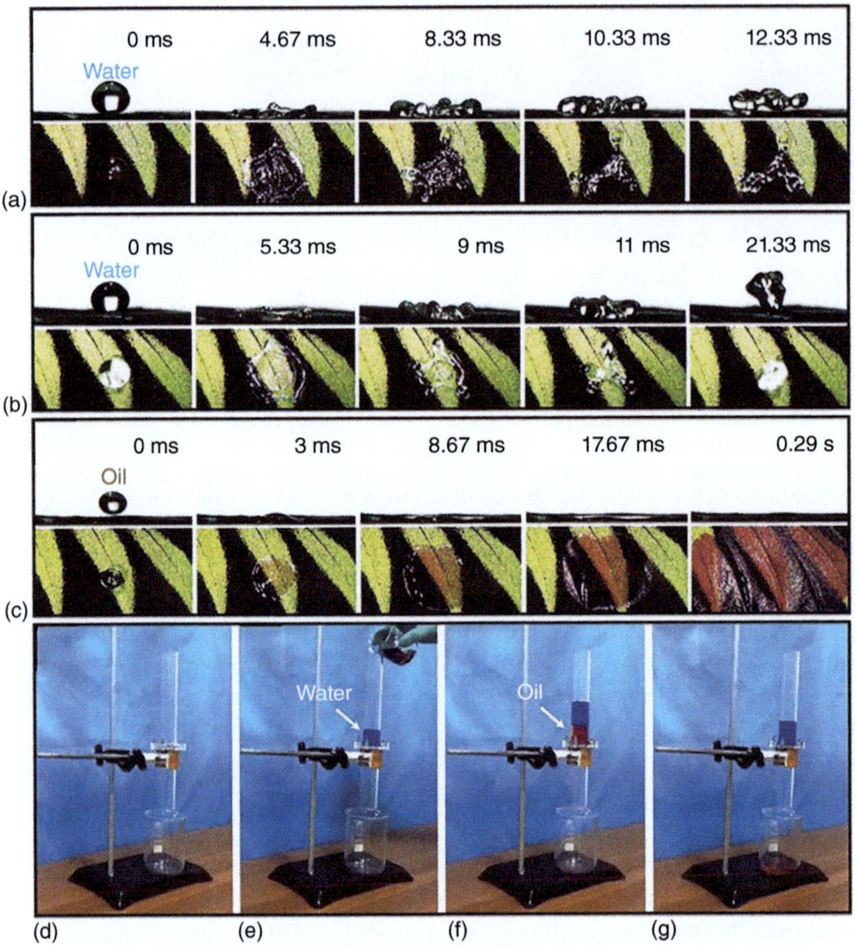

Figure 7.11 Water bouncing dynamics, oil spreading processes, and oil–water separation processes on butterfly wings. (a) Bounce dynamics of a water droplet on green parts. (b) Bounce dynamics of a water droplet on black parts. (c) Spreading processes of an oil droplet on wing surface. (d) Before, (e, f) among, and (g) after the oil–water separation processes. Reprinted with permission from Han et al. [43]. Copyright 2013 WILEY-VCH Verlag GmbH.

that of the green part, which proved better water repellency. Therefore, the black region was chosen as the experimental sample for further study. Synchronized top-view images (bottom images in Figure 7.11a,b) show a faster retraction process along the ridges than in other directions. This variation in retraction speed broke the radial symmetry of the recoiling film, making more film participate in the recoil. This explains the less contact time between butterfly wings and water droplets. The oil wettability on butterfly wings was characterized (Figure 7.11c). When an oil droplet fall on the wing surface, it spread out and the CA reached near 0° quickly. The whole process was completed within 0.3 second, demonstrating a superior property of the surface for oil wetting. Figure 7.11d–g shows that butterfly wings possess oil–water separation ability, which was confirmed by a gravity-driven oil–water separation test. A piece of butterfly wing was placed between two Teflon flanges. Water was colored with methylene blue, and oil (chloroform) was colored with Sudan III to aid visualization, which did not affect the wettability behaviors. A mixture of oil and water was poured into the glass tube connected with Teflon flange. Oil spread and permeated the wing membrane rapidly because of superoleophilicity, whereas water was repelled and kept in the Teflon flange owing to superhydrophobicity. Little amount of water was found in the collected oil, showing high separation efficiency. However, the separation rate was quite low. The possible reason may be that the pores on butterfly wing surface were too tiny.

7.2.2 Design Principle for Butterfly-based Oil–Water Separation Materials

According to the characteristic sizes of wing scales, visual 3D optimized configuration models were constructed to illustrate the nonwetting mechanism in butterfly wing scales (Figure 7.12). On the one hand, microscale ridges on wing scale contributed to water repellency (Figure 7.12b,c). These ridges presented a certain degree of roughness to the scale surface and allowed the trapping of an air layer, which

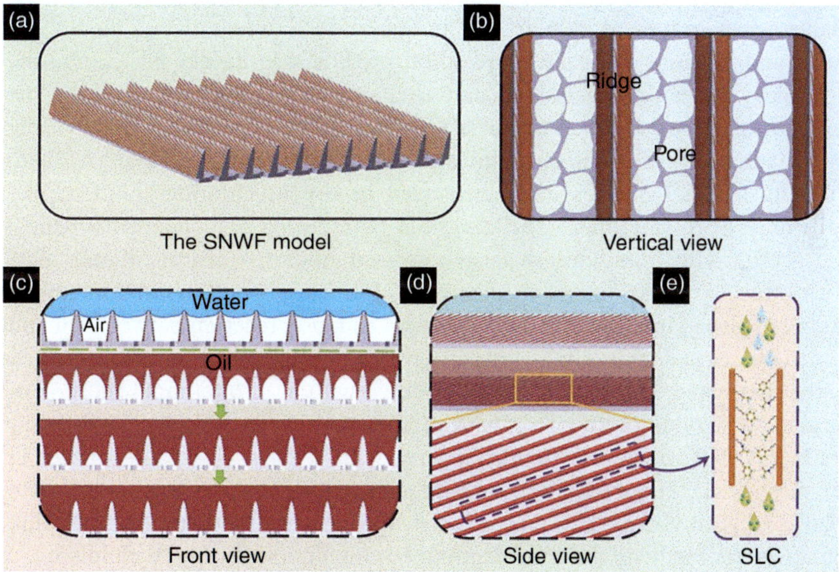

Figure 7.12 Simplified illustrations of the active nonwetting skill of butterfly wings. (a, b) Visual 3D optimized structural models built according to original wing scales. (c) Microscale ridges allow the trapping of a thin air layer, which prevents the penetration of water and enhance the superhydrophobic property. As for oil droplets, penetration of oil into air layer requires less energy due to low surface tension. (d, e) Nanoscale fold stripes distributed on both sides of the ridges constitute a lot of selective liquid channels. Source: Reprinted with permission from Han et al. [43]. Copyright 2013 WILEY-VCH Verlag GmbH.

prevented water penetration into this well-defined space and enhanced superhydrophobic property of the wing scales. As for oil droplets, penetration of oil droplets into the air layer demanded less energy thanks to low surface tension. On the other hand, nanoscale fold stripes were conducive to superhydrophobic, superoleophilic, and self-recovery anti-oil-fouling properties (Figure 7.12d). These fold stripes on both sides of the ridges formed many channels (the space between fold stripes). Because the channels consisted of chitin and water-insoluble organic components, abundant hydrophobic groups were distributed on the jamb wall of the channels. The presence of these groups allowed oil droplets to pass through, while the water molecules were stopped. Due to these features, the space between fold stripes constituted selective liquid channels (SLCs) (Figure 7.12e). The SLCs allowed oil droplets to pass through rapidly along the ridges until oil droplets filled the air layer, and then went through the pores in the scales. In conclusion, butterflies can smartly and efficiently deal with this puzzle via their roughness effect and hydrophobic groups. Based on the SNFW (combination of the ridges and fold stripes) and chemical composition (water-insoluble organic components) of butterfly wings, we can design and implement a biomimetic oil–water separation surface.

7.2.3 Fabrication of Butterfly-inspired Oil–Water Separation Materials

This section introduces the biomimetic fabrication method of SNWF-based monolayer membrane. Stainless steel mesh is a perfect basement membrane due to its affordable prices as well as its porous structures, which are similar to the SNWF. A simple and effective two-step preparation method was developed, combining an electroless deposition technique and a chemical modification process. The manufacturing process is illustrated briefly in Figure 7.13a, and the corresponding internal chemical synthesis mechanism is introduced intuitively in Figure 7.13b–d. When stainless steel meshes were immersed in copper chloride solution, elemental iron (Fe) in steel mesh surface could be replaced by copper ions (Cu^{2+}) (Figure 7.13b). The iron wire surface possessed micro-/nanoscale hierarchical structures due to the self-growing process of Cu. Then, the biomimetic copper meshes were placed in a vacuum-drying oven. Oxidation reaction between Cu and O_2 formed the copper oxide (CuO) under the hydrothermal condition (Figure 7.13c). Then, the pretreated copper films were modified with a 0.1 M ethanol solution of stearic acid for 80 minutes under ambient conditions. The hydrophobic alkyl chains ($CH_3(CH_2)_{16}COO$–) could be chemically combined at the native oxide surface of copper meshes via the double decomposition reaction, forming a self-assembled monolayer (Figure 7.13d). The displacement reaction and double decomposition reaction together made up the abovementioned chemical synthesis mechanism.

7.2.4 Characterizations of Butterfly-inspired Oil–Water Separation Materials

Figure 7.14a is a typical image of the raw stainless-steel mesh with single-layer iron wires, which formed a lot of square pores. The inset shows that the iron wire surface was quite flat. As shown in Figure 7.14b–d, biomimetic monolayer copper

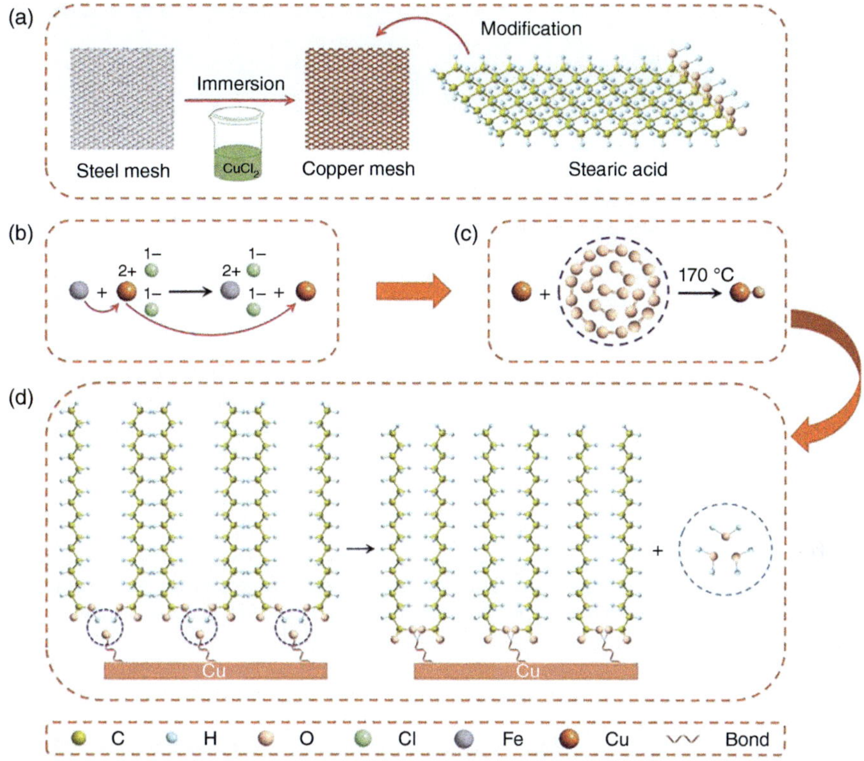

Figure 7.13 Bioinspired fabrication method of the MHDS-based BMCM and the internal chemical synthesis mechanism. (a) The preparation process of MHDS on stainless steel mesh. (b) The ion-exchange process involved in electroless deposition. (c) Copper reacts with oxygen, obtaining copper oxide under hydrothermal condition. (d) Copper stearate 3D networks are gradually constructed and extend along the MHDS-based BMCM via the double decomposition reaction. Source: Reprinted with the permission from Han et al. [43]. Copyright 2013 WILEY-VCH Verlag GmbH.

membranes (BMCMs) were covered with not only the microscale iron wires and copper clusters but also a lot of nanoscale dendritic bumps, forming the typical multiscale hierarchical dendritic structures (MHDSs). The copper clusters covered the iron wires evenly and almost no redundant Cu blocked the mesh pores, which ensured the liquid passed through BMCMs rapidly. The thickness of the copper clusters was about 30 μm. The role of this micrometer structures was equivalent to ridges on butterfly wings, which could prop up the water membrane and form an air layer in the space between the top of copper clusters and the iron wire surface. Besides, high-resolution scanning electron microscope (SEM) images (Figure 7.14d) shows that the copper clusters consisted of a lot of nanoscale projections. Copper clusters and dendritic bumps together made up the SNWF of BMCM. These nanoscale dendritic bumps together with a monolayer of copper stearate 3D networks formed SLCs, which acted as fold stripes on butterfly wings. To examine the average roughness, a 3D microscope system with a superwide depth of field

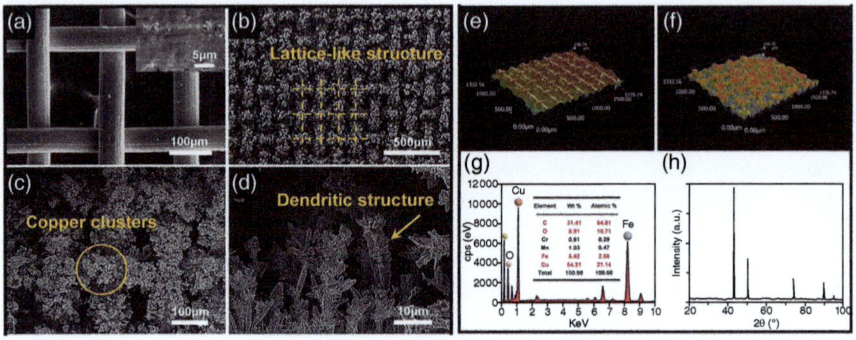

Figure 7.14 Structural and chemical composition analyses of BMCM. (a) Field emission scanning electronic microscopy (FESEM) images of raw stainless steel mesh. The inset shows that the iron wire surface is flat. (b) Low-magnification FESEM images show overall morphology of MHDS-based BMCM, especially the lattice-like structures. (c) Enlarged FESEM images present microstructures of the bioinspired MHDSs, which act as ridges on butterfly wings. (d) FESEM image of bioinspired dendritic structure under high magnification. Images showing the surface roughness of samples (e) stainless steel mesh and (f) BMCM. (g) EDS spectrum demonstrates that the main elements of BMCM are Cu, C, Fe, and O. (h) XRD spectrum of MHDS presents its quasi-PC structures and corresponding grazing angles. Source: Reprinted with permission from Han et al. [43]. Copyright 2013 WILEY-VCH Verlag GmbH.

was employed to measure the surface roughness of the steel mesh (Figure 7.14e) and the BMCM (Figure 7.14f). By comparing the topological differences, it was found that the surface roughness of the BMCM increased obviously after secondary modification. The BMCM possessed an average mean square roughness (R_a) of 25.952 nm, which was much higher than that of stainless steel mesh (17.486 nm). This was due to the existence of dendritic structures on the BMCM surface, which could increase the surface roughness and hierarchy [44, 45]. Chemical composition analysis of BMCM was carried out with EDS and XRD. Figure 7.14e demonstrates the EDS spectra of the BMCM. The peaks of elemental C, O, and Cu indicated that Fe of iron wires displaced the copper ions in the copper chloride solution, and SNWF was treated by stearic acid successfully. The XRD result (Figure 7.14f) further revealed that the BMCM was crystallized. It also indicated that the copper was self-growing on the iron wires surface. The five distinct peaks in the spectra were assigned to copper stearate ($Cu[CH_3(CH_2)_{16}COO]_2$), resulting from the reaction between stearic acid ($CH_3(CH_2)_{16}COOH$) and copper oxide. The appearance of the copper stearate peaks confirmed the presence of low surface energy methylated (—CH_3 and =CH_2) bonds on BMCM.

7.2.5 Performance of Butterfly-inspired Oil–Water Separation Materials

7.2.5.1 The Waterproof Phenomena on Oil–Water Separation Materials

The water bounce processes on BMCM were recorded using a high-speed video camera, when a 16.7 μL water droplet dropped from a needle with a height of

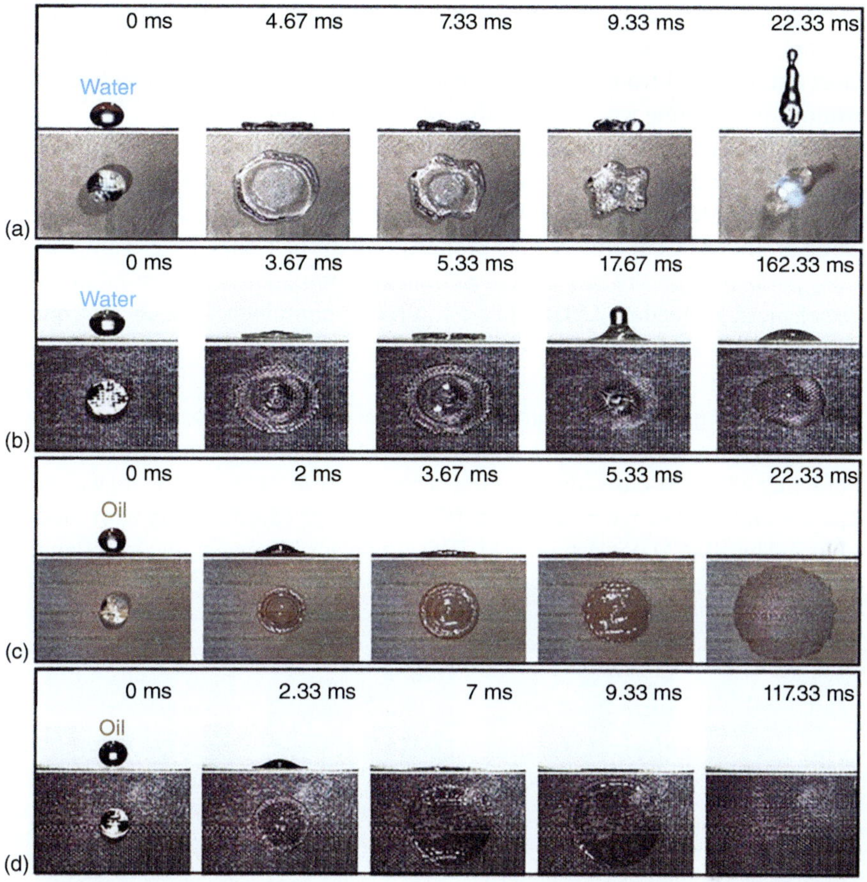

Figure 7.15 Water bouncing dynamics and oil spreading processes on BMCM and raw steel mesh. (a) Bounce dynamics of a water droplet on MHDS-based BMCM. (b) Bounce dynamics of a water droplet on raw steel mesh. (c) Spreading processes of an oil droplet on MHDS-based BMCM. (d) Spreading processes of an oil droplet on raw steel mesh. Source: Reprinted with permission from Han et al. [43]. Copyright 2013 WILEY-VCH Verlag GmbH.

40 mm (Figure 7.15a). The height of the needle was determined by the maximum height avoiding the droplet fragmentation when impacting the BMCM, ensuring maximum droplet momentum. The droplet volume (~17 μL) was optimized as the droplet could be replicated easily and fell down under its own weight. It could be found that water droplets spread out immediately after impacting the BMCM. The initial impact velocity of the droplet was about 0.86 m s^{-1}. At 4.67 ms, the area of the water membrane was maximum. Then, the water droplet shrank quickly and was completely removed from surface of the BMCM within 22.33 ms, which illustrates splendid superhydrophobic and self-cleaning properties. The water droplets impacting examinations were also carried out on the raw steel mesh film for comparison (Figure 7.15b). The raw steel mesh possessed hydrophilic and adhesive properties toward the water, and the water was stuck on the surface without bouncing out.

Impinging behaviors of oil droplets on BMCM were also recorded (Figure 7.15c). The dropping height was 40 mm and the diameter was approximately 2.33 mm. Oil droplets also dropped from a 23-gauge dispensing tip. The above parameters were determined in the same way, just like the water droplet impacting tests. The results showed that oil droplets impacted the BMCM and spread out immediately. The initial impact speed of the droplet was about $0.84\,m\,s^{-1}$. At 22.33 ms, the oil droplets infiltrated the BMCM totally and the oil membrane area reached the maximum, denoting an intrinsic superoleophilic property. Oil droplet impacting examination was also performed on the raw steel mesh film for comparison. (Figure 7.15d). The raw steel mesh needed 117.33 ms to be infiltrated completely, which was much more than that of BMCM.

7.2.5.2 Separating Efficiency of Recycled Oil–Water Separation Experiments

As shown in Figure 7.16a–d, the separation function of water–oil mixtures was realized utilizing a BMCM-based device. First, a piece of BMCM (5 cm × 5 cm) was used to separate oil (dyed in red with Sudan III) from water (dyed in blue with methylene blue). The BMCM consisting of MHDSs was placed between two iron flanges.

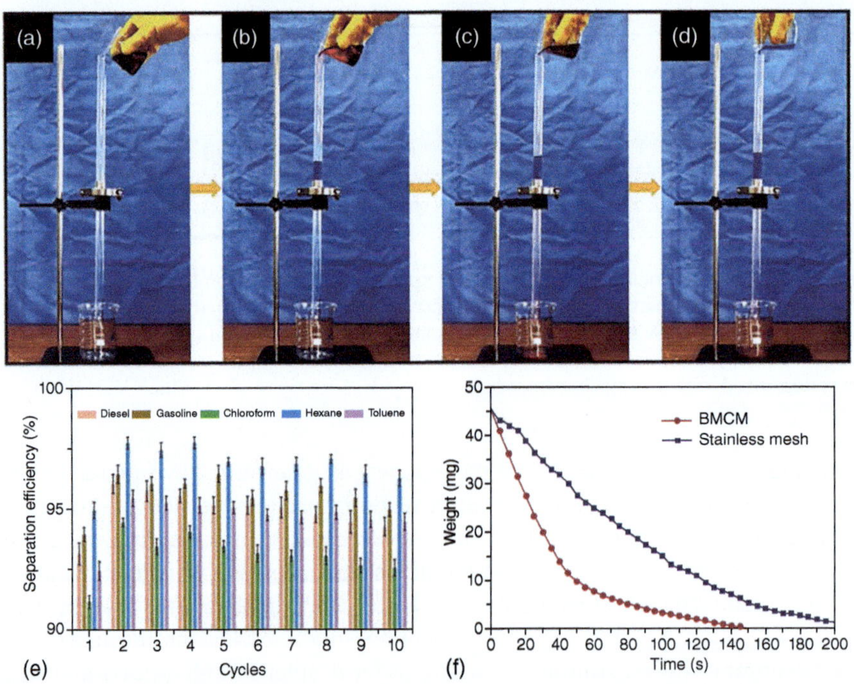

Figure 7.16 Oil–water separation processes and self-recovery anti-oil-fouling property of the BMCM-based device. (a) Before, (b, c) among, and (d) after the oil–water separation processes. (e) Separating efficiency of recycled tests for separating various oils from water. (f) Self-recovery anti-oil-fouling property of BMCM. Source: Reprinted with permission from Han et al. [43]. Copyright 2013 WILEY-VCH Verlag GmbH.

Then, the oil–water mixtures were poured into the glass tube, which was connected with flanges. It could be found that the oil passed through the BMCM quickly and dropped into the beaker underneath, while water was still kept in a glass tube upon BMCM. The experimental results demonstrated the efficient separation capability of BMCM for water–oil mixtures. Oil–water separation efficiency was calculated to estimate the oil–water separation ability. The influence of cycles of oil–water separation processes on the separation efficiency was tested to evaluate reusability. The oil–water separation efficiency (*R*, expressed as a percentage) was then calculated according to the following equation:

$$R(\%) = \left(1 - \frac{C_P}{C_O}\right) \times 100$$

where C_O and C_P are the oil concentration of the initial oil–water mixtures and the collected water after the first separation process, respectively [43]. The oil concentration was measured by an automatic infrared oil analyzer. As shown in Figure 7.16e, BMCM was efficient (>93%) for five various oil–water mixtures after the test was carried out for 10 cycles, indicating the excellent recyclability of BMCM. The deficit in the collected liquid was mainly absorbed by the device itself, and a small amount of oil was still inside the beaker or glass tube. There were slight differences in separation efficiency between various oils. This should be attributed to different densities and adhesion. This device required no external power but only gravity of oil–water mixtures in the separation processes, which means it has the advantage of low-energy consumption. The CA of BMCM before the oil–water separation process was 154.3°. BMCM was still superhydrophobic (CA = 151.5°) after the separation process had been applied for 10 cycles. It also denoted that as-prepared BMCM showed good recyclability for oil–water separation. Then, the recovering ability of BMCM was characterized by a self-built system. The BMCM samples were fixed on a high-sensitivity electronic balance. About 45 mg oil droplets were dropped onto stainless steel mesh and the BMCM evenly through a glass capillary, and a high-speed camera was used to record the changes of readings. Figure 7.16f illustrates the residual oil on the stainless mesh (reference film) and BMCM as a function of time *t*. The oil volatilization rate on the stainless steel mesh was quite constant. It cost more than 200 seconds for the oil to evaporate completely on stainless steel mesh. The evaporation rate of oil on the BMCM was much higher than that on stainless steel mesh. Then, the volatilization slowed down because there was little oil left. At ~148 seconds, oil droplets were completely removed and BMCM returned to the original state, proving that the MHDS-based BMCM was highly repellent to be fouled by oil, which was helpful for its reutilization. The anti-oil-fouling tests of the BMCM contaminated by oil (cyclohexane) are shown graphically in Figure 7.17. Water droplets (Figure 7.17a) maintained a spherical shape at the oil–solid interface without wetting the BMCM surface. Then, the water droplet rolled off from the surface. As shown in Figure 7.17b–d, the water droplets slipped off from the BMCM surface that was contaminated by oil, demonstrating anti-oil-fouling properties were retained even after oil contamination.

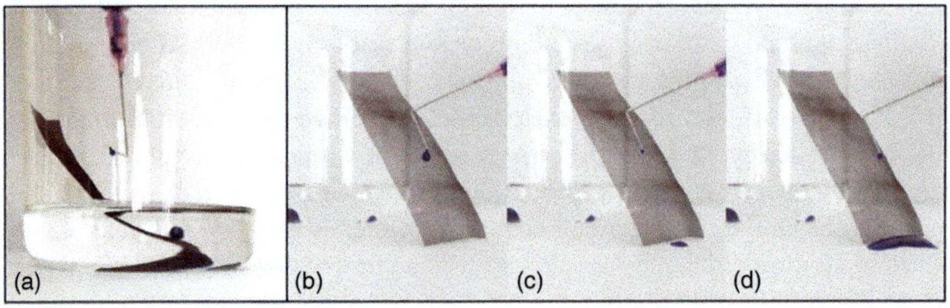

Figure 7.17 Anti-oil-fouling examination after oil contaminations. (a) Water droplet is still repelled by the BMCM immersed in oil. (b, c, and d) The BMCM maintains its water-repellent property even after being contaminated by oil. Source: Reprinted with permission from Han et al. [43]. Copyright 2013 WILEY-VCH Verlag GmbH.

7.2.5.3 Oil–Water Separation Mechanism of Butterfly-inspired Structural Materials

To further explain the internal nonwetting mechanism of the BMCM, one reasonable explanation for the above findings was that the synergistic effect of both chemical compositions and physical structures of BMCM played a key role in realizing the oil–water separation function (Figure 7.18). At the molecular level, copper stearate 3D networks containing hydrophobic alkyl chains ($CH_3(CH_2)_{16}COO$—) had a strong resistance force for water molecules and absorption force for chloroform, which was chemically conducive for repelling water droplets and attracting oil droplets. These copper stearate 3D networks acted like intelligent switches for the passage of various liquids. At the nanoscale level, SLCs were made up of dendritic bumps joining with hydrophobic groups in the space between dendritic-like structures, which corresponded to fold stripes in butterfly wings. At the microscale level, copper clusters consisted of dendritic bumps. Copper clusters had different attitudes toward oil

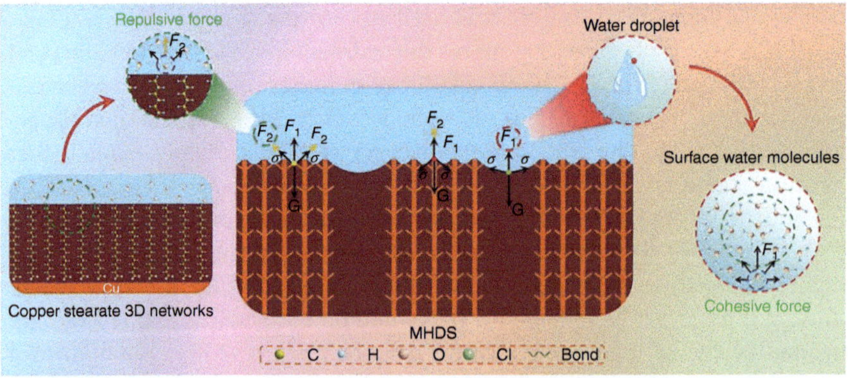

Figure 7.18 Synergistic effect of chemical compositions and MHDSs of the BMCM, realizing the selective transmission of oil–water mixtures. Illustrations show the static microforce analysis of the water and oil molecules, respectively. Here, F_1 is intermolecular force, F_2 is repulsive force, σ is surface tension, and G is gravity. Source: Reprinted with permission from Han et al. [43]. Copyright 2013 WILEY-VCH Verlag GmbH.

and water, which acted as ridges on the wing scales. Microscale copper clusters and nanoscale dendritic bumps together formed the MHDS. Illustrations demonstrated the static microforce analysis of the water molecules and oil molecules. Physically, water molecules were influenced by intermolecular force (F_1), repulsion force of long alkyl chains (F_2), surface tension (σ), and its gravity (G). These forces finally found a balance, so that a stable water membrane appeared on the surface of the MHDS. Chloroforms ran along the SCLs due to the uneven force field and the air layer was finally filled up. It was a common strategy to fabricate superhydrophobic and superoleophilic surfaces by constructing rough structures with proper components. Interestingly, the MHDS played an important role in amplifying the intrinsic wettability of stainless steel film, which dramatically increased the surface roughness. Apparently, it was the hierarchical amplification effect that brought generous rewards for the achievement of superhydrophobic and superoleophilic surfaces.

7.3 Oil–Water Separation Materials Inspired by Fish Scales

7.3.1 Underwater Superoleophobic Performance on Fish Scales

As far as we know, examples from nature in the air rather than in water media have attracted most attention from biomimetic study on self-cleaning effects. For example, seabirds, rather than fish, are polluted by oil during a shipwreck. Boats are contaminated by plankton, while fish can keep their body clean. This interesting phenomenon indicates a new superoleophobic model, which inspires us to develop a novel strategy for designing superoleophobic materials. This bioinspired superoleophobic surface with a low affinity for oil drops is better than traditional methods , which employ surfactant solutions to remove oils from interfaces [46].

7.3.2 Biomimetic Design Strategy for Oil–Water Separation Materials

Biological organisms possessing superhydrophobic properties, such as lotus leaves [47], cicada's wings [48], water strider's legs [40], and desert beetle's back [49] always inspire us to develop novel functional interfaces. Learning from nature, fabrication of multiscale hierarchical structures and modification with low surface free-energy chemical provides efficient methods in developing superhydrophobic interfaces [50, 51]. However, superoleophobic interfaces, whose oil apparent contact angles are greater than 150°, are highly dependent on the modification of fluorinated materials [52], limiting the development of new materials. Previous research mainly focus on the design and preparation of materials, namely, the control of a single factor. In fact, the interfacial properties usually involve the interactions of two or more factors. Therefore, developing new methods to manufacture superoleophobic interfaces remains a challenge (Figure 7.19).

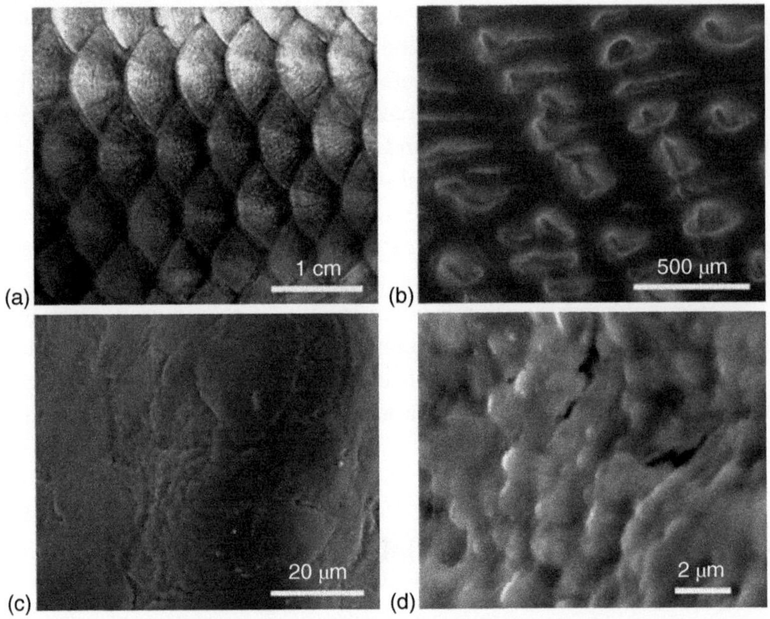

Figure 7.19 Surface structures of fish scales. (a) Optical image of the fish scales. The sector-like scales with diameters of 4–5 mm arrange into an array. (b) SEM image of fish scales. Oriented papillae with 100–300 mm in length and 30–40 mm in width existing on the fish scales are arranged in the radial direction. (c) SEM image of the fish scales at high magnification; the papillae surface is rough. (d) SEM image of the papillae. Nanostructures can be observed on the papillae. Source: Reprinted with permission from Liu et al. [53]. Copyright 2009 WILEY-VCH Verlag GmbH.

7.3.3 Fabrication of Fish-inspired Oil–Water Separation Materials

7.3.3.1 Preparation of Raw Material via a Modified Hummers' Method

A 500 mL reaction flask was charged with 0.5 g of natural flake graphite and a stir bar. About 23 mL concentrated sulfuric acid was added to the reaction system. Next, 0.5 g $NaNO_3$ was added under stirring condition and cooled down by an ice-water bath. The flask was then slowly charged with 3 g $KMnO_4$ to avoid a rapid increase in solution temperature. Once mixed, the flask was transferred to oil bath at 35 °C and stirred for about one hour. Then, 40 mL water was slowly added dropwise, forming a thick paste, and the temperature increased slowly. Meanwhile, the temperature of oil bath rose to 90 °C, and the solution was stirred for 30 minutes. Finally, 100 mL water and 5 mL H_2O_2 (30%) were added slowly. The solution was centrifuged at a speed of 9000 rpm. The obtained solid was then washed with 5% HCl and water, until some flocs were observed above the precipitates. The upper water was carefully removed. The precipitates were then dissolved into 200 mL water, followed by low-power sonication for 30 minutes to form thin graphene oxide (GO).

7.3.3.2 Underwater Superoleophobic Network with Hierarchical Nanostructures

Clean stainless steel mesh was submerged into GO aqueous solution with low-power sonication for 10 mins. After that, as-prepared mesh was dried at 60 °C. Then, the mesh was submerged into GO aqueous solution again for ~10 mins, and dried at 60 °C. Finally, GO was easily coated onto stainless steel mesh surface.

7.3.4 Characterizations of Fish-inspired Oil–Water Separation Materials

Figure 7.20 shows SEM images of uncoated and GO-coated meshes. Figure 7.20a is a typical image of the stainless steel mesh with an average pore diameter of ~38 mm (400 mesh size). The surface of a raw mesh is clean and smooth. After the raw mesh surface was coated with GO, the diameters of uncoated and coated wires were almost unchanged. The magnified view of the joint of two wires proves the GO coatings indeed exist on the mesh surface. In accordance with the transmission electron microscope (TEM) results, the 2D nanosheets have microscopic folds attributed to the increased nanoscale corrugations induced by various oxygen-containing functional groups of GO, forming a multiscale hierarchical rough surface. Figure 7.21a shows the UV–visible absorption spectrum of GO in an aqueous solution with a characteristic peak at 230 nm, which is consistent with the literature data. From the

Figure 7.20 SEM images of raw (a) and GO-coated (b–d) meshes with an average pore diameter of ~38 mm. The inset in (a) shows the magnified view of the smooth surface of a raw mesh. Source: Reprinted with permission from Dong et al. [54]. Copyright 2014 Royal Soc Chemistry.

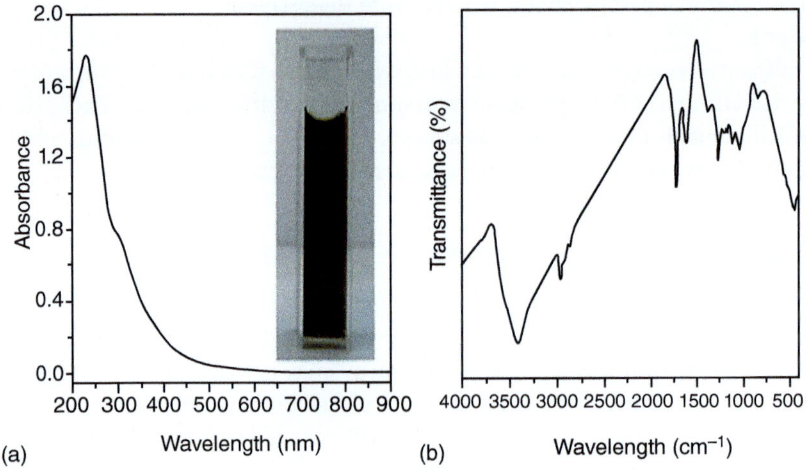

Figure 7.21 (a) UV-visible absorption spectrum and a photograph (inset) of aqueous solution with GO. (b) The FTIR spectrum of GO. Source: Reprinted with permission from Dong et al. [54]. Copyright 2014 Royal Society of Chemistry.

photograph of GO aqueous solution, it is clearly found that GO is well dissolved in water. This corresponds to the presence of its various oxygen-containing functional groups, which was demonstrated by FTIR analysis. As shown in Figure 7.21b, the FTIR spectrum of GO exhibits a series of absorption peaks at about 1730, 1400, and 1050 cm^{-1} attributed to C=O, O—H, and C—O, respectively.

7.3.5 Oil–Water Separation Evaluation of Fish-inspired Structural Materials

Compared to raw meshes, meshes coated with GO become more hydrophilic in air and superoleophobic underwater. Taking advantage of this unique wettability, meshes coated with GO were applied in gravity-driven oil–water separation. The oil–water separation process was carried out as illustrated in Figure 7.22a. A mixture of hexane (dyed with oil Red Dye 24) and water (50%, v/v) was poured onto a pre-wetted mesh coated with GO, which was fixed between two glass tubes and sealed with Teflon tape. Gravity supplies the driving force for the separation process. Because of the excellent underwater superoleophobic and low oil-adhesion properties of the GO coating, hexane was kept above the coated mesh. Afterward, water having higher density than hexane quickly passed through the coated mesh and dropped into the beaker below, which demonstrated the hydrophilicity of the GO coating. Water flux was in the range of 0.2–84 $L\ m^{-2}\ s^{-1}$, which was determined by coating times, GO concentrations, and pore diameters of raw meshes. No oil was found in the collected water. A series of oil–water mixtures were successfully separated using the same device. Heavy oils, such as chloroform and dichloroethane, were separated by slowly pouring oil–water mixtures onto pre-wetted meshes coated with GO. As-prepared meshes retained underwater superoleophobicity after 50 cycles of oil–water separation. In contrast, raw meshes or GO-coated meshes

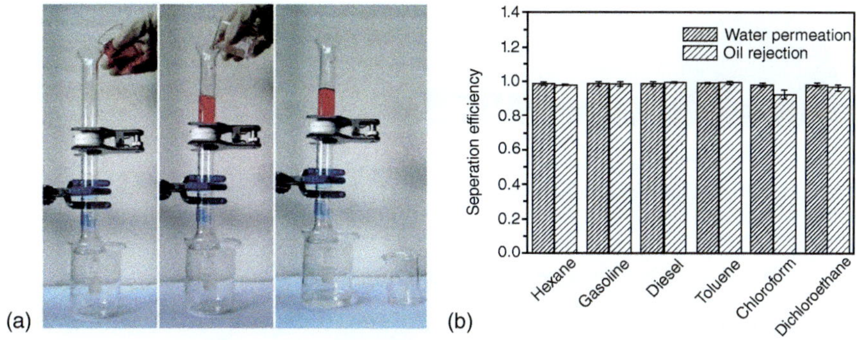

Figure 7.22 (a) Photographs of the oil–water separation process. Hexane was utilized as a typical oil and dyed with oil Red Dye 24 for clear observation. (b) The separation efficiency of the GO-coated mesh with 38 mm pore diameter for various oils in terms of water permeation and the oil rejection coefficient. Source: Reprinted with permission from Dong et al. [54]. Copyright 2014 Royal Society of Chemistry.

with a larger pore diameter, such as ~425 mm (40 mesh size), cannot be utilized to separate oil and water. It is indicated that both the GO coatings and the pore size of stainless steel meshes play a key role in the implementation of oil–water separation. To further investigate the separation efficiency of various oil–water mixtures, $m_{separation}/m_{initial}$ was employed to assess the separation efficiency, where $m_{separation}$ represents the weight of oils or water after oil–water separation, $m_{initial}$ represents the weight of oils or water before being mixed. The separation efficiency of GO-coated meshes for water or light oils is above 98%, and for heavy oils above 90%, as shown in Figure 7.22b. In this situation, the underwater superoleophobic meshes are not an ideal material for heavy oil–water separation because heavy oils will accumulate and form a barrier layer on the film surface, which could avoid water permeation. Nevertheless, underwater superoleophobic and superhydrophobic films can be regarded as complementary materials to separate oil–water mixtures depending upon the situation.

Wettability of a solid surface is mainly determined by its chemical composition and rough structures. A hydrophilic GO coating and the exquisite morphology of coated meshes provide a possibility for the preparation of underwater superoleophobic films. The wettability of water and oil on uncoated and GO-coated meshes was demonstrated by the apparent contact angle measurements. In air, the water contact angle of a raw mesh is ~120°. Compared to raw meshes, the water wetting property of GO-coated meshes is greatly enhanced, with a water contact angle lower than 10° (Figure 7.23a). In addition, GO-coated meshes keep the wet state after being taken out from the water. However, water on the raw mesh surface converges rapidly and then dries in air. As expected, the GO coating has significant influence on the surface wetting behaviors of raw meshes. Moreover, the selective adhering of water on GO nanosheets could be confirmed by testing water contact angles in oil. The water contact angle of a raw mesh is 156.4° attributed to its hydrophobicity. In contrast, the GO coating cannot capture oil on its 2D nanosheets and selectively absorbs water, resulting in a lower water contact angle

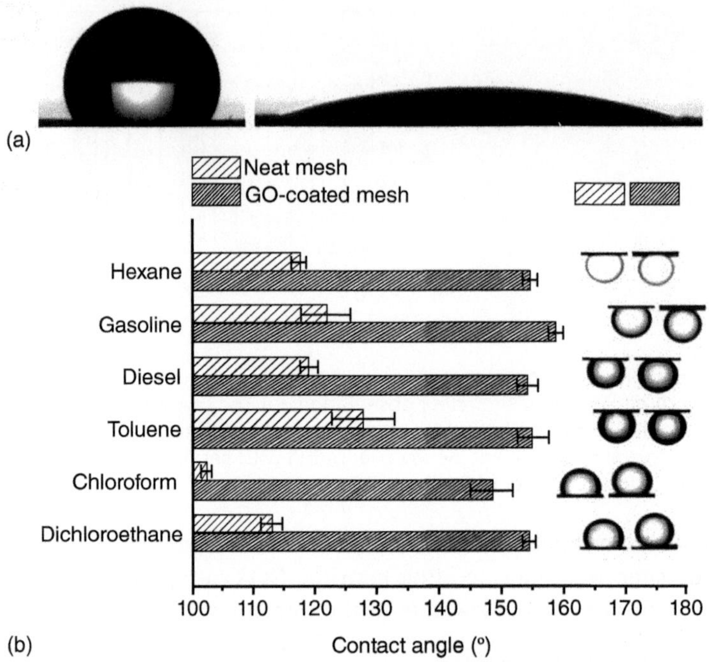

Figure 7.23 (a) Photographs of a water droplet on raw (left) and GO-coated (right) meshes in air. (b) Oil contact angles of raw and GO-coated meshes in aqueous solution. The insets are photographs of oil droplets on raw (left) and GO-coated (right) meshes in aqueous solution. Source: Reprinted with permission from Dong et al. [54]. Copyright 2014 Royal Society of Chemistry.

(89.8°). The underwater oil wettability was tested by submerging the meshes in an aqueous solution. Figure 7.23b shows the apparent contact angles of a few typical oil droplets on uncoated and GO-coated meshes in an aqueous solution. The typical photographs of oil droplets are shown in the right insets. When a GO-coated mesh is pre-wetted by water, it becomes extremely oil-repulsive for various oil droplets, such as hexane, gasoline, diesel, toluene, chloroform, and dichloroethane. The apparent contact angles of all oil droplets are above 150°, illustrating that GO-coated meshes have excellent underwater superoleophobic properties. Without a GO coating, raw meshes possess oleophobic properties. However, the apparent contact angles of all oil droplets are lower than those of GO-coated meshes. Besides, the wettability of uncoated and GO-coated meshes could be affected by the pore size of stainless steel meshes. The dichloroethane contact angles of uncoated and GO-coated meshes with pore diameters in the range of 38–425 mm were examined in an aqueous solution. The experimental results indicate that GO-coated meshes with different pore diameters are all underwater superoleophobic. All of the dichloroethane contact angles are higher than those of raw meshes. It can be found that a dichloroethane droplet is quite unstable on all GO-coated meshes and it easily rolls off the oblique surface. In contrast, a dichloroethane droplet hangs and is fixed on the raw mesh surface and is difficult to fall off from the face-down surface. These results show that GO-coated meshes possess lower oil-adhesion property in oil–water–solid three-phase system than raw meshes. In an oil–gas–solid system, both uncoated and GO-coated meshes show superoleophilic characteristics with a dichloroethane contact angle lower than 101° due to the low surface energy of oils. Different from this, underwater oleophobicity and superoleophobicity on uncoated and GO-coated meshes are realized in oil–water–solid systems thanks to the repellency between polar (water) and nonpolar (oil) molecules. Especially, the GO coatings can absorb adequate amount of water to reach a balanced state. As a consequence, when the GO coatings contact the oil droplets, water trapped in the rough nanostructures supports a strong repulsive force, resulting in higher oil contact angles and lower oil-adhesion force than raw meshes. Therefore, such low adhesion properties can efficiently prevent GO-coated meshes from being polluted or blocked up by oil during oil–water separation processes.

References

1 Quéré, D. (2008). Wetting and roughness. *Annual Review of Materials Research* 38 (1): 71–99.

2 Zhang, Q., Li, Y., Yan, Y. et al. (2020). Highly flexible monolayered porous membrane with superhydrophilicity-hydrophilicity for unidirectional liquid penetration. *ACS Nano* 14 (6): 7287–7296.

3 Zhang, Y., Chen, Y., Shi, L. et al. (2012). Recent progress of double-structural and functional materials with special wettability. *Journal of Materials Chemistry* 22 (3): 799–815.

4 Liu, K., Cao, M., Fujishima, A., and Jiang, L. (2014). Bio-inspired titanium dioxide materials with special wettability and their applications. *Chemical Reviews* 114 (19): 10044–10094.

5 Bhushan, B. and Jung, Y.C. (2011). Natural and biomimetic artificial surfaces for superhydrophobicity, self-cleaning, low adhesion, and drag reduction. *Progress in Materials Science* 56 (1): 1–108.

6 Guo, Z., Zhou, F., Hao, J., and Liu, W. (2005). Stable biomimetic super-hydrophobic engineering materials. *Journal of the American Chemical Society* 127 (45): 15670–15671.

7 Li, Y.X., Shen, J.X., Peng, S.S. et al. (2020). Enhancing oxidation resistance of Cu(I) by tailoring microenvironment in zeolites for efficient adsorptive desulfurization. *Nature Communications* 11 (1): 3206.

8 Feng, L., Zhang, Z., Mai, Z. et al. (2004). A super-hydrophobic and super-oleophilic coating mesh film for the separation of oil and water. *Angewandte Chemie-International Edition* 43 (15): 2012–2014.

9 An, J., Cui, J.-F., Zhu, Z.-Q. et al. (2014). Conductive polymer-coated mesh films with tunable surface wettability for separation of oils and organics from water. *Journal of Applied Polymer Science* 131 (18): 40759.

10 Mahringer, A., Hennemann, M., Clark, T. et al. (2020). Energy efficient, ultrahigh-flux separation of oily pollutants from water with superhydrophilic nanoscale metal-organic framework architectures. *Angewandte Chemie-International Edition* 133 (10): 5579–5586.

11 Cao, Y., Zhang, X., Tao, L. et al. (2013). Mussel-inspired chemistry and Michael addition reaction for efficient oil/water separation. *ACS Applied Materials & Interfaces* 5 (10): 4438–4442.

12 Garcia, B., Saiz-Poseu, J., Gras-Charles, R. et al. (2014). Mussel-inspired hydrophobic coatings for water-repellent textiles and oil removal. *ACS Applied Materials & Interfaces* 6 (20): 17616–17625.

13 Yan, Y., Zhang, Q., Li, Y. et al. (2020). The highly efficient collection of underwater oil droplets on an anisotropic porous cone surface via an electric field. *Journal of Materials Chemistry A* 8 (17): 8605–8611.

14 Gao, C., Sun, Z., Li, K. et al. (2013). Integrated oil separation and water purification by a double-layer TiO_2-based mesh. *Energy & Environmental Science* 6 (4): 1147–1151.

15 Yang, Y., Li, H., Cheng, S. et al. (2014). Robust diamond meshes with unique wettability properties. *Chemical Communications* 50 (22): 2900–2903.

16 Lee, C. and Baik, S. (2010). Vertically-aligned carbon nano-tube membrane filters with superhydrophobicity and superoleophilicity. *Carbon* 48 (8): 2192–2197.

17 Lee, C.H., Johnson, N., Drelich, J., and Yap, Y.K. (2011). The performance of superhydrophobic and superoleophilic carbon nanotube meshes in water-oil filtration. *Carbon* 49 (2): 669–676.

18 Deng, D., Prendergast, D.P., MacFarlane, J. et al. (2013). Hydrophobic meshes for oil spill recovery devices. *ACS Applied Materials & Interfaces* 5 (3): 774–781.

19 Sun, H., Li, A., Qin, X. et al. (2013). Three-dimensional superwetting mesh film based on graphene assembly for liquid transportation and selective absorption. *ChemSusChem* 6 (12): 2377–2381.

20 Xu, L.P., Zhao, J., Su, B. et al. (2013). An ion-induced low-oil-adhesion organic/inorganic hybrid film for stable superoleophobicity in seawater. *Advanced Materials* 25 (4): 606–611.

21 Zhang, L., Zhong, Y., Cha, D., and Wang, P. (2013). A self-cleaning underwater superoleophobic mesh for oil-water separation. *Scientific Reports* 3: 2326.

22 Xue, Z., Wang, S., Lin, L. et al. (2011). A novel superhydrophilic and underwater superoleophobic hydrogel-coated mesh for oil/water separation. *Advanced Materials* 23 (37): 4270–4273.

23 Song, S., Jing, L., Li, S. et al. (2008). Superhydrophilic anatase TiO_2 film with the micro- and nanometer-scale hierarchical surface structure. *Materials Letters* 62 (20): 3503–3505.

24 Darmanin, T. and Guittard, F. (2009). Super oil-repellent surfaces from conductive polymers. *Journal of Materials Chemistry* 19 (38): 7130–7136.

25 Hutton, S.J., Crowther, J.M., and Badyal, J.P.S. (2000). Complexation of fluorosurfactants to functionalized solid surfaces: smart behavior. *Chemistry of Materials* 12 (8): 2282–2286.

26 Yang, J., Zhang, Z., Xu, X. et al. (2012). Superhydrophilic-superoleophobic coatings. *Journal of Materials Chemistry* 22 (7): 2834–2837.

27 Kota, A.K., Kwon, G., Choi, W. et al. (2012). Hygro-responsive membranes for effective oil-water separation. *Nature Communications* 3: 1025.

28 Xin, B. and Hao, J. (2010). Reversibly switchable wettability. *Chemical Society Reviews* 39 (2): 769–782.

29 Feng, X., Feng, L., Jin, M. et al. (2004). Reversible super-hydrophobicity to super-hydrophilicity transition of aligned ZnO nanorod films. *Journal of the American Chemical Society* 126 (1): 62–63.

30 Xu, L., Chen, W., Mulchandani, A., and Yan, Y. (2005). Reversible conversion of conducting polymer films from superhydrophobic to superhydrophilic. *Angewandte Chemie-International Edition* 44 (37): 6009–6012.

31 Yu, X., Wang, Z., Jiang, Y. et al. (2005). Reversible pH-responsive surface: from superhydrophobicity to superhydrophilicity. *Advanced Materials* 17 (10): 1289–1293.

32 Zhang, L., Zhang, Z., and Wang, P. (2012). Smart surfaces with switchable superoleophilicity and superoleophobicity in aqueous media: toward controllable oil/water separation. *NPG Asia Materials* 4 (2): e8.

33 Lim, H.S., Han, J.T., Kwak, D. et al. (2006). Photoreversibly switchable superhydrophobic surface with erasable and rewritable pattern. *Journal of the American Chemical Society* 128 (45): 14458–14459.

34 Tian, D., Zhang, X., Tian, Y. et al. (2012). Photo-induced water-oil separation based on switchable superhydrophobicity-superhydrophilicity and underwater superoleophobicity of the aligned ZnO nanorod array-coated mesh films. *Journal of Materials Chemistry* 22 (37): 19652–19657.

35 Xue, B., Gao, L., Hou, Y. et al. (2013). Temperature controlled water/oil wettability of a surface fabricated by a block copolymer: application as a dual water/oil on-off switch. *Advanced Materials* 25 (2): 273–277.

36 Kwon, G., Kota, A.K., Li, Y. et al. (2012). On-demand separation of oil-water mixtures. *Advanced Materials* 24 (27): 3666–3671.

37 Niu, S., Li, B., Mu, Z. et al. (2015). Excellent structure-based multifunction of morpho butterfly wings: a review. *Journal of Bionic Engineering* 12 (2): 170–189.

38 Xu, S., Wang, Q., and Wang, N. (2020). Chemical fabrication strategies for achieving bioinspired superhydrophobic surfaces with micro and nanostructures: a review. *Advanced Engineering Materials* 23 (3): 2001083.

39 Feng, L., Zhang, Y., Xi, J. et al. (2008). Petal effect: a superhydrophobic state with high adhesive force. *Langmuir* 24 (8): 4114–4119.

40 Gao, X. and Jiang, L. (2004). Biophysics: water-repellent legs of water striders. *Nature* 432: 36.

41 Han, Z., Mu, Z., Li, B. et al. (2016). Active antifogging property of monolayer SiO_2 film with bioinspired multiscale hierarchical pagoda structures. *ACS Nano* 10 (9): 8591–8602.

42 Han, Z., Li, B., Mu, Z. et al. (2015). Fabrication of the replica templated from butterfly wing scales with complex light trapping structures. *Applied Surface Science* 355: 290–297.

43 Han, Z., Li, B., Mu, Z. et al. (2017). Energy-efficient oil-water separation of biomimetic copper membrane with multiscale hierarchical dendritic structures. *Small* 13 (34): 1701121.

44 Yang, S., Chen, L., Mu, L. et al. (2016). Graphene foam with hierarchical structures for the removal of organic pollutants from water. *RSC Advances* 6 (6): 4889–4898.

45 Rana, M., Chen, J.-T., Yang, S., and Ma, P.-C. (2016). Biomimetic superoleophobicity of cotton fabrics for efficient oil-water separation. *Advanced Materials Interfaces* 3 (16): 1600128.

46 Gómez-Sainero, L.M., Cortés, A., Seoane, X.L., and Arcoya, A. (2000). Hydrodechlorination of carbon tetrachloride to chloroform in the liquid phase with metal-supported catalysts effect of the catalyst components. *Industrial & Engineering Chemistry Research* 39 (8): 2849–2854.

47 Kwak, J.M., Kim, S.A., Hong, S.W., and Nam, H.G. (1997). Evaluation of 515 expressed sequence tags obtained from guard cells of Brassica campestris. *Planta* 202 (1): 9–17.

48 Han, Z., Wang, Z., Li, B. et al. (2019). Flexible self-cleaning broadband antireflective film inspired by the transparent cicada wings. *ACS Applied Materials & Interfaces* 11 (18): 17019–17027.

49 Parker, A.R. and Lawrence, C.R. (2001). Water capture by a desert beetle. *Nature* 414: 33–34.

50 Feng, X.J. and Jiang, L. (2006). Design and creation of superwetting/antiwetting surfaces. *Advanced Materials* 18 (23): 3063–3078.

51 Blossey, R. (2003). Self-cleaning surfaces--virtual realities. *Nature Materials* 2 (5): 301–306.

52 Tuteja, A., Choi, W., Ma, M. et al. (2007). Designing superoleophobic surfaces. *Science* 318: 1618–1622.

53 Liu, M., Wang, S., Wei, Z. et al. (2009). Bioinspired design of a superoleophobic and low adhesive water/solid interface. *Advanced Materials* 21 (6): 665–669.

54 Dong, Y., Li, J., Shi, L. et al. (2014). Underwater superoleophobic graphene oxide coated meshes for the separation of oil and water. *Chemical Communications* 50 (42): 5586–5589.

8

Underwater Bioinspired Superhydrophobic Multifunctional Surface

8.1 Underwater Writable and Heat-Insulated Superhydrophobic Paper

8.1.1 The Necessity of Heat-Insulated Superhydrophobic Paper

Conventional paper, as a conventional and indispensable material, has been widely applied in many fields, such as printing journals, books, communications, flexible electronics, packaging engineering, and energy harvesting due to its low production cost, renewability, nontoxicity, good recyclability, and mechanical strength [1–4]. However, the paper can be extremely easily destroyed when it is exposed to an external environment, such as underwater, humid condition, immersed in various liquids, and in the fire. This is mainly caused by hydrophilic hydroxyl groups of cellulose, which is the main component of paper [5]. To overcome such shortcomings, the underwater non-wet and heat-insulation paper has attracted increasingly significant attention in recent years, and substantial efforts have been devoted to improve its comprehensive performance, which greatly broaden its applications [6, 7].

Nature has long been an enormous source of inspiration for humans to design and construct large amounts of biomimetic materials with intriguing functions [8–10]. For instance, fundamental and applied research on lotus-leaf-inspired superhydrophobic materials is one of the hottest topics in materials science. Lotus leaf shows great potential to address some engineering problems due to its fancy properties including self-cleaning [11–13], drag-reduction [14, 15], oil–water separation [16–18], and anti-fogging [19, 20]. It is well known that low surface energy and high surface roughness are two key factors in the construction of superhydrophobic materials, which are prepared by controlling the chemical composition and micro-/nanoscale binary architectures of surfaces [21]. To manipulate these two factors, various advanced techniques have been developed, including electrochemical deposition [22], photolithography [23], and chemical etching [24]. Until now, there are several strategies to prepare superhydrophobic paper. Researchers have used inorganic nanoparticles (NPs) (such as TiO_2 and SiO_2) to construct nanostructures and then treated them with a hydrophobic modification. Inorganic nanoparticles can also be modified by hydrophobic solution in advance and directly applied on the paper surface to manufacture superhydrophobic paper [25–27]. Apart from

Nature-Inspired Structured Functional Surfaces: Design, Fabrication, Characterization, and Applications, First Edition. Zhiwu Han.

nanoparticles, micro-/nanoscale hierarchical structures coated with low surface free energy materials can create hydrophobic surfaces for paper [28–30]. Recently, ultralong hydroxyapatite nanowires were used as building blocks to fabricate superhydrophobic paper [31–34]. Some approaches have also been adopted to fabricate superhydrophobic paper, such as spray-deposition process [35], photopatterning process [36], rapid expansion of supercritical solutions [37], polymerization-induced phase separation [38], plasma treatment [39], and layer-by-layer assembly [40, 41]. Among these methods, the spray-deposition process can not only be scaled up easily but also does not rely on sophisticated devices [42].

In the current research, using different types of inorganic particles to construct nanoscale structures has been a universal strategy to develop superhydrophobic paper [43–45]. In particular, SiO_2 nanoparticles (NPs) have been widely applied to fabricate micro-/nanoscale hierarchical structures owing to their advantages, such as low production costs, high thermal stability, and optical transparency [46]. To prevent agglomeration when inorganic particles blend with a polymer solution, modifiers including surfactants and organic tethers are often employed to functionalize the particles [47, 48]. Among them, most hydrophobic modification materials are costly or toxic (fluorine-containing modified reagents), thus impeding their practical applications. Furthermore, these superhydrophobic coatings suffer from a conventional disadvantage of weak adhesive force with paper. Fortunately, the aforementioned issues can be effectively resolved by improving the surface roughness or incorporating adhesive materials into superhydrophobic materials. It has been indicated that one of the effective ways to increase surface roughness is to achieve roughness at the micro-/nanoscale level for creating hierarchical structures [49]. Such hierarchical binary architectures can not only enhance the fraction of trapped air pockets but also effectively decrease the contact area between the water droplets and corresponding substrates, producing a low slip angle (SA) and adhesive force [50]. On the other hand, to improve the adhesion between the superhydrophobic materials and substrates (copper mesh, glass, cotton cloth), epoxy resin (ER) is one of the considerable candidates due to its good mechanical properties and strong adhesion [51, 52].

In addition to endowing paper-based materials with superhydrophobicity, transparency is another important property with respect to superhydrophobic paper. It is well known that transparency and superhydrophobicity are generally two contradictory characteristics because of their special requirements for surface roughness. Consequently, precise control of surface roughness plays a crucial role in these two characteristics [53–55]. It would be of great practical significance if the paper exhibits multiple liquid repellency, heat resistance, self-cleaning properties, and good chemical stability, as well as mechanical robustness. Currently, investigation on superhydrophobic paper with versatile performances has become a mainstream development trend and a research hotspot [31, 36, 56].

Based on the abovementioned urgent issues, robust, nontoxic, multifunctional superhydrophobic paper with hierarchical micro-/nanostructures are presented in this work, which are inspired by the "lotus effect" in nature [57]. The characteristic of this study is that by introducing the "embedded" strategy and taking advantage of high-adhesive force of ER, SiO_2 microspheres (MPs) can be effectively filled into the gaps between adjacent fiber structures, forming more obvious microstructured surfaces. This can further improve abrasion resistance of the surface. Then SiO_2 NPs are used to construct a binary rough structure on the surface of the microstructures. By optimizing the ratio of materials, bioinspired superhydrophobic paper (BSP) has high transparency, self-cleaning ability, and good aqueous-based liquid repellency. It also retains good mechanical robustness, even under multiple abrasion and origami folding tests. In addition, the BSP shows good heat-insulation property. Apart from these performances, this kind of paper can be used for normal writing and recording even underwater. This study not only fulfills real-life requirements of paper but also expands the potential application fields of functional paper-based materials, especially in complex and harsh conditions such as diversified liquids, high temperature, underwater, and in the fire.

8.1.2 Fabrication of Structural Bioinspired Superhydrophobic Paper (BSP)

The BSP was prepared via a simple two-step process, which is systematically illustrated in Figure 8.1a. It was obtained according to the following procedures: initially, 0.5 g of SiO_2 MPs, 50 ml of absolute ethanol, and 3.5 ml of hexamethyldisilazane (HMDS) were added into a beaker, stirred for 20 minutes with a magnetic stirrer, and subjected to ultrasonic shaking for 30 minutes. The procedure endows SiO_2 MPs with hydrophobicity by immobilizing methyl groups on their surfaces [58, 59].

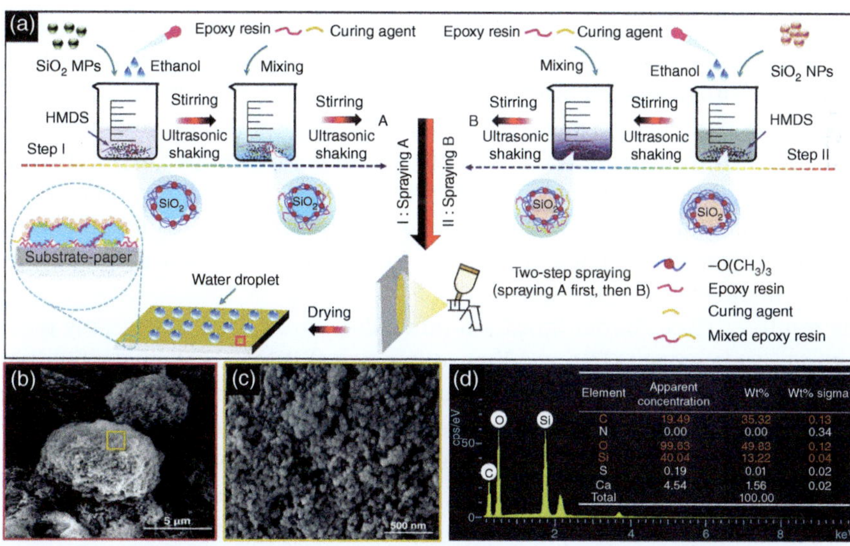

Figure 8.1 Preparation process and characterization of the BSP. (a) Schematic illustration of the preparation procedure of the BSP. (b) SEM images of the BSP presenting the micro-/nanoscale hierarchical structures. (c) Magnified SEM image of (b). (d) EDS spectra of the BSP. Source: Reprinted with permission from Jiao et al. [58] Copyright © 2020 Royal Society of Chemistry.

Next, a transparent ER solution composed of ER (0.3 g) and curing agent (0.1 g) was generated at room temperature. The transparent ER solution was added into the as-prepared solution drop by drop. After that, all of the contents were stirred for 40 minutes and then subjected to ultrasonic shaking for 40 minutes to create suspension A. For suspension B, all other procedures were the same as that of suspension A except for the amount of the silica nanoparticles. 1.0 g of SiO_2 NPs were employed in the synthesis of suspension. Subsequently, the resulting uniform suspensions were manually sprayed onto the standard A4 printer-grade paper via a small glass sprayer. The diameter of the atomizer orifice was about 1 mm. During the spraying process, the atomizing orifice was perpendicular to the paper surface. And the distance between the atomizer orifice and paper was about 10 cm in this study. Suspension A was sprayed for three times and suspension B for four times. One spraying process lasted about 20 seconds. After dried naturally, the paper was sprayed again. Finally, the sample was placed horizontally and dried at 60 °C for 30 minutes. The BSP was successfully obtained.

8.1.3 Characterizations of Bioinspired BSP Surfaces

8.1.3.1 Surface Morphology and Characterization of the BSP

There exist numerous self-assembled micro-/nanoscale hierarchical spherical structures of 6–9 μm diameter, which are similar in structures to lotus leaves. These bioinspired structures are embedded into the gaps of adjacent fibers and uniformly covered on the entire paper surface. Intuitively, the micro-/nanosized SiO_2 particles do not present in the form of uniform dispersion. Instead, they generate cross-linked and aggregative hierarchical structures, which firmly contact the paper. Compared with the conventional paper with a rough surface in the micrometer scale, the BSP possesses more distinct micro-/nanoscale hierarchical surface structures. As shown in the high-magnification scanning electron microscope (SEM) images (Figure 8.1b,c), the hierarchical structures are composed of SiO_2 MPs with a diameter of ~8.5 μm and SiO_2 NPs acting as second-order structures with a diameter of ~25 nm. This dual roughness plays a key role in the unique superhydrophobic properties of the BSP.

The results of EDS, Fourier-transform infrared spectroscope (FTIR), X-ray photoelectron spectroscope (XPS), and X-ray diffraction (XRD) indicate that silica has been successfully functionalized. It also confirms that the modified SiO_2 and ER are successfully covered on the entire paper surface. Figure 8.1d illustrates that the BSP mainly consists of C, O, Si, S, and Ca elements, while the conventional paper is composed of C, O, S, and Ca elements. Si element is obviously increased on the BSP surface. Furthermore, these elements are clearly observed in the field emission scanning electron microscope (FESEM)-mapping of the BSP. FTIR result demonstrates that the composite coating is successfully functionalized by the hydrophobic group. Then, the XPS results of the BSP indicate that the conventional paper is composed of O, Ca, and C elements, while an evident characteristic peak of Si element except for the above elements is detected in the spectrum of the BSP. The Si 2p has one strong characteristic peak centered at about 103.45 eV. One strong characteristic peak appears at $2\theta = 20.62°$, indicating the existence of silica amorphous phase.

8.1.3.2 Surface Wettability, Self-Cleaning, and Optical Transparency Performances of the BSP

To better illustrate the importance of hierarchical micro-/nanostructures, the microstructure and nanostructure surfaces were constructed on paper. SiO_2 MPs (0.5 g) mixed with 0.3 g ER and 0.1 g curing agent were merely sprayed for three times on paper after optimization. The microstructured surface showed a water contact angle (WCA) of $125.4 \pm 1.41°$. In fact, with the increase of the spraying times (e.g. four or five times) of the mixture, the paper surface was still in a hydrophobic state. In this case, it not only increased the weight but also affected the performance (such as writability and high transparency) of the paper. Therefore, from the perspective of considering the multifunctional characteristics of the paper, after optimization, spraying the mixture for three times is chosen as the best. Similarly, SiO_2 NPs (1.0 g) mixed with the same amount of ER and curing agent were sprayed for four times on the paper. Unlike the surface mentioned above, the nanostructured surface displayed the superhydrophobicity with a WCA of $157.16 \pm 1.6°$ and an SA of $8.3 \pm 1.9°$. To get the optimal surface wettability, the micro-/nanoscale hierarchical surface was also prepared. Here, different spraying times and different amounts of SiO_2 NPs were investigated. Results show that the obtained structures appeared to possess optimal superhydrophobicity with a WCA of $162.7 \pm 0.49°$ and an SA of $5.72 \pm 0.62°$. In this case, the amounts of SiO_2 NPs, ER, and curing agent were 1.0, 0.3, and 0.1 g, respectively. The spraying process was repeated for four times. In particular, with the increasing amount of SiO_2 NPs, rougher surface structures were formed. On the surface of the protruding rough structures, the top of the SiO_2 particles were clearly exposed. This is also related to the solvent evaporation process of the composite coating during heat curing. Thus, under the combined effect of the exposed rough structures and hydrophobic groups, the superhydrophobic property of the paper surface was successfully conferred. It can be further explained by the fraction of the air trapped between the micro-/nanoscale hierarchical structures. According to the Cassie–Baxter equation, the surface fraction (f_s) value can be calculated for a non-wetting state through the WCAs on the surface of the BSP (θ_c) and the paper without silica particles (θ_Y). Thus, the f_s value can be expressed by [60]

$$\cos\theta_c = f_s(\cos\theta_Y + 1) - 1 \tag{8.1}$$

The static contact angle (θ_Y) of the paper surface without SiO_2 particles is 118.7°. According to Eq. (8.1), when the amount of SiO_2 NPs increases from 0.25 to 1.0 g, the optimal hierarchical structures are formed on paper surface ($f_s = 8.7\%$). In this case, the fraction of the trapped air between them reaches a maximum ($1 - f_s = 91.3\%$). However, as the amount of SiO_2 NPs increases from 1.0 to 1.5 g, the air fraction between the hierarchical structures decreases gradually (88.75%). It weakens the superhydrophobic properties of the paper. From this point of view, the proper coverage of SiO_2 NPs plays a key role in constructing the hierarchical micro-/nanostructures. This can also be confirmed by the surface roughness that is sprayed different amounts of SiO_2 NPs. Besides, the WCAs and SAs of the BSP with different amounts of ER were also measured under the conditions of spraying SiO_2 MPs (0.5 g) for three times and SiO_2 NPs (1.0 g) for four times. The results show that as the amount of epoxy resin increases, the WCAs decrease and the SAs increase. It confirms that although ER has a bonding effect, different cross-linker amounts affect the superhydrophobic property of the paper surface. Thus, under the premise of a certain amount of SiO_2 particles, it is a good choice to add a small amount of epoxy resin, which is also one of the key factors to obtain multifunctional superhydrophobic paper. To expound the superhydrophobic property of the BSP, a schematic illustration of the paper-based surface with hierarchical micro-/nanostructures was created. Simplified wetting models for three structures were also established. This process can be explained in detail based on the Wenzel and Cassie–Baxter wetting models. Parts of the modified SiO_2 MPs are firstly filled into the gaps of adjacent cellulose structures. Then, these MPs gradually accumulate on the fibers, forming the microstructured surface, as the spraying times increase. The rough surface has the largest contact area with water droplets [61], which makes the paper hydrophobic (WCA = $125.4 \pm 1.41°$). Subsequently, after modified SiO_2 NPs are uniformly deposited onto the microstructured surface, the surface possesses the micro-/nanostructures. It is difficult for water droplets to penetrate into the spaces between the micro-/nanostructures, thus imparting superhydrophobic property. Moreover, the presence of the methyl groups on silica surface can further enhance the superhydrophobic property. Therefore, the water droplets have a minimum contact area with the paper surface under the combined action of the micro-/nanostructures and low surface energy. Compared with the hierarchical surface, the nanostructured surface has a greater contact area with water droplets. It leads to a reduction in the WCA ($157.16 \pm 1.6°$) [61]. To further verify the superhydrophobic properties of the BSP, the bounce dynamics experiment of the water droplet was performed. The water droplet bounces off the BSP quickly without leaving any trace. So, a shorter interaction time at the interface provides good evidence for water repellency of the BSP. In addition, the theoretical contact time of the impacting water droplet was calculated to validate this observation. The theoretical contact time between the droplet and the BSP surface can be expressed by [62]

$$t_c = 2.6\sqrt{\rho D_0^{\,3}/(8\gamma)} \tag{8.2}$$

where ρ is the density of the water droplet, D_0 is the diameter of the water droplet, and γ is the surface tension of the water droplet. In particular, the theoretical contact time (t_c) is irrelevant to the impact velocity but linearly related to the droplet size [63]. Here, the diameter of the water droplet is 2.80 mm, and t_c is 16.05 ms. The result is consistent with the experimental contact time ($t = 16.4$ ms).

Paper, as an indispensable material in daily life, is often exposed to dust and different liquids, and this can affect readability and writability of paper and even lead to its deterioration and destruction. Therefore, the self-cleaning property of the BSP is performed. When the BSP is placed with a slight angle, the water droplets can easily roll off the surface in a spherical shape and clean away fine sand particles.

No droplets are left, showing good self-cleaning abilities of the BSP. Additionally, transparency is another important factor for the BSP, which affects the readability of paper. Importantly, visibility of the BSP does not change significantly after spraying superhydrophobic coating. Thus, it can be concluded that this coating can not only endow paper with water repellency property but also keep its inherent properties, just like conventional paper.

8.1.3.3 Aqueous-Based Liquids Repellency of the BSP

In addition to water-repellency and self-cleaning properties, paper need to repel other different liquids for practical applications. So, six different beverages are used to investigate the repellent property of the BSP. When six different beverages, including green tea, red tea, Coca Cola, coffee, orange juice, and milk, are slowly poured onto the BSP surface, they roll down the surface quickly without wetting the BSP. In order to explain this characteristic, the corresponding relationships between the contact angles and surface tension of different droplets are also measured (Figure 8.2a).

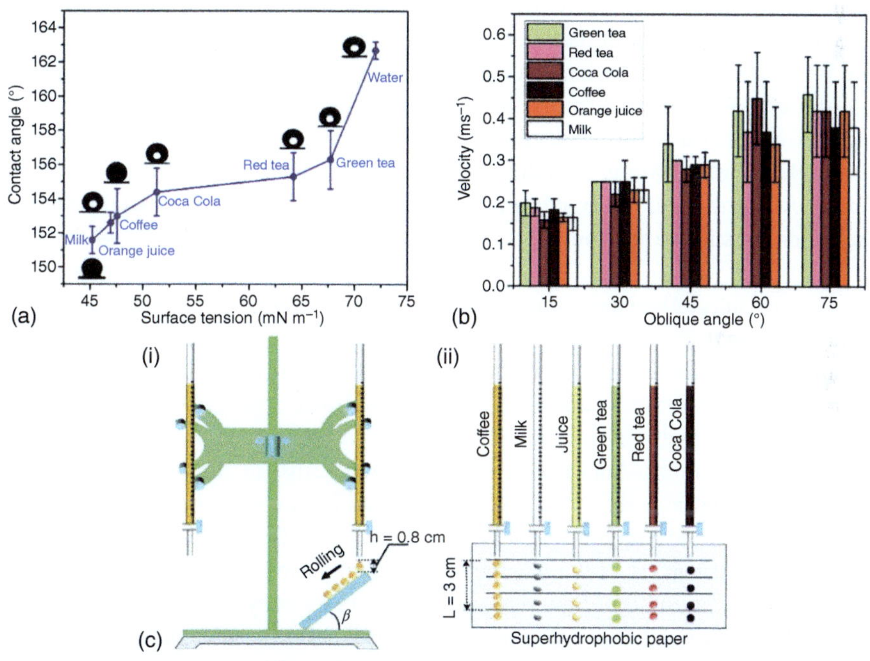

Figure 8.2 Liquid repellent performances of the BSP for different beverages and the effect of different inclination angles on average rolling speed of different beverages. (a) The corresponding relationships between the contact angles and surface tension of different droplets on the surface of the BSP. (b) Schematic illustration of six different beverages, including green tea, red tea, Coca Cola, coffee, orange juice, and milk, were dripped on the surface of BSP to conduct the average rolling velocity test of droplet at different inclination angles. (i) The front view of the schematic illustration. (ii) The side view of (i). (c) The average rolling velocity of six different beverages as a function of oblique angles. Source: Reprinted with permission from Jiao et al. [58] Copyright © 2020 Royal Society of Chemistry.

Results suggest that the contact angles of six different beverages (from milk to green tea) gradually increase as the surface tension increases. Among these six beverages, the BSP possesses the optimal liquid-repellent property for green tea with a contact angle of $156.3 \pm 1.7°$. This is because the smaller the surface tension of the droplets is, the more stable the contact with the solid surface. Thus, in comparison, milk is more firmly in contact with the BSP surface.

To further explore the effect of the BSP on repellency of different droplets, the average rolling velocity of six different beverages on the BSP surface was also evaluated at different tilt angles (Figure 8.2b). In our test, different inclination angles of the BSP were performed ($\beta = 15°, 30°, 40°, 60°$, and $75°$, respectively). The end of the titrator was fixed at 0.8 cm from the BSP surface. The rolling distance was designed to be 3 cm. From the results of the average velocity (Figure 8.2c), the approximate order at the same tilt angle is green tea, red tea, Coca Cola, coffee, orange juice, and milk. Take red tea as an example, when the droplet rolls on the surface with an inclination of 15°, the average rolling velocity is $0.188 \pm 0.02\ \mathrm{m\,s^{-1}}$, while the milk reaches $0.164 \pm 0.03\ \mathrm{m\,s^{-1}}$. On the other hand, the average rolling velocity of the same droplets tends to increase gradually with the increasing tilt angles under different inclination conditions. For example, the average rolling speed is $0.182 \pm 0.03\ \mathrm{m\,s^{-1}}$ when the coffee rolls on the surface with an inclination of 15°. Its speed reaches $0.38 \pm 0.11\ \mathrm{m\,s^{-1}}$ at an inclination of 75°. For six different beverages, as surface tension decreases, the droplets become more firmly in contact with the BSP during rolling. This directly leads to a decrease in the roll-off speed of the droplets. Therefore, the BSP has aqueous-based liquid repellency at any angle. The greater the surface tension, the stronger the performance.

8.1.3.4 Mechanical Abrasion Durability, Chemical Durability, and Boiling Water Resistance of the BSP

Mechanical robustness has been one of the key issues for a long time, which has seriously limited the application of superhydrophobic materials, especially for paper-based superhydrophobic materials. To solve this problem, the robust BSP with hierarchical micro-/nanostructures by introducing ER as an adhesive was prepared in this work. A set of tests were conducted for the BSP, including abrasion test, bending test, and origami folding test. The WCAs with two abrasion cycles are presented in Figure 8.3a. As the number of abrasion cycles increases, the WCAs of the BSP decrease gradually. However, when the number of cycles reaches six times, WCAs still maintain a higher value of $159.4 \pm 1.49°$. After the BSP surface was subjected to wear, the newly exposed SiO_2 NPs mixed with ER began to function again. In the process of wear, a certain layer of SiO_2 NPs and MPs reconstructs the roughness under the action of ER, thereby achieving superhydrophobic property. Thus, by introducing the ER and constructing the hierarchical micro-/nanostructures, abrasion-enhanced BSP can be obtained. The composite coating is flattened and the fiber structure of the paper is clearly visible. However, due to the existence of composite coating between adjacent fiber structures, it plays an important role in maintaining the hydrophobicity of BSP. This is why we have adopted a two-step spraying deposition process to create hierarchical micro-/nanostructures.

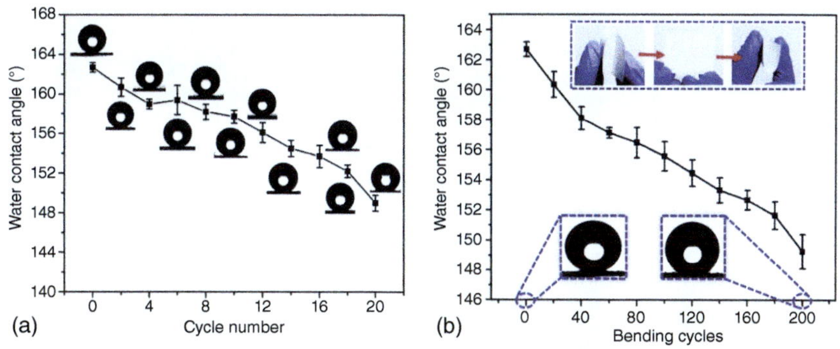

Figure 8.3 Mechanical properties testing of the BSP. (a) WCAs of the BSP after each wear test cycle. (b) WCAs of the BSP as a function of bending cycles. The inset images represent one bending cycle from bending upward and bending in the opposite direction. Source: Reprinted with permission from Jiao et al. [58] Copyright © 2020 Royal Society of Chemistry.

Meanwhile, the bending test of the BSP for 200 times is shown in Figure 8.3b. The process of bending upward and downward is defined as one bending cycle as shown in the inset of the upper-right corner in Figure 8.3b. As the number of the bending cycles increased, the WCAs decreased slightly. When the bending cycles reached 200 times, a WCA of $149.3 \pm 1.1°$ was obtained, exhibiting a critical state of superhydrophobicity and hydrophobicity. Furthermore, an origami folding test on mechanical stability of the BSP was performed. The BSP was folded into three different shapes, including paper crane, boat, and airplane. Obviously, the folded BSP surface did not leave any traces of water during the whole procedure. This indicates that the BSP still exhibited superhydrophobicity after undergoing multiple folds.

Paper-based superhydrophobic materials are also susceptible to damages in extreme environments such as strong acids/bases and boiling water. So, the BSP is immersed in strong acid/alkali and boiling water. When the BSP is immersed in the boiling water for 85 seconds, it still maintains superhydrophobicity with a WCA of $150 \pm 0.6°$. The BSP exhibits good superhydrophobicity with the WCAs of larger than 150° at all pH conditions, especially for pH = 1 and pH = 14 with a WCA of $152.6 \pm 1.6°$ and $158.7 \pm 2.9°$. This indicates that the BSP has stable chemical durability at all pH ranges. In short, the presence of the micro-/nanostructures effectively reduces the chance of contact between the strong acid/alkali, boiling water, and the paper, thereby protecting the paper from damage.

8.1.3.5 Heat-Insulation Performance of the BSP

Paper is extremely flammable when approaching a heat source. This significantly limits its application and causes huge economic losses. It is therefore urgent and important to prepare functional paper that achieves both waterproof and heat insulation. To evaluate the thermal stability of the BSP, thermal gravimetric analysis (TGA) test of the superhydrophobic composite coating was performed as shown in Figure 8.4a.

The curve of the coating decreases slightly before 200 °C, which may be due to evaporation of the water adsorbed on the coating surface, resulting in a small amount

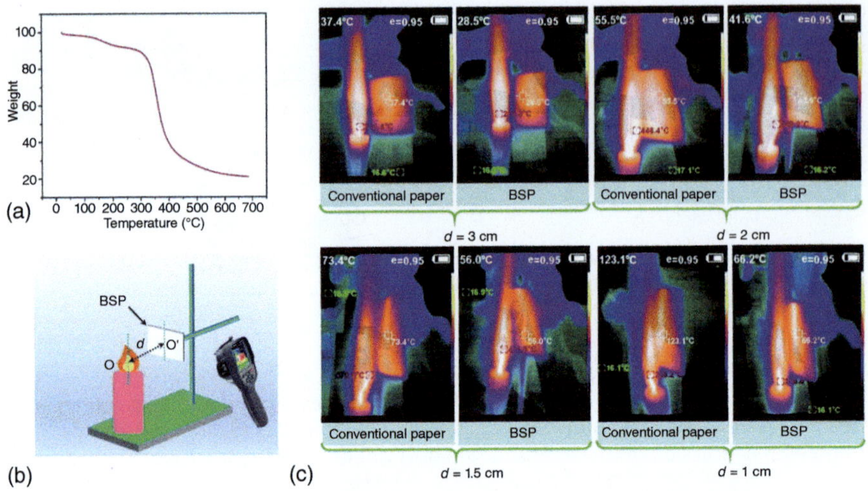

Figure 8.4 Heat-insulation property of the BSP. (a) TGA test of the superhydrophobic composite coating. (b) Schematic diagram of detecting the surface temperature of the conventional paper and the BSP at different distances using thermal imaging camera. (c) Digital photographs to present the surface temperature of the conventional paper and the BSP at different distances at 3, 2, 1.5, and 1 cm, respectively. Source: Reprinted with permission from Jiao et al. [58] Copyright © 2020 Royal Society of Chemistry.

of weight loss. Then, the curve begins to slowly decrease between 200 and 320 °C. This is caused by the burning of the $-OSi(CH_3)_3$ groups in HMDS on the BSP surface that are used to modify SiO_2 MPs and NPs. The decline in this temperature range also demonstrates that HMDS is successfully modified on the silica surface. From 320 to 600 °C, there is a sharp decline in the curve due to the burning of the ER. The curve then remains relatively stable. This is mainly due to the presence of Si—O—Si bonds in the coating. In order to verify the heat-insulation properties of the BSP, the surface temperatures of the conventional paper and the BSP were detected at different distances by thermal imaging camera, as shown in the schematic diagram of Figure 8.4b. As the center distance between the flame and the paper gradually shortens, the surface temperature of the paper increases (Figure 8.4c). It is worth noting that the surface temperature of the BSP is significantly lower than that of the conventional paper. When the center distance reaches 1 cm, the temperature of the conventional paper is 123.1 °C, while only 66.2 °C for the BSP.

8.1.3.6 Underwater Writable Performance of the BSP

In most cases, paper is usually operated in dry conditions for writing or painting. However, paper can easily lose these properties in a humid environment such as underwater. Based on this, an underwater writing test was carried out. Figure 8.5a shows the schematic diagram of the process of underwater writing on the paper surface. The conventional paper and the BSP were immersed in a glass petri dish containing water. Then, a blue pencil and a red pencil were used to write on the BSP and the conventional paper, respectively. The surface wettability of the BSP and the conventional paper was measured immediately after the water was poured

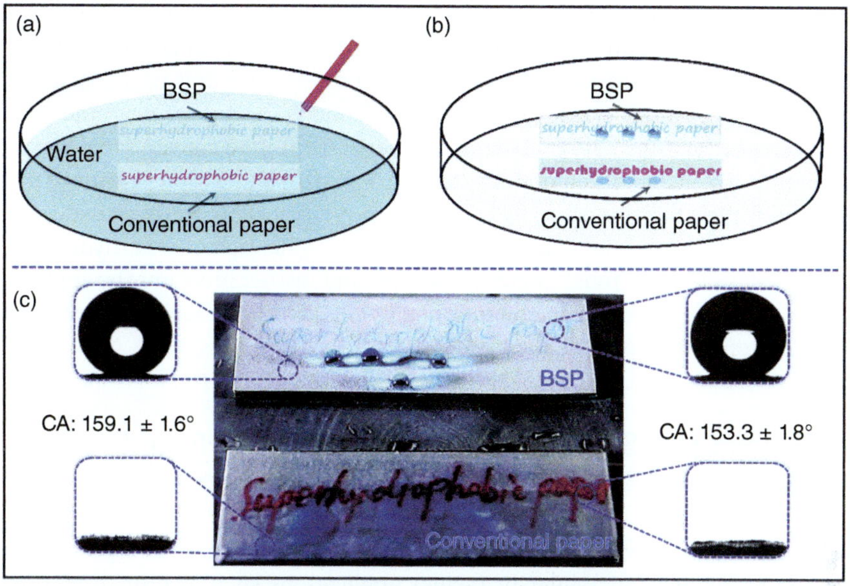

Figure 8.5 Writable underwater performance. (a) The BSP and conventional paper were immersed in glassware with water. (b) Underwater writing on the BSP and conventional paper. (c) WCAs of the BSP and conventional paper after writing. Source: Reprinted with permission from Jiao et al. [58] Copyright © 2020 Royal Society of Chemistry.

from the glass petri dish (Figure 8.5b). As shown in Figure 8.5c, the BSP remains a dry surface after writing. The handwriting is still clearly visible. On the contrary, the conventional paper surface has been completely wetted by water. The handwriting is also slightly damaged. In addition, the droplets exist in spherical shape on the BSP. The word-free region displays good superhydrophobic properties with a WCA of $159 \pm 1.6°$. Surprisingly, in the region with handwriting, the surface also maintains the superhydrophobic properties with a WCA of $153.3 \pm 1.8°$. However, in the word-bearing area and the word-free area of the conventional paper, the water droplets spread out quickly, exhibiting the superhydrophilic properties. The underwater writable property of the BSP provides more favorable and hard evidence for stronger adhesion between the superhydrophobic coating and the paper substrate.

Inspired by the lotus leaf, the biomimetic fluorine-free BSP with hierarchical structures were successfully fabricated through a two-step spray-deposition process. Thanks to the merits of this strategy, the SiO_2 MPs are more firmly filled into the gaps of adjacent fiber structures. And the SiO_2 NPs are distributed on the microstructured surface uniformly. This endows the BSP with robust superhydrophobic and self-cleaning properties, showing a WCA of $162.7 \pm 0.5°$ and an SA of $5.7 \pm 0.6°$. Benefitting from the synergistic effect of SiO_2 and ER, the BSP with fluorine-free coating integrates good aqueous-based liquid-repellency, mechanical robustness, chemical stability, heat insulation, and underwater writable properties. The greater the surface tension of beverages is, the stronger repellent property the BSP performs.

In particular, the hierarchical structures of the BSP exhibit abrasion-enhanced resistance for the mechanical performance. It can bear the abrasion and bending tests for up to 20 and 200 cycles, respectively. Moreover, the BSP presents stable chemical durability at all pH ranges. When the flame is getting closer to the BSP, it also shows the heat-insulation effect in comparison with the conventional paper. More importantly, even written underwater, the area with and without words of the BSP can both maintain the superhydrophobic properties. This study provides an important inspiration for the nonpolluting practical application of superhydrophobic coating on a variety of rough material surfaces and further broadens the application range of paper, especially in complex and harsh conditions such as diversified liquids, high temperature, underwater, and in the fire.

8.2 Bioinspired, Superhydrophobic, and Paper-Based Strain Sensors for Wearable and Underwater Applications

There is currently a growing demand for flexible strain sensors with high performance and water repellency used in various applications such as human motion monitoring, sweat or humidity detection, and certain underwater test fields. Among these strain sensors, paper-based ones have drawn increasing attention because they coincide with the future development trend of environmentally friendly electronic products. However, paper-based electronics are easy to fail when they encounter water and thus cannot be applied under humid or underwater circumstances. Herein, based on a strategy of coupling bionics inspired by lotus leaf and scorpion, which are known for the superhydrophobic characteristic and ultrasensitive vibration-sensing capacity, respectively, a paper-based strain sensor with high performance and water repellency was successfully fabricated. As a result, the strain sensor exhibits a gauge factor of 263.34, a high strain resolution (0.098%), a fast response time (78 ms), excellent stability over 12 000 cycles, and a water contact angle of 164°. Owing to the bioinspired structures and function mechanisms, the paper-based strain sensor can not only serve as regular wearable electronics to monitor human motions in real time but also detect underwater subtle vibration, demonstrating its great potential for numerous applications such as wearable electronics, water environmental protection, and underwater robots.

8.2.1 Preparation and Principle of Bioinspired Strain Sensor

8.2.1.1 Design and Fabrication of Bioinspired Strain Sensor

The design concept of the bioinspired paper-based strain sensor with micro-/nanostructures is illustrated in Figure 8.6. The visual system of scorpions is highly degraded after long-term evolution due to their nocturnal hunting habits. For better survival, scorpions evolved another organ slit sensillum, distributed on the walking legs, to make up the degradation of the visual system and to realize the sensing function of surrounding signals, which helps scorpions sense and locate

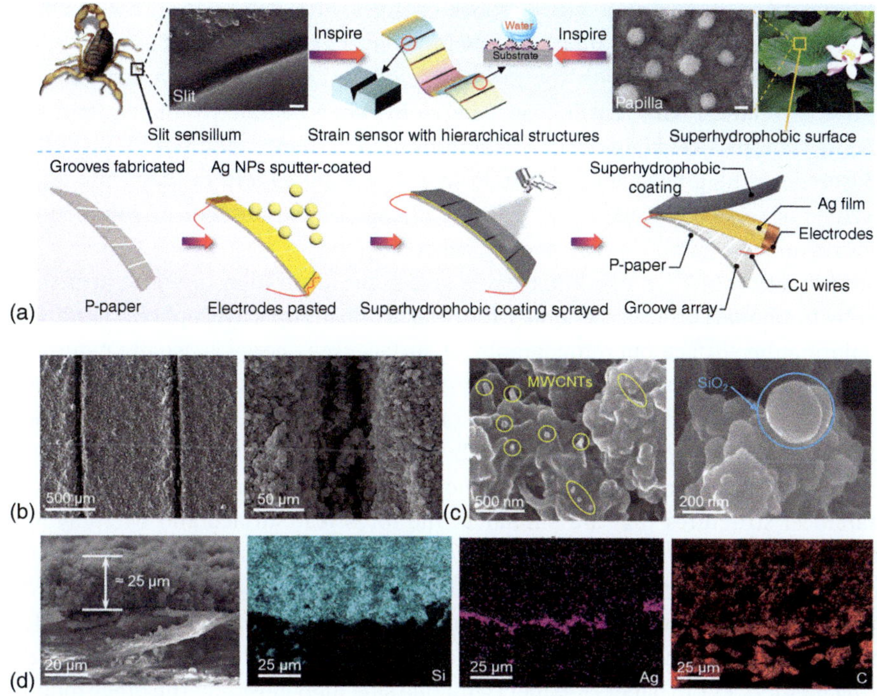

Figure 8.6 A paper-based, superhydrophobic strain sensor inspired by a strategy of coupling bionics. (a) The design concept and fabrication process of the bioinspired strain sensor based on a strategy of coupling bionics. SEM images of the slit unit geometry of scorpion's ultrasensitive vibration-sensing organ and the surface morphology of lotus leaf, respectively. Scale bar: 2 and 5 μm. (b) SEM images of the surface morphology of the strain sensor. (c) Cross-sectional SEM image (far left) and energy dispersive X-ray spectroscopy (EDS) analysis for the cross section of the strain sensor in different elements, showing that the thickness of the coatings is ≈25 μm and the sensor is composed of a paper layer, Ag layer, and superhydrophobic layer, from bottom to top. Source: Reprinted with permission from Liu et al. [64] Copyright © 2020 American Chemical Society.

the vibration sources [65–67]. Basically, silt sensillum of scorpions can be excited by weak stimuli such as an insect's walking movements 20 cm away, or a single grain of sand dropped within 5 cm of the tarsus, reflecting that the slit sensillum has a powerful sensing capacity so that we can learn from it to develop strain sensors with ultrahigh sensitivity [68]. It can be observed that slit sensillum is composed of several slits with a microscale. We further optimized the slit geometry of scorpion's slit sensillum into grooves with V-shaped cross section, for better enhancement of the mechanical–electrical response of piezoresistive strain sensors. A corresponding three-dimensional model was built and set for finite element analysis. The simulation result shows that great deformation occurs at the bottom of the groove when the model is subjected to compressive pressure load. Obviously, the distance between the two walls of the groove decreases, and at the bottom of the groove, the two walls of the groove even touch each other. When functional material is tiled over the substrate surface designed with grooves, the contact area of the functional material at every groove changes with deformation. Obviously, the height of the overlapped functional material layer at the bottom of every groove has a positive correlation with the magnitude of applied compressive load and a negative correlation with tensile load. Therefore, when conductive materials such as Pt, Ag nanoparticles (Ag NPs), and multiwalled carbon nanotubes (MWCNTs) are layered on the predesigned substrate surfaces or doped within them, the total resistance of sensor will vary owing to the change in the number of conducting pathways and the initiation of tunneling effect [69].

Besides, natural plants like lotus leaves give great inspiration to create artificial superhydrophobic surfaces. Generally, creating artificial superhydrophobic surfaces by tailoring the surface microtopography or surface free energy is usually required to endow a material surface with superhydrophobic properties, for example by applying an external force, lithography, or chemical vapor deposition [70]. The most convenient and universally applicable way to build superhydrophobic surfaces for sensors is by employing the natural morphology of nanomaterials to construct hierarchical structures that are similar to the lotus leaf surface and by spraying directly on the target surfaces in a coating way [71]. The proposed bioinspired strain sensor with superhydrophobic characteristics was fabricated by coating as-prepared suspension on a microstructured paper substrate. Firstly, grooves with microscale and V-shaped cross sections were fabricated on the paper substrate by a mechanical cutting method. Photo paper (P-paper) with one side high gloss and the other rough was adopted here due to its higher breaking strength and better wrinkle resistance than other commonly used papers. The paper was cut into strips with a size of 30 mm × 10 mm ($l \times w$), following with the microscale groove array fabricated on the rough side of the paper, as the rough side is porous, which will be more suitable for adsorbability of other materials. The interval between grooves was set as 1 mm. Then Ag NPs were sputtered on top of the paper, for enhancing the conductivity of the strain sensor. To avoid the electrodes of the sensor directly touching the water when used underwater or in a humid environment, the electrodes were inserted on the ends of the Ag layer before spray-coating the superhydrophobic layer.

As for the preparation of superhydrophobic coating, HMDS and n-butyl acetate were used to modify the surface energy of silicon dioxide (SiO_2) NPs and MWCNTs, respectively, while n-butyl acetate was used to disperse the hybrid nanomaterials. Polydimethylsiloxane (PDMS) was then dissolved through magnetic stirring and sonication to form a suspension, acting as the matrix for the SiO_2/MWCNTs layer. Since PDMS has excellent viscoelasticity, it will make the SiO_2/MWCNTs suspension easier to form a layer and also make the coating more flexible. Compared to the heavy use of polymer materials in some reported strain sensors, the usage of PDMS here only accounts for a very small mass ratio of the whole sensor and is not acting as a flexible substrate so that has less damage to the environment. Finally, the superhydrophobic coating was sprayed on top of the Ag layer and totally covered the two electrodes. After a heating process to dry the superhydrophobic coating, the strain sensor was successfully fabricated.

8.2.2 Structure and Morphology Characterization

The surface morphology of the paper-based strain sensor was carefully observed. Parallel grooves can be clearly seen on the sensor surface. Also, MWCNTs with a large aspect ratio and SiO_2 NPs with a spherical shape can be observed too. They built hierarchical structures similar to the surface of the lotus leaf and thus form a relatively rough surface to work as a superhydrophobic layer. Besides, the cross-sectional SEM image of the sensor indicates that the superhydrophobic coating and the Ag layer only have a total thickness of ≈25 μm. Then the cross-sectional element distribution maps of the paper-based strain sensor were analyzed in order to demonstrate the material composition of the sensor matches its fabrication. The results show that the Si element distributes obviously on the top surface of the sensor. More carbon (C) element is distributed in the lower half than in the upper half due to the existence of the paper layer. And a very thin layer formed by Ag element separates the superhydrophobic coating from the paper substrate. In short, in terms of material composition, the fabricated paper-based strain sensor is made up of the paper layer, Ag layer, and superhydrophobic layer, and in terms of the structural framework, the sensor consists of V-shaped grooves inspired from scorpion and papilla-like structures inspired by lotus leaf.

8.2.3 Sensing Performance and Working Mechanism

On the basis of the scorpion's ultrasensitive vibration-sensing organ that consists of many slits, we optimized the slit's structure into V-shaped grooves to promote the sensitivity of the resistance-type sensor. Figure 8.7a,b shows that the fabricated grooves have an expected V-shaped cross section (highlighted in orange dotted line), and the sensor has a thickness of ≈310 μm. In addition, it can be carefully observed that the edges of the groove are slightly higher than the surrounding area due to the accumulation of paper scraps caused by the mechanical cutting method (marked in yellow dotted line in Figure 8.7b). Based on the simulation and theoretical analysis, the grooves with V-shaped cross sections could play an important role in the influence of sensor's resistance variation. When the sensor was subjected to compression stress, more Ag NPs and MWCNTs that distributed on the two walls of each groove would connect with each other, causing an increase in the number of conductive pathways and thus the tunneling effect, in turn leading to a significant decrease in the sensor's resistance [64, 69, 72]. Inversely, if tension force was applied to the sensor, disconnections between the Ag NPs and MWCNTs that distributed on the two walls of each groove would occur and result in an increase in the sensor's resistance.

Besides, micro/nano cracks were generated on the superhydrophobic coating layer due to the mismatches of the superhydrophobic coating and paper substrate in the thermal expansion coefficient and Young's modulus after the drying process (Figure 8.7c). Instantaneous disconnection–reconnection of micro/nano cracks during loading–unloading also endowed the strain sensor with high sensitivity. So,

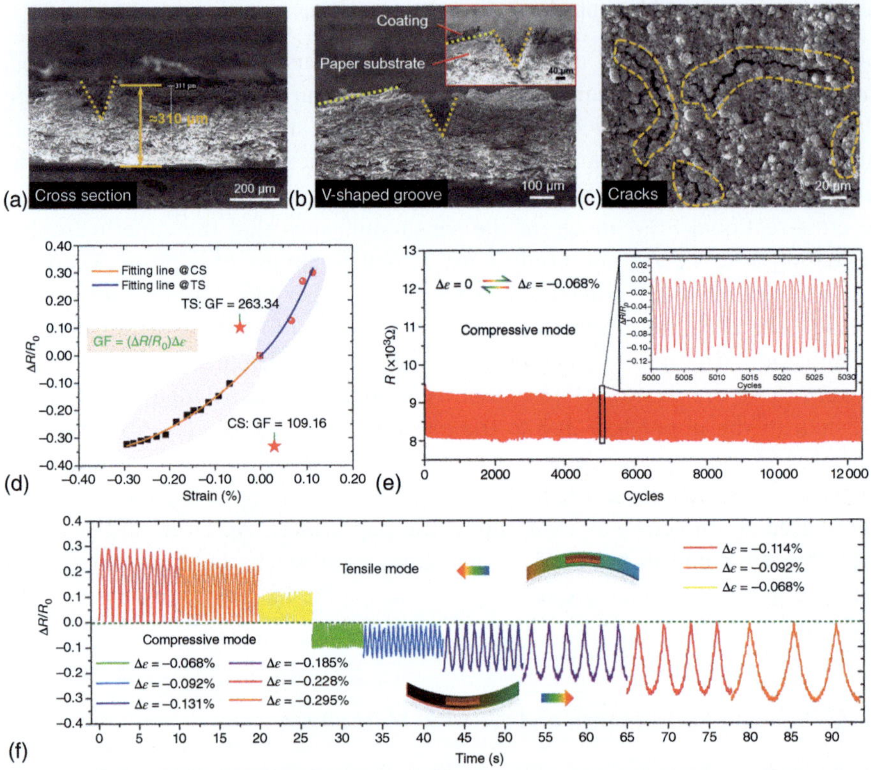

Figure 8.7 Figure 8.7 Morphology and performance characterization of the paper-based strain sensor. (a) SEM image of the cross section of the strain sensor, the orange dotted line representing the profile of the fabricated groove. (b) A magnified view of the groove with a V-shaped cross section. Inset shows the superhydrophobic coating on the paper substrate and the edges of the groove slightly higher than the surrounding area (yellow dotted line). (c) SEM image of generated micro/nano cracks on the conductive superhydrophobic coating. (d) Normalized relative resistance variation as a function of compressive/tensile strain induced by bending. (e) Resistance variation of the strain sensor under cyclic compression from $\Delta\varepsilon = 0\%$ to $\Delta\varepsilon = -0.068\%$ over 12 000 cycles. Inset: close-up of the selected area. (f) Real-time variation in the resistance under repetitive tension or compression from $\Delta\varepsilon = -0.295\%$ to $\Delta\varepsilon = 0.068\%$. Source: Reprinted with permission from Liu et al. [64] Copyright © 2020 American Chemical Society.

the bulk resistance of MWCNTs–polymer composites is mainly determined by the number of carbon nanotubes (CNTs) in the elastic polymer and the spacing between the MWCNTs. As a result, the width of micro/nano cracks would reduce under compression, making more MWCNTs connect with each other and even triggering the tunneling effect, leading to an increase in the number of conductive pathways and a decrease in the sensor's resistance. Conversely, the micro/nano cracks would be enlarged and even propagated upon tension, making more MWCNTs disconnect with each other and weaken the tunneling effect, causing a drop in the number of conducting pathways and an increase in the sensor's resistance. In short, both the microgrooves with V-shaped cross sections and the micro/nano cracks on the surface have a positive impact on the sensor's sensitivity.

After the analysis of the working mechanisms of the bioinspired strain sensor, the sensor's fundamental strain sensing performance was investigated, including its sensitivity, stability as well as responses to different strains. The sensitivity of the sensor is defined by the gauge factor (GF), which is calculated by $GF = (\Delta R/R_0)/\Delta\varepsilon$. Here, $\Delta\varepsilon$ represents the applied strain of the sensor, R_0 is the initial resistance, and ΔR is the difference between the instantaneous resistance and the initial resistance (Figure 8.7d). By measuring the thickness of the sensor (t) and the corresponding curvature radius (r) of the sensor upon bending, the applied strain can be calculated by $\Delta\varepsilon = (\pm t)/2r$ [73–75]. Responses were measured under applied tensile strains (TS) up to 0.12% and compressive strains (CS) from −0.30% to 0%. The sensor exhibits a corresponding GF of 263.34 and 109.16 upon the tensile mode and compressive mode, respectively, which is much higher than other paper-based strain sensors reported recently. Next, to evaluate the durability and stability of the strain sensor, over 12 000 cycles under a cycling compression strain from 0% to 0.068% were conducted (Figure 8.7e). It can be clearly observed that the real-time $\Delta R/R_0$ value varies periodically and no obvious fluctuation or drift of each cycle, except the first ≈200 cycles where the sensor exhibits a slightly downward trend. After that, the sensor shows excellent cycling stability and repeatability during the cyclic loading–unloading process. The origin of this phenomenon is that the micro/nano crack structure evolution on the superhydrophobic layer and the inner CNTs conductive networks turns to be stable gradually after the damage–reconstruction process via the initial several stretching/relaxing cycles. Moreover, the relative change of the resistance for the first 2000–2100, 6000–6100, and 10 000–10 100 cycles, reflects the universal repeatability and durability of the flexible strain sensors (Figure 8.7f). The sensor maintained fast, stable, and highly reproducible responses under tensile or compressive cycles with different strains. As the applied strain increases, the amplitude range of the $\Delta R/R_0$ increases too. When the applied tensile strain or compressive strain is low about 0.068%, the real-time maximum $\Delta R/R_0$ value reaches 0.1 approximately, showing that the sensor is suitable for detecting an extreme small-scale external stimulus and also implying that the sensor indeed has ultrahigh sensitivity. Moreover, as seen in Figure 8.8a,b, when the sensor was subjected to a cyclic compressive strain from 0% to −0.418%, the relative resistance variation of the sensor dropped from 0 to −0.32. However, when the applied cyclic compressive strain varied from 0% to −0.516%, the relative resistance variation of the sensor would change periodically from 0 to −0.37. In other words, the relative resistance variation decreases from −0.32 to −0.37 when the applied compressive strain changes from −0.418% to −0.516%. So, it can be drawn that the sensor has a strain resolution of at least 0.098%. The response time, another crucial index, was also investigated by the application of a quasi-transient stretching to the sensor. As shown in Figure 8.8c, the sensor exhibited a response time of 78 ms, which is much faster than that of other strain sensors reported recently.

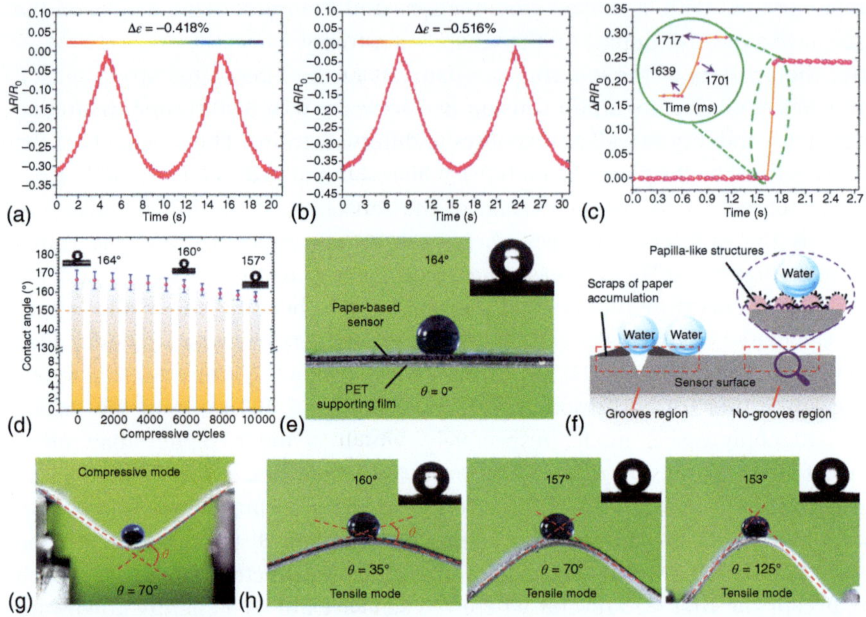

Figure 8.8 Sensing performance and superhydrophobic property of the paper-based strain sensor. (a, b) Real-time relative variation in resistance under repetitive compression at a strain of −0.418% and −0.516%, respectively. (c) Time response of the sensor upon applying a quasi-transient step strain from 0% to 0.1%. Inset: Enlarged figure of the response time. (d) WCAs of the sensor tested after different repeated compressive cycles. (e) Optical photograph of water droplet on the sensor surface. Inset: Enlarged photograph of the WCA. (f) Illustration of the bioinspired structures in the sensor's superhydrophobic characteristic. (g) Optical photograph of water droplet on top of the sensor under compressive mode. (h) Optical photographs of water droplet on top of the sensor under tensile mode with different bending angles. Insets: Enlarged photographs of the WCAs. Source: Reprinted with permission from Liu et al. [64] Copyright © 2020 American Chemical Society.

8.2.4 Characterization of Superhydrophobic Property

Next, we carried out the evaluation of the superhydrophobic function of the strain sensor. The mechanical stability of superhydrophobic coating under repeated loading–unloading cycle is extremely crucial. Although the microcracks generated on the superhydrophobic coating do great enhancement to the sensor's electrical responses, they have a negative influence on the stability of the coating. The WCAs of the sensor were measured after different compressive cycles. The sensor had an initial WCA up to 164° upon the unused state (Figure 8.8d,e), higher than the defined superhydrophobic surfaces that were over 150°. After 6000 compressive cycles, the WCA decreased to 160° and further dropped to 157° when subjected to up to 10 000 compressive cycles. The results demonstrate that the coating has good mechanical stability in repeated cycling working conditions. The paper-based strain sensor was attached to a poly(ethylene terephthalate) (PET) supporting film to help finish the measurement of WCA under different bending status

(Figure 8.8e). Because the sensor has bioinspired micro-/nanostructures including the microgrooves and papilla-like microtopography established by NPs and nanotubes, we need to discuss their effects on the sensor's superhydrophobic property. As none of the materials used for the superhydrophobic coating has superhydrophobicity, even PDMS is just hydrophobic rather than superhydrophobic. Therefore, the superhydrophobic property of the strain sensor mainly depends on the bioinspired structures. According to the raised edges of grooves, the sensor surface is divided into two regions, grooves region and no-grooves region (Figure 8.8f). As mentioned above, scraps of paper accumulated along the edges of grooves, making the edges higher than the surrounding area (Figure 8.8g). When the water droplets fall down within the grooves region, the raised edges play a mechanical supporting role for water droplets. An average WCA of 166° was measured when water droplets dropped on the sensor surface within grooves region, slightly larger than that of within no-grooves region (WCA $\approx$ 158°). Because the papilla-like structures were formed within both the grooves region and the no-grooves region, it can be concluded that the grooves here have an additional effect on improving the sensor's WCA. Figure 8.8g shows an optical photograph of water droplet on top of the sensor under compressive mode with a bending angle of 70°. And in terms of tensile mode, the WCAs of the sensor were measured upon different bending angles (Figure 8.8h). As the sensor was bent up to 125°, its WCA decreased from 164° to 153°. It indicates that the sensor remained superhydrophobic during cycles even within a large bending angle. Beyond that, the sliding angle of the strain sensor was also measured. A water dropping test was carried out with the sensor tilted at an angle of $\approx$ 6°. Water droplets rolled off immediately after being dropped on the no-grooves region of the sensor surface. For the region of the grooves, the critical sliding angle was about 9°. So, the no-grooves region had a smaller sliding angle than that of the region of the grooves. It suggests that the raised edges of grooves have a slightly negative effect on the sensor's sliding angle. In general, the grooves with V-shaped cross sections have a great positive impact on the sensitivity and the WCA, but a negative impact on the sliding angle of the superhydrophobic surface.

8.2.5 Applications of the Bioinspired Strain Sensor

The fabricated strain sensor has great potential applications in human motion monitoring due to its high sensitivity and good flexibility. As shown in Figure 8.9a, the strain sensor was fixed on a volunteer's index finger. The bending angle of the finger is defined as the angle of finger from the horizontal direction. Normally, the motion of finger joints can be divided into clockwise and counterclockwise bending. Ideally, the strain sensor should be able to identify the arbitrary state of the bending finger. More specifically, the output value of the sensor should have a mapping relationship with the bending angle of the finger. Owing to the excellent stability and reliability of the paper-based strain sensor, the value of relative resistance variation under bending clockwise from 0° to 90° is almost the same that of bending anticlockwise from 90° to 0° (Figure 8.9b). An almost same relative change in

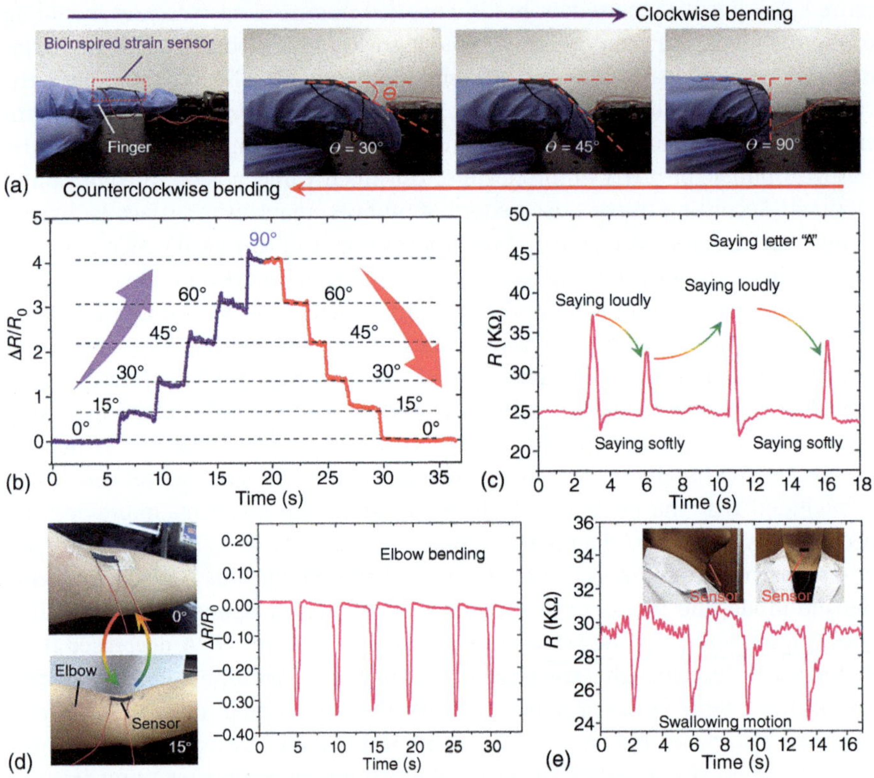

Figure 8.9 Applications of the bioinspired paper-based strain sensor for real-time human motion detection. (a) Photographs of the sensor fixed on the index finger with different bending angles. The finger was bent clockwise from 0° to 90° and then counterclockwise back to the original state. (b) Relative change in resistance under different finger bending angles. (c) Real-time resistance changes of the strain sensor in response to speaking. (d) Relative resistance changes of the strain sensor in response to the bending movement of the elbow joint. (e) Relative resistance changes of the strain sensor in response to the swallowing motion. Source: Reprinted with permission from Liu et al. [64] Copyright © 2020 American Chemical Society.

resistance can be obtained when the finger is bent at an angle of 30°, 45°, and 60° with the horizontal direction, respectively. Further, the plot can also imply the finger bending directions (clockwise or counterclockwise). If the finger is bent clockwise, the paper-based strain sensor will suffer from tensile stress and thus lead to an increase in the sensor's resistance. Accordingly, the increasing $\Delta R/R_0$ of the sensor suggests that the finger is bending clockwise, while the decreasing $\Delta R/R_0$ implies that the finger is bending counterclockwise.

Moreover, the flexible sensor was also attached to a volunteer's (adult male) neck to record the sensor's electrical response when the volunteer was saying the letter "A" in different loudness, including loudly and softly. As seen in Figure 8.9c, higher peaks can be observed when saying the letter "A" loudly than softly. It implies that the sensor may have potential applications in voice recognition in the future. The sensor was also mounted on a volunteer's elbow to monitor joint movement, as shown in Figure 8.9d. The sensor was subjected to compressive stress when the elbow was bent, resulting in a rapid decrease in resistance. During the cyclic bending–straightening movement of the elbow joint, the sensor has a stable and reproducible resistance variation in signal response. The result indicates that the sensor has outstanding reliability for repeated human motion detection. The sensor can also be used for monitoring swallowing behavior in real time by attaching it to the Adam's apple of a volunteer (adult male). A recognizable and repeatable plot was obtained when the volunteer swallowed water repeatedly (Figure 8.9e), showing that the sensor may have a better curative effect on the functional recovery in stroke patients with dysphagia during comprehensive rehabilitation treatment by monitoring swallowing behavior. All these testing results indicate that the bioinspired strain sensor can serve as wearable electronics and has the potential to be used in applications such as in the detection of physical signals for human healthcare and other human–machine interactions.

Considering diversified practical conditions for strain sensors, especially some harsh outdoor conditions such as humid, rainy, or even underwater environments, the waterproof property of the bioinspired paper-based strain sensor should be investigated. As Figure 8.10a shows, the sensor was fixed at one end and suspended at the other; the real-time relative resistance variation was recorded when water droplets (a water droplet of about 0.05 ml) dropped vertically on the strain sensor surface. The plot of $\Delta R/R_0$ of the sensor exhibits regular and significant variation during the test, showing that the sensor is highly sensitive to tiny stimuli generated by a water drop. Similarly, when the applied stimuli changed to a continuous oblique water column, like light reflection, the slanted water column reflected the superhydrophobic surface almost instantaneously (Figure 8.10b), showing that the sensor can work normally upon continuous water column interference. The water did not cause damage to the coating, demonstrating that the coating has strong stability.

Finally, to verify that the sensor can even function normally underwater, the sensor was mounted on the bottom of a container (one end of the sensor was fixed and the other suspended) with water fully immersing the sensor. The water droplets fell on the water surface from a height of $\approx$20 cm, generating vibration waves and then acting as external stimuli to the sensor. The plot of relative resistance variation shows a periodic and synchronous response to the water droplets that intermittently dropped on the water surface (Figure 8.10c). Similar and obvious peaks are observed in the signal response plot, demonstrating that the sensor would not lose function to sense external stimuli even underwater although there is a slight difference between each cycle, which was mainly caused by the slight weight discrepancy of each droplet. Moreover, the water-immersed sensor was introduced into an LED circuitry and strain was applied by a syringe needle, mimicking a scene that the sensor suffered from water flow motion.

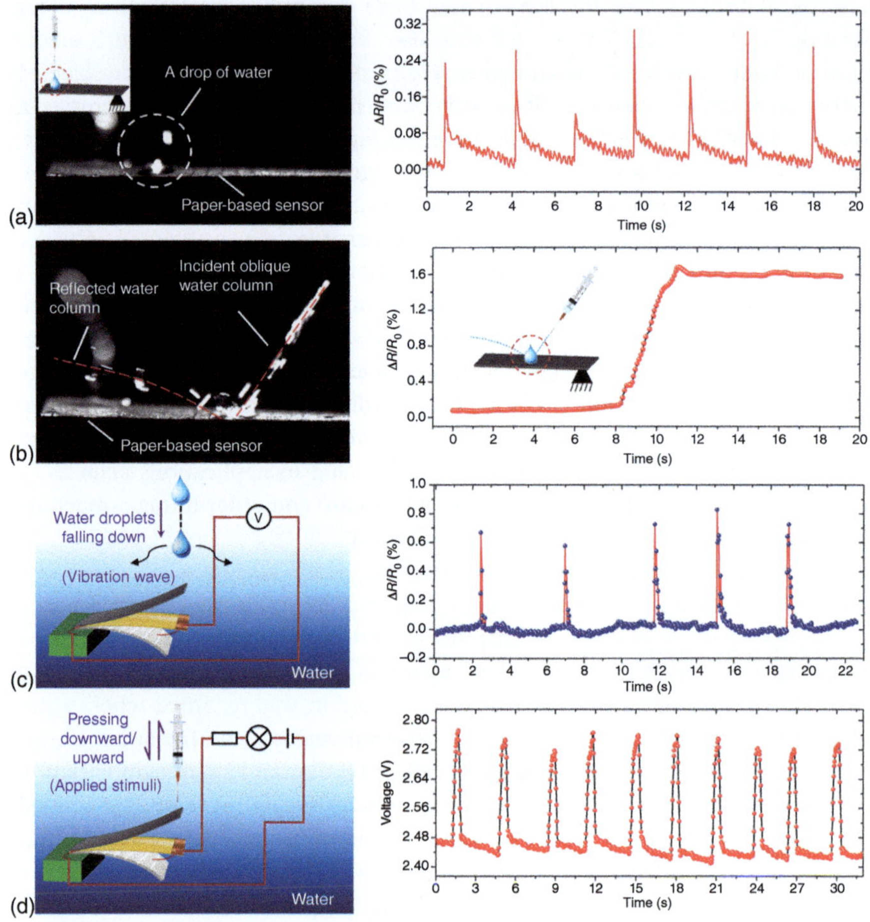

Figure 8.10 Applications of the paper-based strain sensor in real-time detection under humid/underwater harsh conditions. (a, b) Relative resistance changes of the strain sensor in response to vertically and obliquely dropping water droplets, respectively. (c) Relative resistance changes of the sensor fixed underwater toward the vibration wave of water droplets. (d) Real-time voltage changes of the strain sensor fixed underwater toward the loading–unloading cycles of applied stimuli, serving as a variable resistor in the LED circuitry. Source: Reprinted with permission from Liu et al. [64] Copyright © 2020 American Chemical Society.

So, the water-immersed sensor could be regarded as a variable resistor. The voltage of the sensor was measured and used as the output signals. Once the sensor's suspended end was pressed down by the syringe needle, the resistance of the sensor increased sharply, resulting in an increase in the sensor's voltage and a dimming of the brightness of LEDs (Figure 8.10d). Therefore, as the pressure was repeatedly applied and removed, the resistance periodically and synchronously increased and decreased, causing the same voltage change of the sensor and an opposite brightness change of LEDs. All these testing results imply that the bioinspired strain sensor with high sensitivity and the excellent superhydrophobic property also has potential applicability in humid or underwater harsh outdoor conditions.

Based on a strategy of coupling bionics, a paper-based strain sensor inspired by the slit structure of the scorpion's ultrasensitive vibration-sensing organ and the papilla-like structure of the lotus leaf's superhydrophobic surface was successfully fabricated. The optimized scorpion-inspired microgrooves greatly enhanced the sensitivity of the strain sensor, reaching a GF up to 263.34. Furthermore, the sensor displayed excellent stability, reproducibility, and durability during over 12 000 compression cycles. Based on the micro-/nanoscale hierarchical structures constructed by nanomaterials that are similar to the surface papilla structure of the lotus leaf, the bioinspired paper-based strain sensor has a WCA of 164° and remained a WCA of 157° after a long time of repeated usage. Besides, the strain sensor also exhibited an ultralow strain detecting resolution of 0.098%. In terms of applications, the strain sensor can serve as a flexible, highly sensitive wearable electronics to detect various human bodily motions from subtle deformations to vigorous movements. Notably, it also has great potential for applications in humid, rainy, or underwater harsh conditions due to its outstanding superhydrophobicity.

8.3 Underwater Superhydrophobic Air Film Reduction Resistance

After laser machining, the surface of the aluminum alloy was distributed with a pit-shaped micro-/nanostructure with machining spacing of 120, 140, and 160 μm. Then, a silica coating in the previous section was sprayed on the surface. The surface exhibits excellent superhydrophobic and superaerophilic properties. In Figure 8.11 showed the optimal superhydrophobic performance with a WCA of 155.7 ± 0.1° in

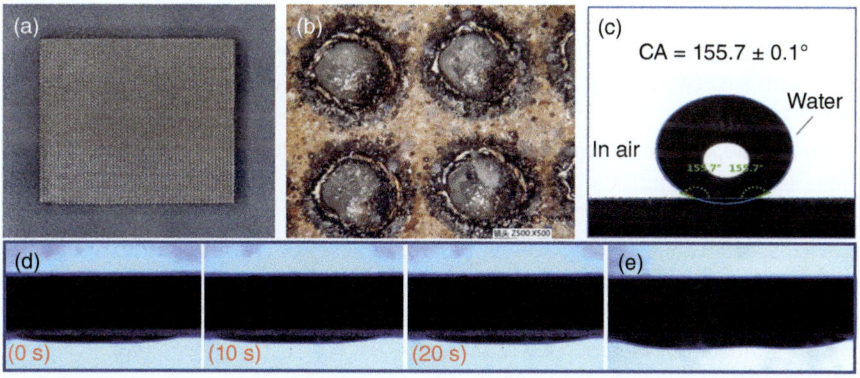

Figure 8.11 Biomimetic superhydrophobic and underwater superaerophilic aluminum alloy surfaces were fabricated by laser processing combined with water bath heating and modification. (a) Morphology of the substrate of surfaces with microcavity arrays prepared by the method of laser marking. (b) Micromorphology of surfaces. (c) The water contact angle of the surfaces. (d) Capturing behaviors for one underwater bubble on the surface. (e) Capturing behaviors for two underwater bubbles on the surface.

air. When the bionic superhydrophobic sample is placed underwater, an air film will form on the sample surface. The air film can spread out quickly in 5 ms. And the air film can be maintained for quite a long time, thus resistance by underwater air film can be reduced.

References

1 Barhoum, A., Samyn, P., Öhlund, T., and Dufresne, A. (2017). Review of recent research on flexible multifunctional nanopapers. *Nanoscale* 9 (40): 15181–15205.
2 Jiang, D., Guo, F., Xu, M. et al. (2019). Conformal fluorine coated carbon paper for an energy harvesting water wheel. *Nano Energy* 58 (5): 842–851.
3 Li, Z., Yang, X., Li, W., and Liu, H. (2019). Stimuli-responsive cellulose paper materials. *Carbohydrate Polymers* 210: 350–363.
4 Li, H., He, Y., Yang, J. et al. (2019). Fabrication of food-safe superhydrophobic cellulose paper with improved moisture and air barrier properties. *Carbohydrate Polymers* 211: 22–30.
5 Liu, K., Li, P., Zhang, Y. et al. (2019). Laboratory filter paper from superhydrophobic to quasi-superamphiphobicity: facile fabrication, simplified patterning and smart application. *Cellulose* 26 (6): 3859–3872.
6 Khanjani, P., King, A.W.T., Partl, G.J. et al. (2018). Superhydrophobic paper from nanostructured fluorinated cellulose esters. *ACS Applied Materials & Interfaces* 10 (13): 11280–11288.
7 Koşak Söz, Ç., Trosien, S., and Biesalski, M. (2018). Superhydrophobic hybrid paper sheets with Janus-type wettability. *ACS Applied Materials & Interfaces* 10 (43): 37478–37488.
8 Wang, S., Liu, K., Yao, X., and Jiang, L. (2015). Bioinspired surfaces with superwettability: new insight on theory, design, and applications. *Chemical Reviews* 115 (16): 8230–8293.
9 Han, Z., Feng, X., Guo, Z. et al. (2018). Flourishing bioinspired antifogging materials with superwettability: progresses and challenges. *Advanced Materials* 30 (13): 1–32.
10 Han, Z., Jiao, Z., Niu, S., and Ren, L. (2019). Ascendant bioinspired antireflective materials: opportunities and challenges coexist. *Progress in Materials Science* 103: 1–68.
11 Lu, Y., Sathasivam, S., Song, J. et al. (2015). Robust self-cleaning surfaces that function when exposed to either air or oil. *Science* 347 (6226): 1132–1135.
12 Min, W.L., Jiang, B., and Jiang, P. (2008). Bioinspired self-cleaning antireflection coatings. *Advanced Materials* 20 (20): 3914–3918.
13 Han, Z., Wang, Z., Li, B. et al. (2019). Flexible self-cleaning broadband antireflective film inspired by the transparent cicada wings. *ACS Applied Materials & Interfaces* 11 (18): 17019–17027.
14 Rastegari, A. and Akhavan, R. (2019). On drag reduction scaling and sustainability bounds of superhydrophobic surfaces in high Reynolds number turbulent flows. *Journal of Fluid Mechanics* 864: 327–347.

15 Lee, C. and Kim, C.J. (2011). Underwater restoration and retention of gases on superhydrophobic surfaces for drag reduction. *Physical Review Letters* 106 (1): 1–4.

16 Zhou, C., Chen, Z., Yang, H. et al. (2017). Nature-inspired strategy toward superhydrophobic fabrics for versatile oil/water separation. *ACS Applied Materials & Interfaces* 9 (10): 9184–9194.

17 Li, J., Xu, C., Zhang, Y. et al. (2016). Robust superhydrophobic attapulgite coated polyurethane sponge for efficient immiscible oil/water mixture and emulsion separation. *Journal of Materials Chemistry A* 4 (40): 15546–15553.

18 Long, M., Peng, S., Deng, W. et al. (2018). Highly efficient separation of surfactant stabilized water-in-oil emulsion based on surface energy gradient and flame retardancy. *Journal of Colloid and Interface Science* 520: 1–10.

19 Cao, M., Guo, D., Yu, C. et al. (2016). Water-repellent properties of superhydrophobic and lubricant-infused 'slippery' surfaces: a brief study on the functions and applications. *ACS Applied Materials & Interfaces* 8 (6): 3615–3623.

20 Han, Z., Feng, X., Jiao, Z. et al. (2018). Bio-inspired antifogging PDMS coupled micro-pillared superhydrophobic arrays and SiO_2 coatings. *RSC Advances* 8 (47): 26497–26505.

21 Bhushan, B. and Jung, Y.C. (2011). Natural and biomimetic artificial surfaces for superhydrophobicity, self-cleaning, low adhesion, and drag reduction. *Progress in Materials Science* 56 (1): 1–108.

22 Darmanin, T., De Givenchy, E.T., Amigoni, S., and Guittard, F. (2013). Superhydrophobic surfaces by electrochemical processes. *Advanced Materials* 25 (10): 1378–1394.

23 Jiang, H., Liu, Y., Zhang, Y. et al. (2018). Reed leaf-inspired graphene films with anisotropic superhydrophobicity. *ACS Applied Materials & Interfaces* 10 (21): 18416–18425.

24 Wenxuan, Y., Yuan, Y., Guoyong, L. et al. (2018). The anti-icing/frosting aluminum surface with hydrangea-like micro/nano structure prepared by chemical etching. *Materials Letters* 226: 4–7.

25 Huang, L., Chen, K., Lin, C. et al. (2011). Fabrication and characterization of superhydrophobic high opacity paper with titanium dioxide nanoparticles. *Journal of Materials Science* 46 (8): 2600–2605.

26 Gao, Z., Zhai, X., Liu, F. et al. (2015). Fabrication of TiO_2/EP super-hydrophobic thin film on filter paper surface. *Carbohydrate Polymers* 128: 24–31.

27 Kong, X., Zhang, J., Xuan, Q. et al. (2018). Superhydrophobic coating for antifouling of Chinese paintings. *Langmuir* 34 (28): 8294–8301.

28 Zhang, W., Lu, P., Qian, L., and Xiao, H. (2014). Fabrication of superhydrophobic paper surface via wax mixture coating. *Chemical Engineering Journal* 250: 431–436.

29 Zhang, L., Kwok, H., Li, X., and Yu, H. (2017). Superhydrophobic substrates from off-the-shelf laboratory filter paper: simplified preparation, patterning, and assay application. *ACS Applied Materials & Interfaces* 9 (45): 39728–39735.

30 Balu, B., Breedveld, V., and Hess, D.W. (2008). Fabrication of 'roll-off' and 'sticky' superhydrophobic cellulose surfaces-via plasma processing. *Langmuir* 24 (9): 4785–4790.

31 Wen, G., Gao, X., Tian, P. et al. (2018). Modifier-free fabrication of durable and multifunctional superhydrophobic paper with thermostability and anti-microbial property. *Chemical Engineering Journal* 346: 94–103.

32 Wen, G. and Guo, Z. (2018). Nonflammable superhydrophobic paper with biomimetic layered structure exhibiting boiling-water resistance and repairable properties for emulsion separation. *Journal of Materials Chemistry A* 6 (16): 7042–7052.

33 Chen, F.F., Zhu, Y.J., Xiong, Z.C. et al. (2016). Highly flexible superhydrophobic and fire-resistant layered inorganic paper. *ACS Applied Materials & Interfaces* 8 (50): 34715–34724.

34 Yang, R., Zhu, Y., Chen, F. et al. (2018). Recyclable, fire-resistant, superhydrophobic, and magnetic paper based on ultralong hydroxyapatite nanowires for continuous oil/water separation and oil collection. *ACS Sustainable Chemistry & Engineering* 6 (8): 10140–10150.

35 Huang, J., Lyu, S., Chen, Z. et al. (2019). A facile method for fabricating robust cellulose nanocrystal/SiO_2 superhydrophobic coatings. *Journal of Colloid and Interface Science* 536: 349–362.

36 Wang, M., Wang, Y., Gao, B. et al. (2019). Fast strategy to functional paper surfaces. *ACS Applied Materials & Interfaces* 11 (15): 14445–14456.

37 Werner, O., Quan, C., Turner, C. et al. (2010). Properties of superhydrophobic paper treated with rapid expansion of supercritical CO_2 containing a crystallizing wax. *Cellulose* 17 (1): 187–198.

38 Zhang, Y., Li, P., Liu, P. et al. (2019). Fast and simple fabrication of superhydrophobic coating by polymer induced phase separation. *Nanomaterials* 9 (3): 1–10.

39 Liu, S., Zhou, H., Wang, H. et al. (2017). Argon plasma treatment of fluorine-free silane coatings: a facile, environment-friendly method to prepare durable, superhydrophobic fabrics. *Advanced Materials Interfaces* 4 (11): 1700027.

40 Yang, J., Li, H., Lan, T. et al. (2017). Preparation, characterization, and properties of fluorine-free superhydrophobic paper based on layer-by-layer assembly. *Carbohydrate Polymers* 178: 228–237.

41 Peng, L., Meng, Y., and Li, H. (2016). Facile fabrication of superhydrophobic paper with improved physical strength by a novel layer-by-layer assembly of polyelectrolytes and lignosulfonates-amine. *Cellulose* 23 (3): 2073–2085.

42 Hwang, H., Kim, N., Lee, S. et al. (2011). Facile fabrication of transparent superhydrophobic surfaces by spray deposition. *ACS Applied Materials & Interfaces* 3 (7): 2179–2183.

43 Wei, J., Zhang, G., Jing, D. et al. (2018). Facile, scalable spray-coating of stable emulsion for transparent self-cleaning surface of cellulose-based materials. *ACS Sustainable Chemistry & Engineering* 6 (9): 11335–11344.

44 Chen, G., Zhu, P., Kuang, Y. et al. (2017). Durable superhydrophobic paper enabled by surface sizing of starch-based composite films. *Applied Surface Science* 409: 45–51.

45 Ogihara, H., Xie, J., and Saji, T. (2013). Factors determining wettability of superhydrophobic paper prepared by spraying nanoparticle suspensions. *Colloids and Surfaces A: Physicochemical and Engineering Aspects* 434: 35–41.

46 Wang, Q., Xiong, J., Chen, G. et al. (2019). Facile approach to develop hierarchical roughness fiber@SiO_2 blocks for superhydrophobic paper. *Materials* 12 (9): 1–9.

47 Kango, S., Kalia, S., Celli, A. et al. (2013). Surface modification of inorganic nanoparticles for development of organic–inorganic nanocomposites – a review. *Progress in Polymer Science* 38 (8): 1232–1261.

48 Sperling, R.A. and Parak, W.J. (2010). Surface modification, functionalization and bioconjugation of colloidal inorganic nanoparticles. *Philosophical Transactions of the Royal Society A – Mathematical Physical and Engineering Sciences* 368: 1333–1383.

49 Song, K.M., Ahn, S.H., and Cho, Y.S. (2015). Fabrication of a hierarchical aluminum oxide surface with micro/nanostructures via a single process and its application as a superhydrophobic surface: simple sintering method with an aluminum microsized powder. *Surface & Coatings Technology* 282: 68–77.

50 Ebert, D. and Bhushan, B. (2012). Durable lotus-effect surfaces with hierarchical structure using micro- and nanosized hydrophobic silica particles. *Journal of Colloid and Interface Science* 368 (1): 584–591.

51 Si, Y., Guo, Z., and Liu, W. (2016). A robust epoxy resins @ stearic acid-$Mg(OH)_2$ micronanosheet superhydrophobic omnipotent protective coating for real-life applications. *ACS Applied Materials & Interfaces* 8 (25): 16511–16520.

52 Hsu, C.P., Chang, L., Chiu, C.W. et al. (2013). Facile fabrication of robust superhydrophobic epoxy film with polyamine dispersed carbon nanotubes. *ACS Applied Materials & Interfaces* 5 (3): 538–545.

53 Ogihara, H., Xie, J., Okagaki, J., and Saji, T. (2012). Simple method for preparing superhydrophobic paper: spray-deposited hydrophobic silica nanoparticle coatings exhibit high water-repellency and transparency. *Langmuir* 28 (10): 4605–4608.

54 Peng, S., Meng, W., Guo, J. et al. (2019). Photocatalytically stable superhydrophobic and translucent coatings generated from PDMS-grafted-SiO_2/TiO_2@PDMS with multiple applications. *Langmuir* 35 (7): 2760–2771.

55 Tang, X., Nan, S., Wang, T. et al. (2013). Facile strategy for fabrication of transparent superhydrophobic coatings on the surface of paper. *RSC Advances* 3 (36): 15571–15575.

56 Baidya, A., Ganayee, M.A., Ravindran, S.J. et al. (2017). Organic solvent-free fabrication of durable and multifunctional superhydrophobic paper from waterborne fluorinated cellulose nanofiber building blocks. *ACS Nano* 11 (11): 11091–11099.

57 Teisala, H. and Butt, H.J. (2019). Hierarchical structures for superhydrophobic and superoleophobic surfaces. *Langmuir* 35 (33): 10689–10703.

58 Jiao, Z.B., Chu, W.C., Liu, L.P. et al. (2020). Underwater writable and heat-insulated paper with robust fluorine-free superhydrophobic coatings. *Nanoscale* 12: 8536–8545.

59 Yuan, Y., Chen, Y., Chen, W., and Hong, R. (2015). Preparation, durability and thermostability of hydrophobic antireflective coatings for solar glass covers. *Solar Energy* 118: 222–231.

60 Kim, J., Mirzaei, A., Woo, H., and Sub, S. (2018). Novel superamphiphobic surfaces based on micro–nano hierarchical fluorinated Ag/SiO_2 structures. *Applied Surface Science* 445: 262–271.

61 Jung, Y.C. and Bhushan, B. (2009). Mechanically durable carbon nanotube–composite hierarchical structures with superhydrophobicity, self-cleaning, and low-drag. *ACS Nano* 3 (12): 4155–4163.

62 Zhang, C., Zheng, Y., Wu, Z. et al. (2019). Non-wet kingfisher flying in the rain: the water-repellent mechanism of elastic feathers. *Journal of Colloid and Interface Science* 541: 56–64.

63 Richard, D., Clanet, C., and Quéré, D. (2002). Surface phenomena: contact time of a bouncing drop. *Nature* 417: 811.

64 Liu, L.P., Jiao, Z.B., Zhang, J.Q. et al. (2021). Bioinspired, superhydrophobic, and paper-based strain sensors for wearable and underwater applications. *ACS Applied Materials & Interfaces* 13: 1967–1978.

65 Brownell, P. and Farley, R.D. (1979). Orientation to vibrations in sand by the nocturnal scorpion paruroctonus mesaensis: mechanism of target localization. *Journal of Comparative Physiology A* 131: 31–38.

66 Brownell, P.H. and Hemmen, J.V.L. (2001). Vibration sensitivity and a computational theory for prey-localizing behavior in sand scorpions. *American Zoologist* 41 (5): 1229–1240.

67 Brownell, P. and Farley, R.D. (1979). Detection of vibrations in sand by tarsal sense organs of the nocturnal scorpion, *Paruroctonus mesaensis*. *Journal of Comparative Physiology A* 131: 23–30.

68 Stürzl, W., Kempter, R., and Van Hemmen, J.L. (2000). Theory of arachnid prey localization. *Physical Review Letters* 84: 5668.

69 Hu, N., Karube, Y., Yan, C. et al. (2008). Tunneling effect in a polymer/carbon nanotube nanocomposite strain sensor. *Acta Materialia* 56 (13): 2929–2936.

70 Li, L., Bai, Y., Li, L. et al. (2017). A superhydrophobic smart coating for flexible and wearable sensing electronics. *Advanced Materials* 29 (43): 1702517.

71 Li, Q., Liu, H., Zhang, S. et al. (2019). Superhydrophobic electrically conductive paper for ultrasensitive strain sensor with excellent anticorrosion and self-cleaning property. *ACS Applied Materials & Interfaces* 11 (24): 21904.

72 Hu, L., Li, Q., Zhang, S. et al. (2018). Electrically conductive polymer composites for smart flexible strain sensors: a critical review. *Journal of Materials Chemistry C* 6: 12121–12141.

73 Amjadi, M. and Sitti, M. (2018). Self-sensing paper actuators based on graphite-carbon nanotube hybrid films. *Advanced Science* 5 (7): 1800239.

74 Liao, X., Liao, Q., Yan, X. et al. (2015). Flexible and highly sensitive strain sensors fabricated by pencil drawn for wearable monitor. *Advanced Functional Materials* 25 (16): 2395–2401.

75 Liao, X., Zhang, Z., Liao, Q. et al. (2016). Flexible and printable paper-based strain sensors for wearable and large-area green electronics. *Nanoscale* 8 (26): 13025–13032.

9

Bioinspired Responsive Surfaces Toward Multiple Organic Vapors

Fancy physicochemical phenomena in the biological world that occur on natural surfaces with evolved optimal structures are ubiquitous and have attracted extensive interest for a long time [1–8]. Notably, diverse natural structures serving as a sustainable source of inspiration provide abundant clues for scientists and engineers to design new biomimetic materials and bioinspired devices [9–13]. These versatile innovations inspired by natural structures hold great potentials for desirable applications in a broad range of fields [14]. For instance, 3D artificial compound eyes for distortion-free imaging [15], gyroids in butterfly wing scales as bio-photonic material [16], liquid-crystalline nanostructures for photonic applications [17], directional surfaces for adhesion, wetting, and transport [18], structured films for antifogging [19], and mechanical stable coating for underwater superoleophobicity [20]. It is well known that butterfly has evolved a remarkable diversity of wing scale structures exhibiting inherent multifunctional integration for survival in harsh environments [21], such as robust ice-phobic/anti-icing properties [22], structural colors [23–25], omnidirectional antireflection properties [26], directional adhesion [27], fluorescence emission [28], and chemical sensing capacity [29]. Particularly, due to its multifunction integration properties, *Morpho* butterfly is of great interest as a typical biological prototype, which is becoming a research hot spot in both fundamental research and practical applications [30–32].

In fact, the recent decade has seen more and more *Morpho*-inspired materials designed and fabricated by many research groups all over the world [33–44]. Nevertheless, most theoretical and experimental studies are limited in optical properties of the original butterfly and its replicas. To date, only a few cases referring to vapor response have been explored from an application perspective since the highly selective vapor response of *Morpho* butterfly wing scales was firstly reported [29, 31, 45]. This partly results from the insufficient capacity of conventional fabrication methods for accurate reproduction of complex 3D elaborate natural structures. On the other hand, specialized instruments employed to characterize vapor responsive properties of biomimetic samples are severely lacking. Thus, fast vapor identification using a biomimetic surface with complex 3D hierarchical structures is still an open challenge. The underlying vapor responsive mechanism of this biomimetic surface needs deeper understanding and further exploration.

Nature-Inspired Structured Functional Surfaces: Design, Fabrication, Characterization, and Applications, First Edition. Zhiwu Han.

In previous research, a facile bio-template method combining sol–gel and acid etching procedures to mimic various ultrafine 3D structures of butterfly wing scales was developed [19, 46, 47]. It has been proved that this method is especially suitable for the biomimetic fabrication of complex natural structures. Here, inspired by the vapor responsive properties of *Morpho* butterfly, we report a biomimetic film with vapor-trapping structures for real-time *in situ* fast detection of multiple organic vapors (MeOH, EtOH, acetone, ether, methylbenzene, and xylene vapors). Such inorganic biomimetic ultrathin films (BUFs) can be obtained through the modified bio-template method without any post process. The 3D hierarchical vapor-trapping structures of the BUF can allow the target vapor to enter the middle pagoda structure supported by a SiO_2-based substrate through the top porous plate. Then, the refractive index of the BUF is determined by the microenvironment composed of a solid sandwich-like structure system and the entrapped target vapor. Due to the variable refractive indices of different vapors, the responsive performance of the BUF is diverse and recognizable. Through locating feature-positions on the reflectance curves, six different vapors can be qualitatively distinguished rapidly at room temperature. This work is anticipated to overcome the weaknesses of conventional micro-/nanofabrication and produce a highly applicable biomimetic film for real-time detection of multiple organic vapors. The cost-effective attempt of mimicking complex 3D structures from original natural species in this work will also inspire researchers to design new biomimetic materials with novel structures for a wide range of practical applications.

9.1 Responsive Performance of *Morpho* Butterfly Wings

9.1.1 Vapor Responsive Platform for Responsive Measurements

9.1.1.1 Build-Up of Vapor Responsive Platform

For real-time detection of various vapors *in situ* using the as-fabricated nature-inspired functional structured surfaces (NIFSS), a homemade vapor detection platform was built up (Figure 9.1a). This platform includes three main subsystems: vapor-generating subsystem, vapor-detecting subsystem, and vapor post-processing subsystem. In detail, the vapor-generating system consists of a nitrogen (N_2) tank and target vapor reactor on a hot plate. They are linked by a flexible gas-guide tube with adjustable valves. Here, N_2 serves as a carrier gas to transport the target vapors. Since common organic reagents are in liquid state, heating and stirring the organic reagents on the hot plate could offer a simple but effective approach to generate the desirable target vapors. Vapor-detecting system is the core part of the detecting platform. It is made up of a self-built vapor chamber, optical fiber with a metal probe, spectrograph, and computer. The vapor post-processing system, including an exhaust vapor reactor, can handle redundant vapors to avoid pollution.

In this system, a vapor chamber with a gas-guide tube and metal probe is the core device, where the NIFSS will be set in. On one side, a three-way tube is inserted into the chamber, which can allow the carrier gas N_2 and target vapor to enter

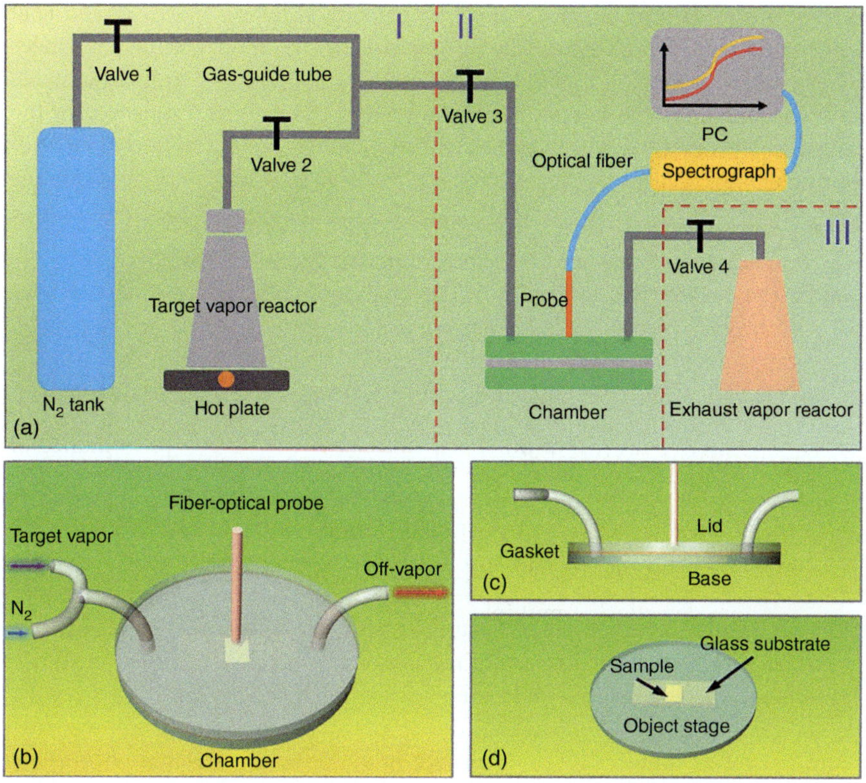

Figure 9.1 *In situ* vapor fast identification platform and its core device. (a) Schematic illustration of the built-up fast identification platform including vapor-generating subsystem (I), vapor-detecting subsystem (II), and vapor post-processing subsystem (III). (b) Test chamber as core device with gas-guide tubes and a fiber-optic probe outside. (c) Front view of the chamber showing its PMMA lid, gasket, and PMMA bottom. (d) Section view of the chamber inside demonstrating its internal construction.

the chamber inside. On the other side, off-vapor can go out of the vapor chamber through a gas-guide tube (Figure 9.1b). The vapor chamber is composed of a polymethyl methacrylate (PMMA) lid, flexible gasket, and PMMA base (Figure 9.1c). By adjusting the gasket thickness, the vapor thickness in the chamber can be controlled. The PMMA base serves as an object stage to support the BUF on a glass substrate (Figure 9.1d). The whole platform offers a continuous system to detect target vapors in real time.

9.1.1.2 Operating Principle of Vapor Responsive Platform

When the system starts to work, the NIFSS are first put in the chamber with all valves closed. Then, open valves 1 and 3 and the chamber will be gradually filled with N_2. After a while, the air in this system will be cleared away. At this time, open valve 4 and bubbles can be observed rising in the exhaust vapor reactor. Next, open valve 2, the target vapor generated in the reactor will enter the chamber. The optical response toward different target vapors can be obtained by the spectrograph. The experimental results can be intuitively displayed on the computer, which can help us detect multiple vapors easily. In the last step, open valves 1, 3, and 4, the used vapor in the chamber will be completely blown into the exhaust vapor reactor by carrier gas N_2. Hence, the vapor chamber will gradually recover to its initial state for the next detection.

9.1.1.3 Responsive Performance of Original Structures and NIFSS Toward N_2

To explore the responsive properties of the NIFSS toward different vapors, the optical performance of the NIFSS under a gas environment is first characterized using the vapor-detecting platform described earlier. The sensitivity of the samples in this work is defined as reflectance intensity (R %) toward the selected wave range (380–1045 nm). Previously, the optical performance of structured materials largely depends on their surface morphologies. Based on such insight, the reflectance measurements of the biomimetic SiO_2 structure are carried out. The reflectance spectra in the air serve as the reference or blank. For biomimetic structures, due to the fact that the periodic pagoda structure is partly covered by the porous plate, the reflectance spectra exhibit quite different features from that of *Morpho* structure, especially in the visible region (380–780 nm). Reflectance trends of the NIFSS in air and in/out N_2 are almost the same in a monotonically increasing situation. With the increasing wavelength, the reflectance dramatically increases in the whole visible wavelength range. In particular, the reflectance peaks appearing over the near-infrared (NIR) range (850–1045 nm) are much broader than that of *Morpho* structure. Interestingly, the NIFSS with biomimetic structures demonstrates a more stable response recovery property ($\Delta R < 5\%$) than the original bio-template, which offers recyclable potential of the NIFSS for vapor detection.

9.1.2 Responsive Mechanism of Butterfly Wings Toward Incident Angles

To take advantage of the elaborate structures of original biological prototypes, a male *Morpho* butterfly individual was first investigated in detail. The wingspan of the selected butterfly with light blue color is c. 14 cm from the top view (Figure 9.2a). From the side view, the wing length is c. 12 cm. Intriguingly, a much deeper blue color can be observed when the wing surface is slightly tilted (Figure 9.2b), which is the result of the relative change in the observation angle. It also provides additional proof that the blue color is structure-based rather than pigments. It also indicates the existence of elaborate structures on the wing surface. The bottom view of the entire butterfly shows more recognizable biological features, especially symmetrically distributed eyespots on the reverse side of the wing surface (Figure 9.2c). To confirm the optical response properties of the wing surface, a series of *in situ* reflectance spectra with respect to variable incident angles were obtained using a fiber-optic spectrometer. No matter how incident angles change from normal (0°) to large tilting degree, strong feature peaks always appear near 460 nm with reflectance varying from 28%

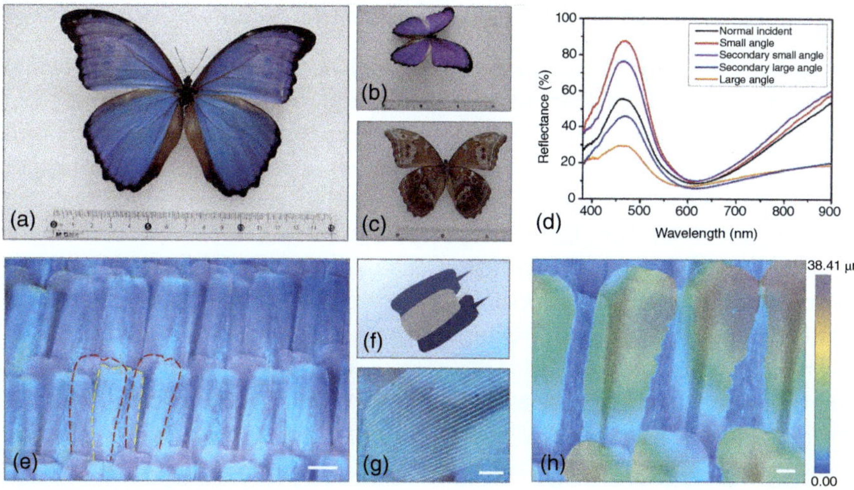

Figure 9.2 Optical response performance and wing surface morphologies of original *Morpho* butterfly. Digital photos of an intact male *Morpho* butterfly from (a) top view, (b) side view, and (c) bottom view. (d) Reflectance spectra of wing surface toward different incident angles. (e) Stereomicroscope image of semitransparent glass scales (red dashed) and blue ground scales (yellow dashed) on wing surface in good order. (f) Schematic illustration of overlapping wing scales. (g) Single glass scale with periodic parallel ridges. (h) Roughness of wing surface demonstrating clear boundaries of two type scales. Scale bars: (e) 50 μm, (f) 10 μm, (h) 20 μm.

to 90% (Figure 9.2d). These results correspond to a wavelength range of blue color (450–475 nm) in the visible region (380–780 nm).

The optical responses with variable reflectance Intensities are determined by the physical interactions between incident light and surface structures of butterfly wings. With the 3D stereoscopic microscope, the original state of wing scales with ultrafine structures was observed *in situ*. Leaf-like wing scales can be divided into two groups: upper semitransparent glass scales and lower blue ground scales (Figure 9.2e). Glass scales and ground scales arrange in line. Meanwhile, they overlap each other in highly organized ways (Figure 9.2f). Periodic parallel ridges of high density are well distributed along the scale length direction on the scale surface (Figure 9.2g). It further confirms that the optical response properties of the wing surface are structure-based. The remarkable structure-based optical response properties lay the foundation for the development of other extended functions. However, the wing surface is not smooth but rough since original scales are slightly bent. The highest altitude difference is about 36.41 μm from the ground scale surface to glass scale surface (Figure 9.2h). Although surface roughness is essential for special functionalities, such as directional adhesion, hydrophobicity [48], and self-cleaning [49], high roughness makes it hard to fabricate ultrathin structured surfaces, limiting its further practical applications. Moreover, these scales are just inserted into the wing surface and can easily come off. It suggests that using original wing scales as responsive materials directly is not feasible in a real environment.

9.1.3 3D Visible Parameterized Models of Butterfly Feature Structures

It is necessary to fully characterize the original structural features of original wing scales at the micro-/nanoscale before designing a solution for artificial analogous structures. Next, the elaborate structures of original wing scales are well investigated. At the microscale, the scanning electron microscope (SEM) images reveal that leaf-like wing scales tidily are distributed on the wing surface from the top view (Figure 9.3a). For glass scales, they have a width of c. 82 μm and length of c. 200 μm. This is consistent with previous observation results obtained using the stereoscopic microscope. Surprisingly, when cutting off a single scale along the width direction, a novel pagoda structure array appears on the scale surface, which offers abundant superficial areas for the narrow scale (Figure 9.3b). Every pagoda monomer is of c. 1.2 μm high. The distance between two adjacent ridges with the profile of the pagoda structure is less than 1 μm. Interestingly, these ridges are composed of five to nine stacked lamella with a width of c. 133 nm (Figure 9.3c). At the nanoscale, the side view of a single ridge indicates two adjacent lamellas are linked by numerous nanoribs to form window-like grids (Figure 9.3d). It is noted that these ridges are supported by a thin substrate composed of fiber-like materials (Figure 9.3e). In consideration of the material components of the whole wing scales, it has been proved that the original hierarchical structures, including pagoda-like ridges with stacked lamella and thin substrate with cross-linked fibers, are formed by chitin, a natural polysaccharide.

To easily understand the original hierarchical structures from an intuitive view, a visual 3D structure model is built according to the structure features presented by SEM images. In brief, a pagoda array composed of parallel ridges and a pagoda profile is supported by a thin substrate (Figure 9.3f). Obviously, the accurate reproduction of such complex 3D hierarchical structures is challenging for most conventional micro-/nanofabrication methods, such as laser processing, photolithography, and

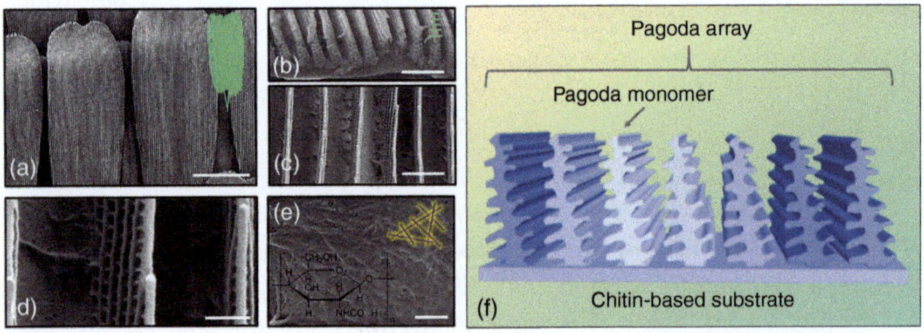

Figure 9.3 SEM images and 3D visual model of original hierarchical structures. (a) Leaf-like wing scales with furcate tails. (b) Cross section of single wing scale showing pagoda structure array. (c) Periodic parallel ridges with stacked lamellae. (d) Side view of a single ridge with window-like ribs. (e) Chitin-based scale substrate composed of cross-linked fibers. (f) Structure model of typical pagoda array with periodic pagoda monomers on thin chitin-based substrate. Scale bars: (a) 50 μm, (b) 2 μm, (c) 1 μm, (d) 200 nm, (e) 200 nm.

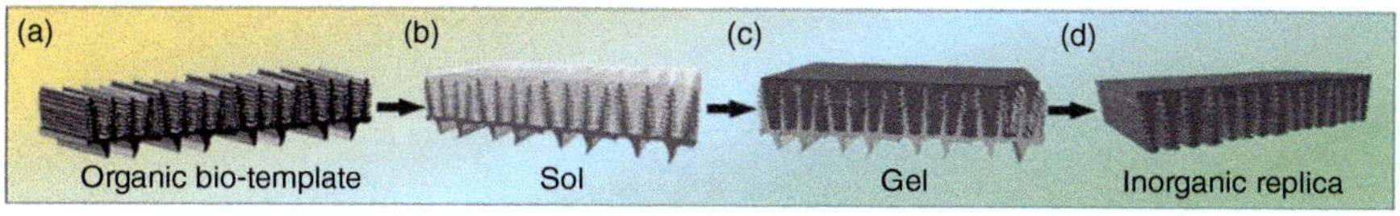

Figure 9.4 Biomimetic fabrication procedure and structure characterization of the biomimetic film. (a) Pagoda structure array in original bio-template. (b) Precursor sol is immersed into the nanogaps between adjacent pagoda pairs. (c) Gel is solidified and inlaid into the bio-template after calcination. (d) Inorganic replica with similar pagoda structure array is obtained after removal of bio-template via mixed acid etching.

vapor deposition. It is urgent to develop a facile method to mimic these 3D natural structures, which is a critical step for a wide range of fundamental research and practical applications.

9.2 Fabrication of Butterfly-Inspired Structural Responsive Surfaces

Due to the chitin's intrinsic nature of low density, chitin-based structures contribute to lightweight wings for butterflies, which can also help butterflies to fly for a longer time in the real world. However, in complex practical situations, it is insufficient for chitin-based structures due to their weak mechanical strength. Thus, turning chitin-based organisms into inorganic materials would be a wise choice to overcome the weaknesses of the original organisms.

Here, based on previous studies, a modified facile one-step bio-template method combining sol–gel and selective acid etching procedures was introduced to realize the transformation from organic bio-templates to inorganic replicas. To be brief, a wing slice was selected as an organic bio-template after simple pretreatment using EtOH and diethyl ether (Figure 9.4a). Then, the precursor solution with the products from hydrolysis of tetraethyl orthosilicate (TEOS) was injected into the bio-template. The generated sol would spontaneously fill into the nanogaps between pagoda monomers due to capillarity action (Figure 9.4b). The sol with good liquidity was gradually solidified under heating conditions to form the gel embedded itself in the bio-template. By this time, a combination of organic bio-template and inorganic matter was obtained (Figure 9.4c). At last, the organic bio-template was gradually etched by mixed nitric acid (HNO_3) and perchloric acid ($HClO_4$). Finally, the inorganic replica on a glass substrate was obtained as expected (Figure 9.4d).

9.3 Characterizations of Butterfly-Inspired Structural Responsive Surfaces

9.3.1 Morphology Characterizations

The detailed structure features were characterized by SEM. The results clearly indicate that the parallel ridges are inherited from the organic bio-template by

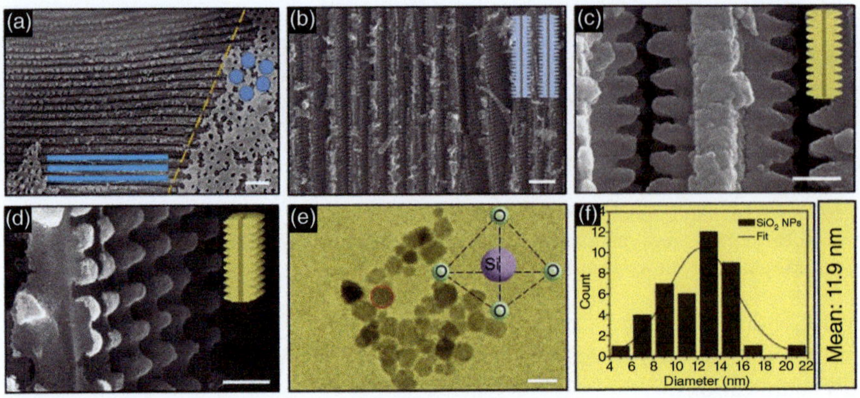

Figure 9.5 (a) Flat plate with dense random pores on the top of periodic parallel structure array. (b) Pagoda structures arranged in parallel. (c) Single pagoda with raised ridges and papilla-like edges. (d) Side view of single pagoda demonstrating stacked lamellae. (e) TEM image of crushed SiO_2-based pagoda structures as nanoparticles. (f) Size distribution analysis of the nanoparticles. Scale bar: (a) 2 μm, (b) 1 μm, (c) 200 nm, (d) 200 nm, (e) 20 nm.

the inorganic replica. Surprisingly, the ridge array is covered with a porous plate (Figure 9.5a). The pores are distributed throughout the whole plate in a random way. In addition, the pore sizes vary from 200 to 520 nm. No other *Morpho*-inspired replicas have displayed analogous composite structures. Thus, it is essential to find out the underlying reason how these pores form. From the outset, we know that the original pagoda array is standing on a chitin-based substrate. It is hypothesized that firstly the chitin-based substrate is partly mineralized in bio-template removal process under heating conditions to form an organic–inorganic composite plate. Then, with the organic bio-template gradually charring, some volatile gases would come out to form bubbles in the hot gel. Finally, the bubbles would fracture and release the gases, leading to the formation of abundant pores on the mineralized plate. Thus, a porous plate on the top of the ridge array is obtained to constitute a sandwich-like structure system with a pagoda structure array on a glass substrate.

As the magnifications increase, more structure details of the as-fabricated ridges are observed. Numerous papilla-like structures are found, which are distributed symmetrically along both sides of the ridge. These ridges are close to each other with a distance of c. 750 nm, which leads to the narrow gaps between two adjacent ridges (Figure 9.5b). Single ridges are composed of a pillar in the middle and papilla-like structures on both sides (Figure 9.5c). The width of the pillar is determined by SEM to be c. 733 nm. Similarly, the regularly spaced papillae around the pillar have a radius of c. 50 nm. In fact, these papillae have regular geometries and high degree of symmetry, resulting from window-like grids in the original bio-template as described above. More accurately, during the biomimetic fabrication process, the sol is filled into the window-like grids and becomes solidified by heating to form these papilla-like structures. To fully reveal the morphology of biomimetic ridges, a side view of the inorganic ridge presents lamellae with papilla-like structures stacked layer by layer, which is consistent with the original structures of the bio-template

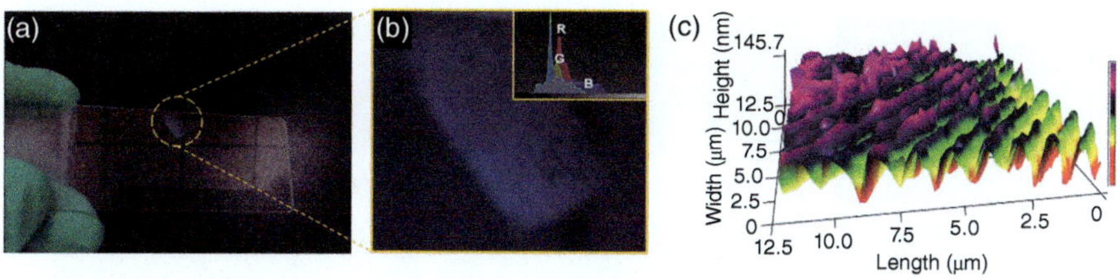

Figure 9.6 (a) Digital photo of the biomimetic film on a glass substrate with vivid blue structural color. (b) Enlarged image and its RGB composition (inset). (c) AFM image of the local surface of the NIFSS under 3D visual model.

(Figure 9.5d). According to the involved chemical reaction of the sol–gel process, the inorganic ridges should be composed of SiO_2-based materials. Hence, the crushed ridges are characterized by transmission electron microscope (TEM). The morphology of the obtained particles is close to spherical, indicating their regular shape (Figure 9.5e). The size of these SiO_2 nanoparticles is calculated with the distribution varying from 4 to 22 nm (Figure 9.5f). It suggests that the inorganic ridges result from highly controllable organization of SiO_2 nanoparticles with the bio-template assistant. The results further indicate that 3D complex structures inspired by *Morpho* butterfly are reproduced successfully in a time-saving way.

It is well known that material structure is very important for its functions. As a demonstration, the biomimetic film on a glass substrate shows bright blue structural colors, which has more similarities to the original blue butterfly wing surface (Figure 9.6a). The RGB analysis also confirms the existence of the structural blue color (Figure 9.6b). The atomic force microscope (AFM) image of the selected biomimetic film (c. 12.5 µm × 12.5 µm) under 3D model demonstrates the fluctuant surface morphology, which agrees well with the raised ridge structures in the biomimetic film (Figure 9.6c). The thickness of the biomimetic film determined by AFM is 145.7 nm, which is thin enough for a film with 3D complex structures along the vertical direction.

9.3.2 Composition Characterizations

To examine its material composition, energy disperse spectroscopy (EDS) spectrum and element maps of the porous plate are obtained. The EDS spectrum shows that the main chemical elements include silicon (Si), oxygen (O), and carbon (C) in trace amounts. Moreover, the colorful maps demonstrate the element distribution of Si, O, and C in the ultrathin film surface, indicating that all the main chemical elements are well distributed. In addition, element composition and distribution of the smashed powders from the solidified gel are also characterized by using the EDS (Figure 9.7). The experimental results provide additional proof to confirm that the NIFSS is indeed SiO_2-based inorganic materials, which also offers a verification of the complete removal of original organic bio-template. The ultrathin structured SiO_2-based film offers a great possibility for extending its applications in extreme conditions, such as narrow space, high temperature (c. 100 °C), or strong acidity environment.

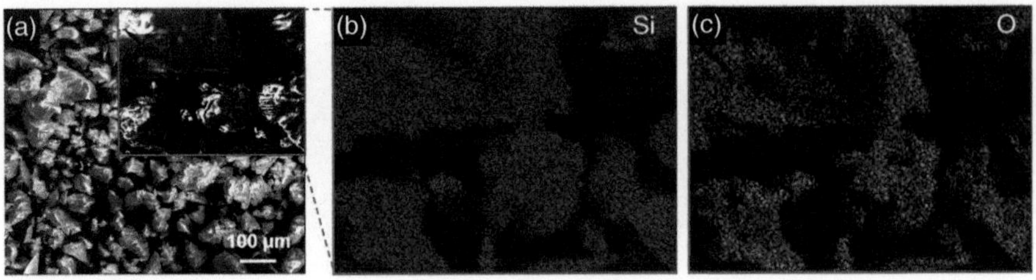

Figure 9.7 (a) Smashed powders of the solidified gel. (b, c) Element maps of two main elements: Si and O.

9.4 Responsive Performance of NIFSS Toward Multiple Organic Vapors

9.4.1 Reflectance Spectra of NIFSS Toward Six Organic Vapors

The response changes of the NIFSS toward different common organic vapors are monitored in the same way as above. In detail, the NIFSS are first exposed to target vapors in the chamber. Then, the optical responses with respect to six randomly selected organic vapors are recorded; in this case, MeOH, EtOH, acetone, ether, methylbenzene, and xylene vapors are selected. Similarly, vapor sensitivity is expressed as reflectance intensity in corresponding spectra. As expected, the experimental reflectance spectra of these organic vapors exhibit remarkable different features in the whole wavelength range including the visible region (380–780 nm) and NIR region (780–1045 nm) (Figure 9.8). To compare the response differences of these organic vapors, four characteristic positions in corresponding reflectance spectra are selected: starting point (SP), inflection point (IP), demarcation point (DP), and peak region (PR). Generally, it is found that reflectance intensity increases sharply with the increment of c. 20% from SP to IP over a narrow wavelength range of c. 50 nm. The IPs of all these six target vapors appear near 460 nm, which is highly consistent with the reflectance peak position of the original wing surface with the pagoda structure. It also provides additional strong evidence for the existence of the biomimetic structure. Then, the reflectance changes gradually level off the wavelengths increasing from IP to DP. Moreover, the reflectance trends from IP to DP are almost sustained in NIR region. Interestingly, their reflectance peaks all appear near 1000 nm in the corresponding PR.

9.4.2 Responsive Evaluation Based on Introduced Sensitive Corner

Apart from these general similarities owing to the biomimetic structure, there exist some critical features to distinguish these target vapors. Among six organic vapors, the NIFSS toward EtOH vapor and methylbenzene vapor exhibit a much stronger response in a short wavelength (c. 380 nm). Accordingly, their SPs are much higher than that of other organic vapors. Here, θ is defined as the tuning angle, which is

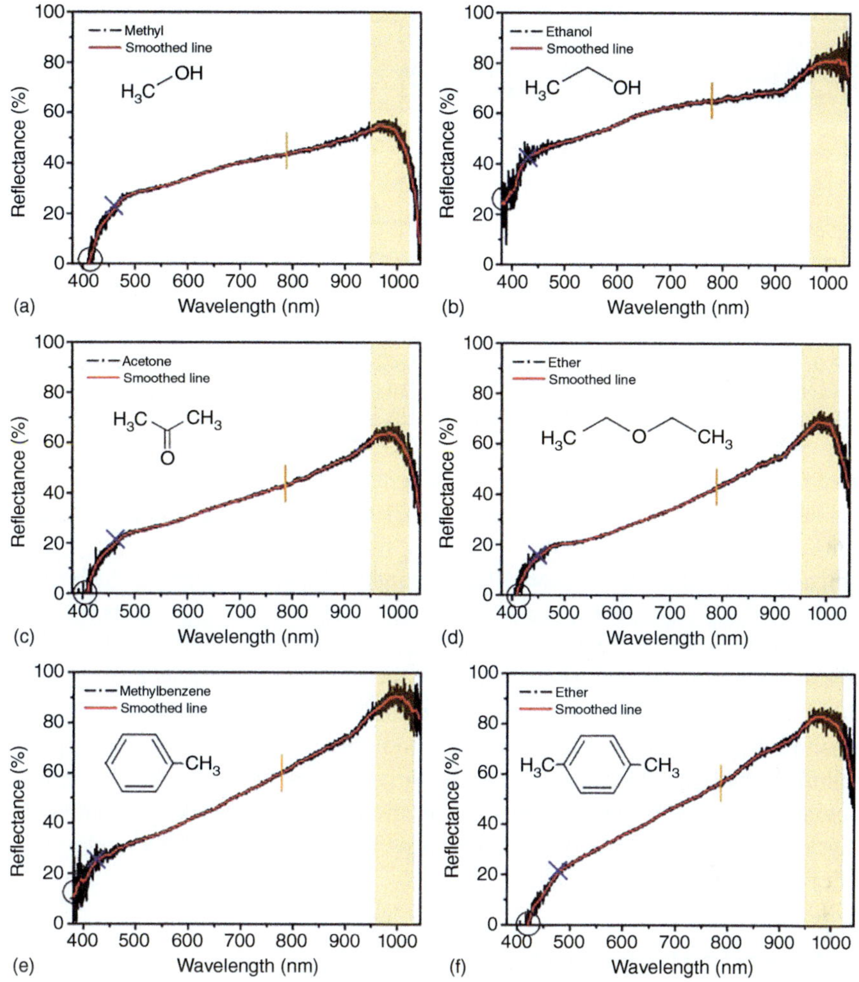

Figure 9.8 Experimental reflectance spectra of the NIFSS toward different organic vapors. (a) MeOH vapor. (b) EtOH vapor. (c) Acetone vapor. (d) Ether vapor. (e) Methylbenzene vapor. (f) Xylene vapor. Here, SP (black circle), IP (blue cross), DP (yellow line), and PR (pink shelter).

introduced to describe the change in the degree of the reflectance curves. If θ is bigger, the vapor response will change more dramatically.

$$\theta = \arctan \frac{\Delta R(\%)}{\Delta \lambda(\text{nm})} = \arctan \frac{R_{\text{IP}} - R_{\text{SP}}}{\lambda_{\text{IP}} - \lambda_{\text{SP}}} \tag{9.1}$$

where R_{IP} and R_{SP} are reflectance intensity of target vapor at IP and SP, respectively. λ_{IP} and λ_{SP} are corresponding wavelengths at IP and SP, respectively. A simple calculation reveals that the vapor response of the NIFSS toward MeOH vapor is stronger than the other five organic vapors. The response order, in this case, can be described as $\theta_{\text{MeOH}} > \theta_{\text{EtOH}} > \theta_{\text{xylene}} > \theta_{\text{acetone}} > \theta_{\text{ether}} > \theta_{\text{methylbenzene}}$. From the responsive intensity aspect, more obvious distinguished features can be

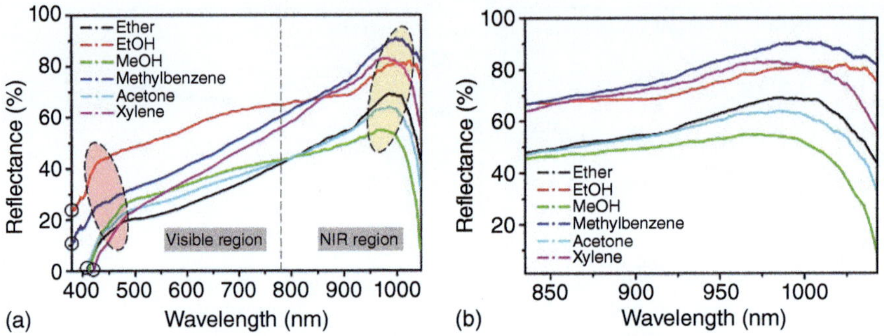

Figure 9.9 Responsive comparison of the target vapors. (a) The reflectance spectra in the whole wavelength range from 380 to 1045 nm. (b) Enlarged peak regions in the NIR range. Here, SP (black circle), IP (pink oval), DP (gray dashed), and NIR peak position (yellow oval).

obtained by the overall comparisons of the reflectance spectra (Figure 9.9a). In the visible region, the responsive intensities all reach their local maxima at DP (780 nm). It is noted that $R_{\mathrm{EtOH}} > R_{\mathrm{methylbenzene}} > R_{\mathrm{xylene}} > R_{\mathrm{MeOH}} > R_{\mathrm{acetone}} > R_{\mathrm{ether}}$. In comparison, in the NIR region, the responsive intensities are in the order (Figure 9.9b) $R_{\mathrm{methylbenzene}} > R_{\mathrm{EtOH}} > R_{\mathrm{xylene}} > R_{\mathrm{ether}} > R_{\mathrm{acetone}} > R_{\mathrm{MeOH}}$. Thus, in the whole wavelength, the NIFSS always demonstrate excellent ability to discriminate these randomly selected six organic vapors by four characteristic positions. Considering all these target vapors are detected in the same system under the same conditions, it is possible to infer that the response differences are in part determined by the intrinsic properties of the target vapors, such as refractive index (RI), molecular weight, chain length, or specific chemical groups. Alternatively, the physicochemical interactions between target vapors and the NIFSS might also account for the final vapor responses.

9.4.3 Responsive Mechanism of NIFSS

9.4.3.1 Responsive Mechanism Based on Sandwich-Like Structures

The underlying response mechanism of the NIFSS is not only a fundamental issue but also a critical step to put forward such responsive materials in practical applications. According to the above experimental results, a structure-based vapor-trapping mechanism has been proposed to explain the fast vapor identification performance of the NIFSS. First, a sandwich-like structure system is built up for the following illustration, which includes top porous plate, middle pagoda structure, and bottom SiO_2-based substrate. Then, the target vapor molecules in the chamber can be divided into four states (Figure 9.10a). For free state, most vapor molecules are randomly dispersed around the sandwich-like structure system. Molecular diffusion will dominate the refractive index of the vapor layer above the surface. For adsorptive state, part of vapor molecules are adsorbed on the surface of the porous plate. For the inlaid state, only a minority of vapor molecules will be inlaid in the randomly distributed pores. For the entrapped state, quite a few vapor molecules could pass through the porous plate and enter the nanogaps between

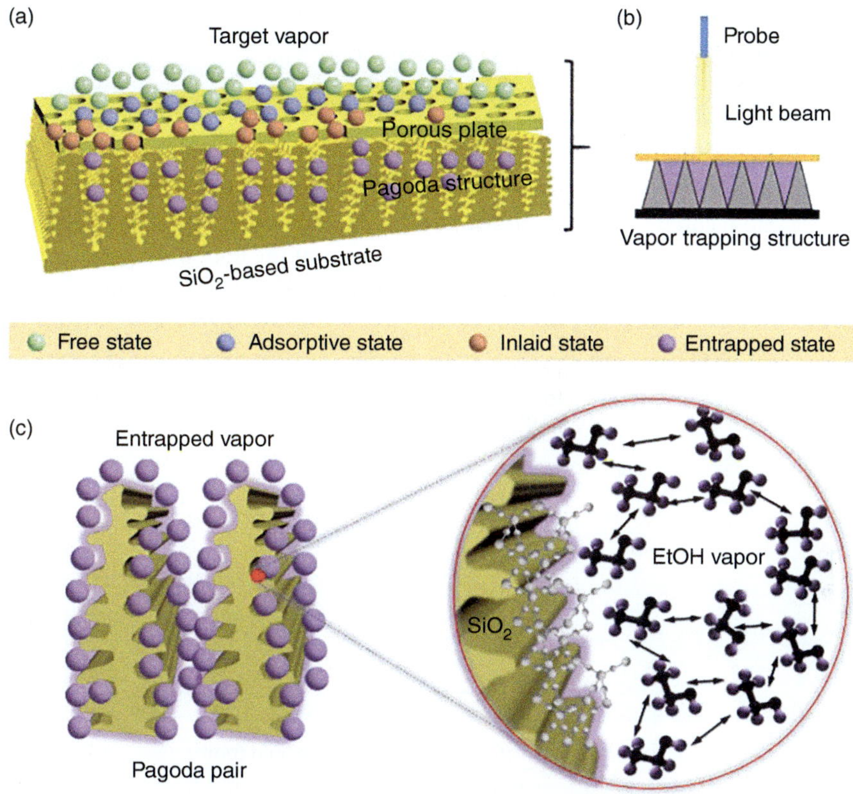

Figure 9.10 Schematic of vapor-trapping mechanism of the sandwich-like structure system. (a) Four vapor molecule states around the gas-trapping structure: free state, adsorptive state, inlaid state, and entrapped state. (b) Sandwich-like structure system with top porous plate (yellow), middle pagoda structure (gray), and bottom SiO_2-based substrate (black). (c) Entrapped vapor behaviors between adjacent SiO_2-based pagoda pairs.

adjacent pagoda structures. Owing to the abundant wrinkle-like surface in the pagoda structure array, it will allow a considerable amount of vapor molecules to be retained on the surface. In other words, these target vapor molecules could be entrapped on the NIFSS with unique vapor-trapping structures. Thus, the vapor detection scene at the micro level can be well understood (Figure 9.10b).

Particularly, we focus on the possible interaction between entrapped vapor and pagoda structure. As a demonstration, EtOH vapor is selected as the example in this case due to its outstanding vapor response performance. EtOH molecules entrapped in the sandwich-like structure system could gradually gather together through van der Waals force. Then the vapor would eventually fill all of the nanogaps. The SiO_2 3D networks might also be helpful to offer sufficient space to attract these vapor molecules, especially residual hydroxyls (–OH) on the SiO_2 surface would form hydrogen bonds with free vapor molecules around the SiO_2-based pagoda pairs (Figure 9.10c). All these possible physicochemical effects work together to endow the vapor trapping structure system with a vapor–solid heterogeneous microenvironment.

9.4.3.2 FDTD Simulation of Reflectance Spectra

Because the solid phase (SiO_2) is definite here, the RI of this system mainly depends on the intrinsic properties of various entrapped vapors. The whole sandwich-like structure system can be ideally regarded as three thin layers, namely, SiO_2 layer on the top, SiO_2/vapor layer in the middle, and SiO_2 layer at the bottom (Figure 9.11a). The RI value of the middle layer might vary between that of air (1.00) and SiO_2 (1.46),

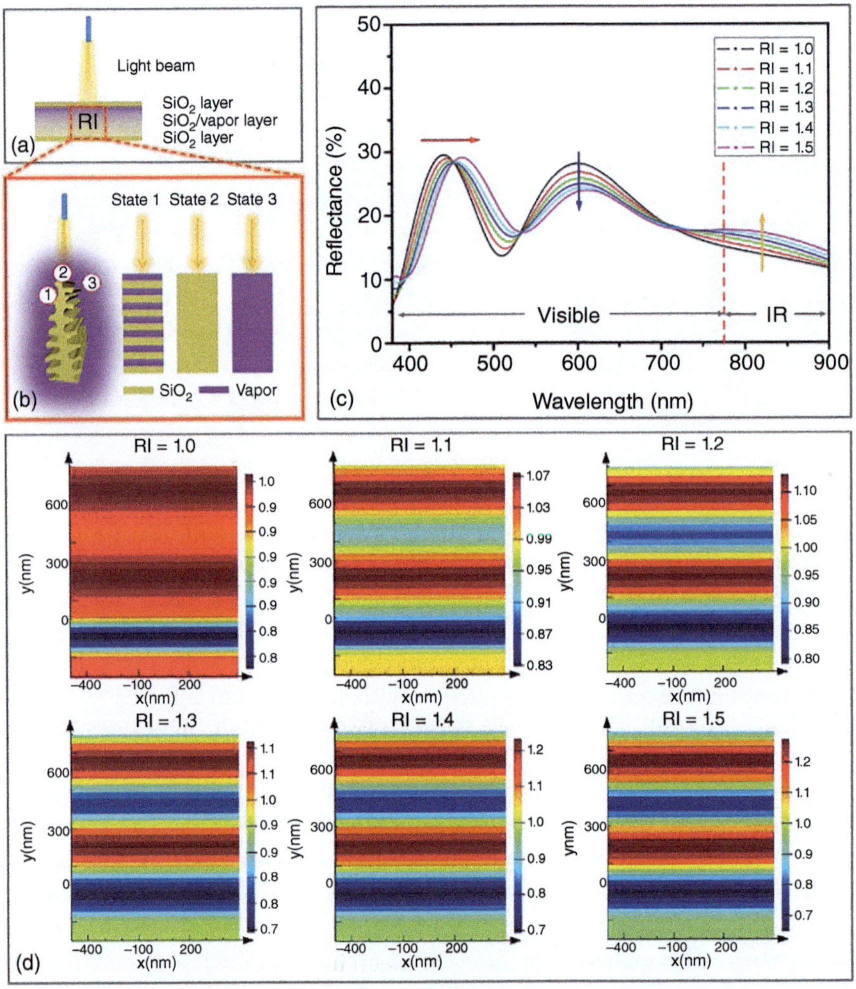

Figure 9.11 Theoretical simulations of the sandwich-like structure system with variable RIs. (a) Simplified sandwich-like SiO_2–SiO_2/vapor–SiO_2 multilayer. (b) Three states in the middle layer with entrapped vapor. (c) Simulated reflectance spectra of the sandwich-like structure system with RI varying from 1.0 to 1.5. (d) Electric field profiles of the NIFSS at the $z = 0$ plane toward variable RI from 1.0 to 1.5.

which partly depends on the amount of vapor entrapped in the middle layer. When a single pagoda structure of the middle layer is exposed to incident light, there would be three states in the vapor-filled middle layer (Figure 9.11b). For state 1, partial light irradiates across the multilayer composed of SiO_2 lamellae and entrapped vapor. For state 2, some light irradiates the raised SiO_2 ridge. For state 3, other light would irradiate the surrounding vapor. Based on such insights, we built a sandwich-like multilayer model to simulate the responsive properties of variable RIs (Figure 9.11c).

The simulation results indicate that there are two feature peak series in the visible region where they display different spectrometric behaviors. The feature peaks in the range of 400–500 nm demonstrate a clear redshift (red arrow) with the RI of the middle layer changing from 1.0 to 1.5. When combining the results of experimental spectra with the simulated ones, we can easily infer that the RI approximates the value of the sandwich-like structure system entrapped certain vapor. For instance, the feature point (IP) of the structure system entrapped EtOH vapor appears near 440 nm in the experimental spectrum. Then, a possible feature peak near 440 nm in the simulated spectrum can be observed. Thus, the corresponding RI value should be between 1.2 and 1.3. Notably, if more RIs are simulated, the obtained RI value will be more accurate. In contrast, the feature peaks in the range of 550–700 nm display another different behavior. The peak positions are all relatively stable near 600 nm, while the reflectance intensity (blue arrow) decreases gradually as the RIs increase. The underlying reason may be that when incident light enters an optically denser medium (NIFSS) from an optically thinner medium (air/vapor), optical phase would reverse 180°. In this case, based on the following optical equation of thin-film interference,

$$2nd = (i + 1/2)\lambda \tag{9.2}$$

where n and i are integers, d (c. 150 nm here) is the thickness of the thin film. Thus, it is easy to accurately calculate the characteristic wavelength $\lambda = 600$ nm, which is completely consistent with the simulated results. It indicates that the characteristic wavelength is determined by the thickness of the sandwich-like structure system itself in this case. More interestingly, the reflectance intensity gradually increases as RI increases in the NIR region. In other words, the reflectance intensity of such a sandwich-like structure system is positively correlated with RI in the NIR region. Actually, as more target vapor is entrapped in the sandwich-like structure system, the RI of the middle layer will gradually increase. Then, the reflectance of NIR will accordingly increase, which can keep the system at a relatively low temperature to some degree. This will reduce the negative effect of evaporation in the system resulting from temperature variations. It can also make a relatively stable inner microenvironment for robust vapor identification.

Furthermore, light–matter interactions between incident light and the NIFSS are also intuitively described by the local electric field profiles in 2D (Figure 9.11d), which provides additional insight into the variable RI effect on vapor responsive properties of the NIFSS at a selected wavelength range of 380–900 nm. The results demonstrate the 2D spatial distribution of the electric field with RI changing from 1.0 to 1.5. It indicates that the strongest electric field enhancement occurs at RI = 1.4

and 1.5, while the weakest one occurs at RI = 1.0. It is also found that the electric field profiles display similar alternative strip patterns irrespective of the RI changes. Notably, intensity changes in the electric field become greater as the RI increases. The typical strong (red)–weak (blue) alternation of the electric field is in accordance with the thin-film interference theory. When RI varies from 1.4 to 1.5, both the distribution and intensity of the electric field tend to be stable. It is possible because the RI value of the middle layer is close to that of SiO_2 (1.46). In other words, when the RI of the middle layer (SiO_2/vapor) is much close to that of the top layer or bottom layer (SiO_2 only), the intensity difference of the electric field is smaller. For instance, when RI = 1.0 and 1.5, the intensity of the electric field around the structure group shows a great difference. The reason might be that the sandwich-like structure system with a RI-variable middle layer can adjust the RI gradient between SiO_2/vapor layer and pure SiO_2 layer. The RI gradient is mainly determined by the entrapped vapors. Thus, when vapors gradually gather in the vapor-trapping structure, their corresponding reflectance spectra would change significantly due to the variable RI gradient between interlayers. It further confirms the rationality of simulating the responsive properties of the NIFSS toward multiple organic vapors with variable RIs.

References

1 Chen, H., Zhang, P., Zhang, L. et al. (2016). Continuous directional water transport on the peristome surface of *Nepenthes alata*. *Nature* 532: 85–89.
2 Liu, K., Cao, M., Fujishima, A., and Jiang, L. (2014). Bio-inspired titanium dioxide materials with special wettability and their applications. *Chemical Reviews* 114 (19): 10044–10094.
3 Ju, J., Bai, H., Zheng, Y. et al. (2012). A multi-structural and multi-functional integrated fog collection system in cactus. *Nature Communications* 3: 1247.
4 Gao, X. and Jiang, L. (2004). Water-repellent legs of water striders. *Nature* 432: 36.
5 Han, Y., Li, M., Yang, Q. et al. (2017). Collective wetting of a natural fibrous system and its application in pump-free droplet transfer. *Advanced Functional Materials* 27 (22): 1606607.
6 Yong, J., Chen, F., Yang, Q. et al. (2017). Superoleophobic surfaces. *Chemical Society Reviews* 46: 4168–4217.
7 Meng, Q., Wang, Q., Liu, H., and Jiang, L. (2014). A bio-inspired flexible fiber array with an open radial geometry for highly efficient liquid transfer. *npg Asia Materials* 6: e125.
8 Yang, E., Son, J.H., Lee, S.I. et al. (2016). Water striders adjust leg movement speed to optimize takeoff velocity for their morphology. *Nature Communications* 7: 13698.
9 Aizenberg, J. and Fratzl, P. (2009). Biological and biomimetic materials. *Advanced Materials* 21 (4): 387–388.
10 Kolle, M., Lethbridge, A., Kreysing, M. et al. (2013). Bio-inspired band-gap tunable elastic optical multilayer fibers. *Advanced Materials* 25 (15): 2239–2245.

11 Pokroy, B., Epstein, A.K., Persson-Gulda, M., and Aizenberg, J. (2009). Fabrication of bioinspired actuated nanostructures with arbitrary geometry and stiffness. *Advanced Materials* 21 (4): 463–469.

12 Li, K., Ju, J., Xue, Z. et al. (2013). Structured cone arrays for continuous and effective collection of micron-sized oil droplets from water. *Nature Communications* 4: 2276.

13 Liu, K., Tian, Y., and Jiang, L. (2013). Bio-inspired superoleophobic and smart materials: design, fabrication, and application. *Progress in Materials Science* 58 (4): 503–564.

14 Zhou, H., Fan, T., and Zhang, D. (2011). Biotemplated materials for sustainable energy and environment: current status and challenges. *ChemSusChem* 4 (10): 1344–1387.

15 Wu, D., Wang, J., Niu, L. et al. (2014). Bioinspired fabrication of high-quality 3D artificial compound eyes by voxel-modulation femtosecond laser writing for distortion-free wide-field-of-view imaging. *Advanced Optical Materials* 2 (8): 751–758.

16 Wilts, B.D., Zubiri, B.A., Klatt, M.A. et al. (2017). Butterfly gyroid nanostructures as a time-frozen glimpse of intracellular membrane development. *Science Advances* 3 (4): e1603119.

17 Wang, L. and Li, Q. (2016). Stimuli-directing self-organized 3D liquid-crystalline nanostructures: from materials design to photonic applications. *Advanced Functional Materials* 26 (1): 10–28.

18 Hancock, M.J., Sekeroglu, K., and Demirel, M.C. (2012). Bioinspired directional surfaces for adhesion, wetting and transport. *Advanced Functional Materials* 22 (11): 2223–2234.

19 Han, Z., Mu, Z., Li, B. et al. (2016). Active antifogging property of monolayer SiO_2 film with bioinspired multiscale hierarchical pagoda structures. *ACS Nano* 10 (9): 8591–8602.

20 Xu, L., Peng, J., Liu, Y. et al. (2013). Nacre-inspired design of mechanical stable coating with underwater superoleophobicity. *ACS Nano* 7 (6): 5077–5083.

21 Liu, C., Ju, J., Zheng, Y., and Jiang, L. (2014). Asymmetric ratchet effect for directional transport of fog drops on static and dynamic butterfly wings. *ACS Nano* 8 (2): 1321–1329.

22 Guo, P., Zheng, Y., Wen, M. et al. (2012). Icephobic/anti-icing properties of micro/nanostructured surfaces. *Advanced Materials* 24 (19): 2642–2648.

23 Kolle, M., Salgard-Cunha, P.M., Scherer, M.R.J. et al. (2010). Mimicking the colourful wing scale structure of the *Papilio blumei* butterfly. *Nature Nanotechnology* 5: 511–515.

24 Vukusic, P. and Sambles, J.R. (2003). Photonic structures in biology. *Nature* 424: 852–855.

25 Sato, O., Kubo, S., and Gu, Z.Z. (2008). Structural color films with lotus effects, superhydrophilicity, and tunable stop-bands. *Accounts of Chemical Research* 42 (1): 1–10.

26 Siddique, R.H., Gomard, G., and Hölscher, H. (2015). The role of random nanostructures for the omnidirectional anti-reflection properties of the glasswing butterfly. *Nature Communications* 6: 6909.

27 Zheng, Y., Gao, X., and Jiang, L. (2007). Directional adhesion of superhydrophobic butterfly wings. *Soft Matter* 3 (2): 178.

28 Vukusic, P. and Hooper, I. (2005). Directionally controlled fluorescence emission in butterflies. *Science* 310 (5751): 1151.

29 Potyrailo, R.A., Ghiradella, H., Vertiatchikh, A. et al. (2007). *Morpho* butterfly wing scales demonstrate highly selective vapour response. *Nature Photonics* 1: 123–128.

30 Niu, S., Li, B., Mu, Z. et al. (2015). Excellent structure-based multifunction of *Morpho* butterfly wings: a review. *Journal of Bionic Engineering* 12: 170–189.

31 Potyrailo, R.A., Starkey, T.A., Vukusic, P. et al. (2013). Discovery of the surface polarity gradient on iridescent *Morpho* butterfly scales reveals a mechanism of their selective vapor response. *Proceedings of the National Academy of Sciences of the United States of America* 110 (39): 15567–15572.

32 Sambles, J.R. (2012). Biophotonics: blue butterflies feel the heat. *Nature Photonics* 6: 141–142.

33 Peng, W., Zhu, S., Wang, W. et al. (2012). 3D network magnetophotonic crystals fabricated on *Morpho* butterfly wing templates. *Advanced Functional Materials* 22 (10): 2072–2080.

34 Mohri, Y., Kobashi, J., Yoshida, H., and Ozaki, M. (2017). *Morpho*-butterfly-inspired patterning of helical photonic structures for circular-polarization-sensitive, wide-angle diffuse reflection. *Advanced Optical Materials* 5 (7): 1601071.

35 Song, B., Eom, S.C., and Shin, J.H. (2014). Disorder and broad-angle iridescence from *Morpho*-inspired structures. *Optics Express* 22 (16): 19386–19400.

36 Song, B., Johansen, V.E., Sigmund, O., and Shin, J.H. (2017). Reproducing the hierarchy of disorder for *Morpho*-inspired, broad-angle color reflection. *Scientific Reports* 7: 46023.

37 Butt, H., Yetisen, A.K., Mistry, D. et al. (2016). *Morpho* butterfly-inspired nanostructures. *Advanced Optical Materials* 4 (4): 497–504.

38 Zhang, S. and Chen, Y. (2015). Nanofabrication and coloration study of artificial *Morpho* butterfly wings with aligned lamellae layers. *Scientific Reports* 5: 16637.

39 Zhang, F., Shen, Q., Shi, X. et al. (2015). Infrared detection based on localized modification of *Morpho* butterfly wings. *Advanced Materials* 27 (6): 1077–1082.

40 Shen, Q., He, J., Ni, M. et al. (2015). Subtractive structural modification of *Morpho* butterfly wings. *Small* 11 (42): 5705–5711.

41 Wang, W., Zhang, W., Gu, J. et al. (2013). Design of a structure with low incident and viewing angle dependence inspired by *Morpho* butterflies. *Scientific Reports* 3: 3427.

42 Liu, Y., Huang, L., and Shi, W. (2012). Structural color bio-engineering by replicating *Morpho* wings. *Advanced Materials Research* 391: 409–417.

43 Chung, K., Yu, S., Heo, C. et al. (2012). Flexible, angle-independent, structural color reflectors inspired by *Morpho* butterfly wings. *Advanced Materials* 24 (18): 2375–2379.

44 Weatherspoon, M.R., Cai, Y., Crne, M. et al. (2008). 3D rutile titania-based structures with *Morpho* butterfly wing scale morphologies. *Angewandte Chemie International Edition* 120 (41): 8039–8041.

45 Potyrailo, R.A., Bonam, R.K., Hartley, J.G. et al. (2015). Towards outperforming conventional sensor arrays with fabricated individual photonic vapour sensors inspired by *Morpho* butterflies. *Nature Communications* 6: 7959.

46 Han, Z., Niu, S., Yang, M. et al. (2013). An ingenious replica templated from the light trapping structure in butterfly wing scales. *Nanoscale* 5 (18): 8500–8506.

47 Han, Z., Mu, Z., Li, B. et al. (2016). A high-transmission, multiple antireflective surface inspired from bilayer 3D ultrafine hierarchical structures in butterfly wing scales. *Small* 12 (6): 713–720.

48 Wang, L., Gong, Q., Zhan, S. et al. (2016). Robust anti-icing performance of a flexible superhydrophobic surface. *Advanced Materials* 28 (35): 7729–7735.

49 Zhu, J., Hsu, C.M., Yu, Z. et al. (2009). Nanodome solar cells with efficient light management and self-cleaning. *Nano Letters* 10 (6): 1979–1984.

10

Prospects and Outlook

For nature-inspired structured functional surfaces (NISFSs), in this book, we introduced the basic ideology of bionics and typical biological prototypes with natural structural surfaces. Then the definition of NISFS was given, and the NISFS was classified into two types based on the origin of structural surfaces. Before biomimetic fabrication of NISFS, basic design principles were introduced and commonly used fabrication methods were explained and enumerated one by one. Also, synthetic design and fabrication strategies were discussed to guide the functional integration of NISFS. Considering the study of natural biological surfaces is of great importance, the collection, preservation, and pretreatment of biological prototypes were also mentioned. Then, relatively advanced characterization methods and composition analysis techniques of both biological prototypes and NISFS were introduced with specific study cases for better understanding. As the main body of the book, combining our group's recent research achievements on NISFS, a few kinds of NISFS with different structural features were emphatically discussed in the chapters. They are structural antireflective surfaces inspired by butterfly and cicada wings, structural antifogging surfaces inspired by butterfly wings and mayfly compound eyes, structurally colored surfaces inspired by butterfly wings, oil–water separation materials inspired by butterfly wings and fish scales, and structural responsive surfaces toward multiple gases/vapors inspired by butterfly wings. For each NISFS, the basic concept, brief classification, classic theory models, the structural characteristics and chemical compositions of original biological prototypes, biomimetic fabrication with various chemical and physical methods, and surface performance of the fabricated NISFS were investigated specifically. However, limited by the research time, technical means, and theoretical level, follow-up research can be carried out from the following three aspects discussed in what follows.

For structures and functions of typical biological prototypes, in the case of butterflies, we can continue to investigate other butterfly species with unique wing scale structures and excellent surface characteristics and explore their structure–function relationship and design accordingly. In addition, the scope of research objectives can be extended from butterflies to other species in nature, and typical biological prototypes with other excellent functional properties are preferred. The research flow path of these work can be described as "macroscopic observation-structural characterization-bionic manufacturing-performance

Nature-Inspired Structured Functional Surfaces: Design, Fabrication, Characterization, and Applications, First Edition. Zhiwu Han.

testing-performance reproduction-mechanism reveal-simulation demonstration." Research and analysis of biological prototypes' functional characteristics, reveal their inherent mechanism, design and manufacture corresponding with multifunctional integrated bionic new materials.

On the other hand, for raw materials, biomimetic design, and manufacturing methods, in this book, biomimetic functional surfaces with different micro-/nanostructures were fabricated by wet chemical methods. The methods and materials involved also have many limitations. For example, due to the inherent defects of the sol–gel technology, the surface with some micro-/nanostructures may shrink or expand, or even crack, which makes it difficult to carry out the secondary template process. Afterward, the strong acid in the selective acid etching process is a potential danger for the human body. Besides, the material has a certain brittleness, which makes the toughness of the bionic functional surface insufficient. The area of the bionic functional surface is on the centimeter scale, and it is difficult to realize large-area controllable production under laboratory conditions. Therefore, follow-up research can deal with the abovementioned defects in the manufacturing and material selection processes of biomimetic functional surfaces, to obtain a cleaner, environmentally friendly, energy-saving, and efficient biomimetic manufacturing method, and excellent material properties, such as lightweight, high strength, flexibility, and high elasticity.

Besides, for theoretical and simulation of characteristics of bionic functional surfaces, although the research work is based on classical optical theory and wettability model, the theoretical analysis and numerical simulation of the mapping relationship between biomimetic micro-/nanostructure and its surface properties are carried out. Both of them have been simplified to a certain extent because the complexity of the 3D multiscale hierarchical micro-/nanostructure makes the classical theoretical model too difficult to fully apply. Therefore, how to extract the empirical formula, basic theory, and accurate model with universality and accuracy for such complex and fine natural 3D structures will be one of the important research contents in this field, which should be carried out in the future.

Index

Nature-Inspired Structured Functional Surfaces: Design, Fabrication, Characterization, and Applications, First Edition. Zhiwu Han.